Lecture Notes on

GEOMETRICAL ASPECTS of PARTIAL DIFFERENTIAL EQUATIONS

Series on
Soviet and East European Mathematics
Volume 9

Lecture Notes on

GEOMETRICAL ASPECTS of PARTIAL DIFFERENTIAL EQUATIONS

V V Zharinov
Steklov Mathematical Institute

World Scientific
Singapore • New Jersey • London • Hong Kong

Published by

World Scientific Publishing Co. Pte. Ltd.

P O Box 128, Farrer Road, Singapore 9128

USA office: Suite 1B, 1060 Main Street, River Edge, NJ 07661

UK office: 73 Lynton Mead, Totteridge, London N20 8DH

LECTURE NOTES ON GEOMETRICAL ASPECTS OF PARTIAL DIFFERENTIAL EQUATIONS

ISBN 981-02-0753-0

Printed in Singapore by Utopia Press.

TABLE OF CONTENTS

CHAPTER II. EXTERNAL GEOMETRY OF PDE

CHAPTER III. FURTHER STUDIES

APPENDIX

INTRODUCTION

Geometrical approach is a natural powerful language to study symmetries, conservation laws, Backlund transformations and other similar properties of nonlinear systems of partial differential equations (PDE). Let us explain our point in more detail. A given system of PDE can be studied from different sides and with different methods. In this sense the main and generally accepted direction (especially for linear systems) is connected with the study of existence, uniqueness, smoothness, singularities and continuation of solutions of a given PDE. This direction is concerned with the description of the solution set of a given PDE. But who told us that the solution set is the only thing we should be interested in. Indeed, there is another wide spectrum of objects associated with a given system of PDE, characterizing the system from a quite different angle. For example, when PDE are used as mathematical models in the theoretical physics, object of this class describe such fundamental phenomena as invariance, conservation of energy, momentum and so on. Here we get direct characteristics of the original physical problem, not of its mathematical model. In the general situation these objects can be considered as functionals over the solution set, i.e., dual elements, carrying a specific information about the system under study. The appropriate frames for their description are given by the differential geometry.

A geometrical counterpart of a nonlinear system of PDE is a *differential manifold*, i.e., a pair *M=(M,CTM)*, consisting of a smooth manifold *M* and an involutive distribution (called the *Cartan distribution*) *CTM* over it. The dimension of the Cartan distribution is equal to the number of the independent variables of the background system of PDE, and its integral submanifolds correspond to solutions of the system. The manifold *M* is usually infinite-dimensional, and this leads to a nontrivial local (i.e., in one chart) geometry, in contrast with the finite-dimensional situation, where the Frobenius theorem ensures the existence of a standard foliating atlas.

Differential manifolds are the objects of the *category of differential manifolds*. Morphisms of this category are *Lie-Backlund mappings*, i.e., smooth mappings compatible with the corresponding Cartan distributions. They map integral submanifolds of the source manifold into integral submanifolds of the target manifold, i.e., solutions are transformed into solutions. In particular, symmetries of a background PDE are described by Lie-Backlund automorphisms of the corresponding differential manifold, while infinitesimal symmetries are described by a special class of vector fields, called *Lie-Backlund fields*.

The de Rham complex together with the Cartan distribution give rise to a spectral sequence (*Vinogradov C-spectral sequence*), associated with a given differential manifold. Elements of this sequence describe conservation laws, functionals, the characteristic mapping, the Euler-Lagrange operator, and other similar objects over the background system of PDE. In contrast with the de Rham complex the spectral sequence usually is not exact even locally, this corresponds to existence of conservation laws for PDE determined in star-shaped domains, where due to the Poincare lemma all cohomology spaces are trivial.

In practice PDE are realized as *reductions* of more simple (mostly trivial) ambient PDE, like manifolds usually are represented as surfaces in Euclidean spaces. This leads to *external* geometry of the corresponding differential manifolds (besides the *internal* one that we discussed before). The main component of the latter is the relation between conservation laws and infinitesimal symmetries, generalizing the classical Green's formula for the Laplace equation. The other ones are the characteristic mapping, and a generalization of Lie-Backlund morphisms, known as *Backlund transformations*.

All these objects are widely used and studied, especially in mathematical physics, in different settings and in different languages. The geometrical approach we present here goes back to pioneer's works of Lie, Backlund, Noether, E.Cartan. In recent times it was developed by Ovsyannikov [1], Ibragimov [1],

Vinogradov [4] (see also Vinogradov, Krasil'shchik [1]), Olver [1] and many others. The purpose of these lectures is to give a short simple introduction into the fascinating domain of infinite-dimensional differential geometry, providing natural frames for the study of specific properties of PDE, and explaining them in a clear precise language. Even if this approach does not help much in actual calculations, it gives a better understanding of the whole picture, and as a result allows to do these calculations in a neat unified way.

0. Infinite-Dimensional Differential Geometry

0.1. Manifolds

Let I be a finite or countable set, whose elements we shall call *indices*. We denote by $\#I$ the number of elements of I, so $\#I$ is a natural number or ∞. Let $\mathbb{R}^I$ be the linear space of all real functions $u=(u^i)$ on I. The space $\mathbb{R}^I$ has the natural topology of the Euclidean space if I is finite, and of the projective limit of such spaces (i.e., Fréchet space topology) if I is countable. The elements $u^i\in\mathbb{R}$, $i\in I$, we shall call *coordinates*. The dual space, i.e., the space of all linear continuous functionals on $\mathbb{R}^I$, we denote by $\mathbb{R}^I_0$. It consists of all real *finite* functions on I, i.e. functions $\omega=(\omega_i)$, where only a finite number of components $\omega_i\neq 0$, $i\in I$.

For an open set $U\subset\mathbb{R}^I$ we denote by $\mathcal{C}^\infty(U)$ the space of all real functions on U that depend on finitely many coordinates and are *smooth* (i.e., of $\mathcal{C}^\infty$-class) as functions of a finite number of variables. Namely, a function $f\in\mathcal{C}^\infty(U)$ iff f depends on coordinates u^i, $i\in K$, where $K=K(f)$ is a finite subset of I, and $f\in\mathcal{C}^\infty(U_K)$, where U_K is the natural projection of $U\subset\mathbb{R}^I$ onto $\mathbb{R}^K$. The set of all spaces $\mathcal{C}^\infty(U)$, $U\subset\mathbb{R}^I$, gives rise to a sheaf - *the sheaf of smooth functions on* $\mathbb{R}^I$ *of a locally finite number of variables.* If I is a finite set this sheaf is the usual sheaf of smooth functions on the Euclidean space $\mathbb{R}^I$.

Let M be a set. A *chart* on M is a triple $(U,\varphi,\mathbb{R}^I)$, where U is a subset of M, φ is a bijection of U onto an open subset $\varphi(U)$ of $\mathbb{R}^I$. A chart $(U,\varphi,\mathbb{R}^I)$ is called *smoothly compatible* with a chart $(V,\psi,\mathbb{R}^J)$ if the image $\psi(U\cap V)$ is open in $\mathbb{R}^J$, and the transition function $\varphi\circ\psi^{-1}$ has smooth components $u^i=(\varphi\circ\psi^{-1})^i(v)$ on $\psi(U\cap V)$, where u^i and v^j are coordinates on $\mathbb{R}^I$ and $\mathbb{R}^J$. Note that linear spaces $\mathbb{R}^I$ and $\mathbb{R}^J$ are isomorphic and $\#I=\#J$ in this case. A family of pairwise smoothly compatible charts, covering the whole set M, is called a *smooth atlas* on M. Two smooth atlases on M are called *equivalent* if their union is the smooth atlas on M. The set M, equipped with an equivalence class of smooth atlases, is called the *smooth manifold*. The number $\dim M=\#I$ does not depend on a chart $(U,\varphi,\mathbb{R}^I)$ and is called the *dimension* of a smooth manifold M.

Let $\mathcal{O}$ be an open subset of a smooth manifold M. A function f on $\mathcal{O}$ is called *smooth* if the composition $f\circ\varphi^{-1}$ is a smooth function on $\varphi(\mathcal{O}\cap U)$ for every chart $(U,\varphi,\mathbb{R}^I)$. We denote by $\mathcal{C}^\infty(\mathcal{O})$ the algebra of all smooth functions on $\mathcal{O}$ with pointwise operations. In a chart $(U,\varphi,\mathbb{R}^I)$ on M with coordinates u^i, $i\in I$, a function $f\in\mathcal{C}^\infty(U)$ has the form

$$f(a) = \tilde{f}(\varphi(a)) \text{ for all } a \in U,$$

where $\tilde{f}\in\mathcal{C}^{\infty}(\varphi(U))$. In this case we shall identify f and $\tilde{f}$, i.e., we shall write $f=f(u)$, $u\in U$.

For sets A and B we denote by $A\times B$ their *direct product*, i.e. the set of all pairs (a,b), $a\in A$ and $b\in B$. Let M and N be smooth manifolds with atlases $\mathfrak{M}$ and $\mathfrak{N}$, respectively. Then the direct product $M\times N$ will be the smooth manifold with the atlas $\mathfrak{M}\times\mathfrak{N}$, consisting of all charts $(U\times V,\varphi\times\psi,\mathbb{R}^{I+J})$, where $(U,\varphi,\mathbb{R}^{I})\in\mathfrak{M}$, $(V,\psi,\mathbb{R}^{J})\in\mathfrak{N}$, $\varphi\times\psi(a,b)=(\varphi(a),\psi(b))$ for all $a\in U$, $b\in V$, and $\mathbb{R}^{I+J}=\mathbb{R}^{I}\times\mathbb{R}^{J}$.

0.2. Vector Fields and Differential Forms

Let M be a smooth manifold and a point $a\in M$. Let us denote by $\mathcal{C}^{\infty}(\{a\})$ the algebra of all smooth functions defined in a neighborhood of a, i.e.,

$$\mathcal{C}^{\infty}(\{a\}) = \bigcup_{\mathcal{O}\ni a} \mathcal{C}^{\infty}(\mathcal{O}).$$

A linear mapping $\zeta:\mathcal{C}^{\infty}(\{a\})\longrightarrow\mathbb{R}$ is called a *tangent vector* at a point a if

$$\zeta(fg) = \zeta(f)g(a) + f(a)\zeta(g)$$

for all $f,g\in\mathcal{C}^{\infty}(\{a\})$. The linear space T_aM of all tangent vectors at the point a is called the *tangent space* to M at a. The space T_aM can be identified with the space $\mathbb{R}^{I}$ if the point a belongs to a chart $(U,\varphi,\mathbb{R}^{I})$. The collection of all tangent spaces T_aM corresponding to all points $a\in M$ gives rise to the *tangent bundle* of M denoted by TM. Local smooth sections of TM are called *vector fields* on M, i.e., a vector field ζ is a smooth mapping from $\mathcal{O}$ into TM, where $\mathcal{O}$ is an open subset of M (the domain of ζ), $\zeta(a)\in T_aM$ for all $a\in\mathcal{O}$. In other words, a vector field ζ on $\mathcal{O}$ is a *differentiation* of the algebra $\mathcal{C}^\infty(\mathcal{O})$, i.e., a linear mapping $\zeta:\mathcal{C}^\infty(\mathcal{O})\longrightarrow\mathcal{C}^\infty(\mathcal{O})$ such that the Leibniz' rule

$$\zeta(fg) = \zeta(f)g + f\zeta(g)$$

is valid for all $f,g\in\mathcal{C}^\infty(\mathcal{O})$. In a chart $(U,\varphi,\mathbb{R}^I)$ on M with coordinates u^i a vector field ζ on U has the representation

$$\zeta = \sum_{i\in I} \zeta^i(u)\,\frac{\partial}{\partial u^i},$$

where components $\zeta^i(u)$ are smooth functions on U. Note that the action $\zeta(f)$ is well-defined for any $f\in\mathcal{C}^\infty(U)$, because $\partial f\big/\partial u^i\neq 0$ only for $i\in K(f)$ and the set $K(f)$ is finite.

Let us denote by $\mathcal{T}(\mathcal{O})$ the set of all vector fields on M with a common domain $\mathcal{O}$. For vector fields $\zeta,\eta\in\mathcal{T}(\mathcal{O})$ their *Lie bracket* $[\zeta,\eta]$ is defined by the rule

$$[\zeta,\eta](f) = \zeta(\eta(f)) - \eta(\zeta(f))$$

for all $f\in\mathcal{C}^\infty(\mathcal{O})$. It is easy to check that $[\zeta,\eta]\in\mathcal{T}(\mathcal{O})$. Moreover, $[\zeta,\eta]+[\eta,\zeta]=0$ and the *Jacoby identity*

$$[\xi,[\zeta,\eta]] + [\zeta,[\eta,\xi]] + [\eta,[\xi,\zeta]] = 0$$

is valid for all vector fields $\xi,\zeta,\eta\in\mathcal{T}(\mathcal{O})$. Hence, $\mathcal{T}(\mathcal{O})$ is the Lie algebra.

Let M be a smooth manifold and a point $a\in M$. Linear continuous functionals on the tangent space T_aM are called *cotangent vectors* at a. The linear space of all cotangent vectors at a is denoted by T_a^*M. It is called the *cotangent space* at a and can be identified with the dual space $\mathbb{R}_0^I$ if the point a belongs to a chart $(U,\varphi,\mathbb{R}^I)$. The collection of all cotangent spaces T_a^*M corresponding to all points $a\in M$ gives rise to the *cotangent bundle* of M denoted by T^*M. Local smooth sections of T^*M are called *1-forms*, i.e., a *1*-form ω is a smooth mapping from $\mathcal{O}$ into T^*M, where $\mathcal{O}$ is an open subset of M (the domain of ω), $\omega(a)\in T_a^*M$ for all $a\in\mathcal{O}$. The linear space of all *1*-forms on an open set $\mathcal{O}$ is denoted by $\Lambda(\mathcal{O})$. In a chart $(U,\varphi,\mathbb{R}^I)$ on M with coordinates u^i a *1*-form $\omega\in\Lambda(U)$ has the representation

$$\omega = \sum_{i\in I} \omega_i(u)\, du^i,$$

where coefficients $\omega_i(u)$ are smooth functions on U, only a finite number of them differ from zero, so the action

$$\omega(\zeta)(u) = \sum_{i\in I} \omega_i(u)\, \zeta^i(u)$$

is well-defined for all $u\in U$ and $\zeta\in\mathcal{T}(U)$.

In the same way, for a point $a\in M$ and a natural number $q\in\mathbb{N}$ let us denote by Λ_a^qM the linear space of all *q-covectors* at a, i.e., q-linear continuous alternating functionals on T_aM, where $\Lambda_a^0M=\mathbb{R}$, $\Lambda_a^1M=T_a^*M$, and $\Lambda_a^qM=0$ if $q>\dim M$ (the last condition takes place only when $\dim M$ is finite). The collection of all spaces Λ_a^qM corresponding to all points $a\in M$ gives rise to the bundle Λ^qM on M. Local smooth sections of Λ^qM are called *q-forms*, i.e., a q-form ω is a smooth mapping from $\mathcal{O}$ into $\Lambda^q M$, where $\mathcal{O}$ is an open subset of M (the domain of ω), $\omega(a)\in\Lambda^q_a M$ for all $a\in\mathcal{O}$. The linear space of all q-forms on an open set $\mathcal{O}$ is denoted by $\Lambda^q(\mathcal{O})$. By the definition, $\Lambda^0(\mathcal{O})=\mathcal{C}^\infty(\mathcal{O})$, $\Lambda^1(\mathcal{O})=\Lambda(\mathcal{O})$, and $\Lambda^q(\mathcal{O})=0$ if $q>\dim M$. In a chart $(U,\varphi,\mathbb{R}^I)$ on M with coordinates u^i a q-form $\omega\in\Lambda^q(U)$ has the representation

$$\omega = \sum_{i_1,\dots,i_q} \omega_{i_1\dots i_q}(u)\, du^{i_1}\wedge\dots\wedge du^{i_q}$$

$$= \sum_{\langle i\rangle} \omega_{\langle i\rangle}(u) \left(\wedge du\right)^{\langle i\rangle},$$

where coefficients $\omega_{i_1\dots i_q}(u)$ are smooth functions on U alternating by indices $i_1,\dots,i_q\in I$, only a finite number of these coefficients differ from zero, $\wedge$ denotes the alternating tensor product. Hence,

$$\omega(\zeta_1,\dots,\zeta_q)(u) = \sum_{i_1,\dots,i_q} \omega_{i_1\dots i_q}(u)\ \text{A.P.} \prod_{\lambda=1}^{q} \zeta_\lambda^{i_\lambda}(u)$$

for all fields $\zeta_1,\dots,\zeta_q\in\mathcal{T}(U)$, where A.P. denotes the alternating part of the product. For an open $\mathcal{O}\subset M$ we denote by $\Lambda^*(\mathcal{O})=\Sigma\Lambda^q(\mathcal{O})$ the direct sum of all linear spaces $\Lambda^q(\mathcal{O})$, where $q=0,1,2,\dots$, i.e., $\omega\in\Lambda^*(\mathcal{O})$ iff $\omega=\omega^0+\omega^1+\omega^2+\dots$, where only a finite number of forms $\omega^q\in\Lambda^q(\mathcal{O})$ differ from zero. Elements of $\Lambda(\mathcal{O})$ are called *differential forms*.

For forms $\omega\in\Lambda^{q}(\mathcal{O})$ and $\theta\in\Lambda^{p}(\mathcal{O})$ their *wedge product* $\omega\wedge\theta\in\Lambda^{q+p}(\mathcal{O})$ is defined by the rule

$$\omega\wedge\theta(\zeta_1,\dots,\zeta_{q+p}) = \text{A.P.}\Big(\omega(\zeta_1,\dots,\zeta_q)\theta(\zeta_{q+1},\dots,\zeta_{q+p})\Big)$$

for all fields $\zeta_1,\dots,\zeta_{q+p}\in\mathcal{T}(\mathcal{O})$, where A.P. denotes the alternating part of the product. With this definition the linear space $\Lambda^{*}(\mathcal{O})$ becomes the algebra.

The *exterior derivative* $d\omega\in\Lambda^{q+1}(\mathcal{O})$ of a q-form $\omega\in\Lambda^{q}(\mathcal{O})$ is defined by the rule

$$d\omega(\zeta_0,\dots,\zeta_q) = \frac{1}{q+1}\sum_{\lambda=0}^{q}(-1)^{\lambda}\zeta_{\lambda}\Big(\omega(\zeta_0,\dots,\hat{\zeta}_{\lambda},\dots,\zeta_q)\Big)$$

$$+\frac{1}{q+1}\sum_{\lambda<\mu}(-1)^{\lambda+\mu}\omega\Big([\zeta_{\lambda},\zeta_{\mu}],\zeta_0,\dots,\hat{\zeta}_{\lambda},\dots,\hat{\zeta}_{\mu},\dots,\zeta_q\Big)$$

for all fields $\zeta_0,\zeta_1,\dots,\zeta_q\in\mathcal{T}(\mathcal{O})$, where the terms $\hat{\zeta}_{\lambda}$ and $\hat{\zeta}_{\mu}$ are omitted. In particular, $df(\zeta)=\zeta(f)$ if $f\in\Lambda^{0}(\mathcal{O})=\mathcal{C}^{\infty}(\mathcal{O})$. It is clear that the composition $d\circ d=0$, so we have the *de Rham complex*

$$0\longrightarrow\mathbb{R}\longrightarrow\Lambda^{0}(\mathcal{O})\xrightarrow{d}\Lambda^{1}(\mathcal{O})\xrightarrow{d}\Lambda^{2}(\mathcal{O})\xrightarrow{d}\dots$$

for every open subset $\mathcal{O}$ of the manifold M. This complex is finite iff $\dim M$ is finite. By the Poincaré lemma, it is locally exact when $\dim M$ is finite. From the proof of this lemma one can see that the same is true also when $\dim M$ is infinite. Hence, *every point $a\in M$ has an open neighbourhood $\mathcal{O}\subset M$ such that a form $\omega\in\Lambda^{*}(\mathcal{O})$ has a representation $\omega=d\theta$, $\theta\in\Lambda^{*}(\mathcal{O})$, iff $d\omega=0$*. The exterior differentiation acts as an *anti-differentiation* on the algebra $\Lambda^{*}(\mathcal{O})$, i.e.

$$d(\omega\wedge\theta) = (d\omega)\wedge\theta + (-1)^q\omega\wedge(d\theta)$$

for all forms $\omega\in\Lambda^q(\mathcal{O})$ and $\theta\in\Lambda^p(\mathcal{O})$.

The *interior product* $\zeta\rfloor\omega$ of a vector field $\zeta\in\mathcal{T}(\mathcal{O})$ and a q-form $\omega\in\Lambda^q(\mathcal{O})$, where $\mathcal{O}$ is an open subset of a smooth manifold M, $q=1,2,3,\ldots,$ is the $(q-1)$-form defined by the rule

$$(\zeta\rfloor\omega)(\zeta_1,\ldots,\zeta_{q-1}) = \omega(\zeta,\zeta_1,\ldots,\zeta_{q-1})$$

for all fields $\zeta_1,\ldots,\zeta_{q-1}\in\mathcal{T}(\mathcal{O})$. The interior product acts as an anti-differentiation on the algebra $\Lambda^*(\mathcal{O})$, i.e.

$$\zeta\rfloor(\omega\wedge\theta) = (\zeta\rfloor\omega)\wedge\theta + (-1)^q\omega\wedge(\zeta\rfloor\theta)$$

for all forms $\omega\in\Lambda^q(\mathcal{O})$ and $\theta\in\Lambda^p(\mathcal{O})$. Moreover,

$$\zeta\rfloor(\eta\rfloor\omega) + \eta\rfloor(\zeta\rfloor\omega) = 0$$

for all fields $\zeta,\eta\in\mathcal{T}(\mathcal{O})$ and forms $\omega\in\Lambda^q(\mathcal{O})$, $q=2,3,4,\ldots$.

The *Lie derivative* $L_\zeta\omega$ of a q-form $\omega\in\Lambda^q(\mathcal{O})$ with respect to a vector field $\zeta\in\mathcal{T}(\mathcal{O})$, where $\mathcal{O}$ is an open subset of a smooth manifold M, $q=0,1,2,\ldots,$ is the q-form defined by the rule

$$(L_\zeta\omega)(\zeta_1,\ldots,\zeta_q) = \zeta\Big(\omega(\zeta_1,\ldots,\zeta_q)\Big) - \sum_{\lambda=1}^{q}\omega(\zeta_1,\ldots,[\zeta,\zeta_\lambda],\ldots,\zeta_q)$$

for all fields $\zeta_1,\ldots,\zeta_q\in\mathcal{T}(\mathcal{O})$. In particular, $L_\zeta f=\zeta(f)$ if $f\in\Lambda^0(\mathcal{O})$ $(=\mathcal{C}^\infty(\mathcal{O}))$. The Lie derivative acts as a differentiation on the algebra $\Lambda^*(\mathcal{O})$, i.e.

$$L_\zeta(\omega\wedge\theta) = (L_\zeta\omega)\wedge\theta + \omega\wedge(L_\zeta\theta)$$

for all forms $\omega\in\Lambda^q(\mathcal{O})$ and $\theta\in\Lambda^p(\mathcal{O})$. Moreover,

$$L_\zeta(L_\eta\omega) - L_\eta(L_\zeta\omega) = L_{[\zeta,\eta]}\omega$$

for all fields $\zeta,\eta\in\mathcal{T}(\mathcal{O})$ and forms $\omega\in\Lambda^q(\mathcal{O})$, $q=0,1,2,\dots$.

There are the following relations between the operations d, $\rfloor$ and L_ζ:

$$L_\zeta(d\omega) = d(L_\zeta\omega)$$

for all fields $\zeta\in\mathcal{T}(\mathcal{O})$ and forms $\omega\in\Lambda^q(\mathcal{O})$, $q=0,1,2,\dots,$

$$L_\zeta(\eta\rfloor\omega) = [\zeta,\eta]\rfloor\omega + \eta\rfloor(L_\zeta\omega)$$

for all fields $\zeta,\eta\in\mathcal{T}(\mathcal{O})$ and forms $\omega\in\Lambda^q(\mathcal{O})$, $q=1,2,3,\dots,$

$$L_\zeta\omega = d(\zeta\rfloor\omega) + \zeta\rfloor(d\omega)$$

for all fields $\zeta\in\mathcal{T}(\mathcal{O})$ and forms $\omega\in\Lambda^q(\mathcal{O})$, $q=0,1,2,\dots,$ where by the definition, $\zeta\rfloor\omega=0$ if $\omega\in\Lambda^0(\mathcal{O})$.

0.3. Mappings

Let M and N be smooth manifolds. A mapping $f:M\longrightarrow N$ is called *smooth* if the mappings $\psi\circ f\circ\varphi^{-1}:\varphi(U)\longrightarrow\psi(V)$ have smooth components $v^j=(\psi\circ f\circ\varphi^{-1})^j(u)$, $j\in J$, $u\in U$, for all charts $(U,\varphi,\mathbb{R}^I)$ on M and $(V,\psi,\mathbb{R}^J)$ on N such that $f(U)\subset V$. Hence, we have the *category of smooth manifolds*. Objects of this category are smooth manifolds, and morphisms are smooth mappings. In particular, a one-to-one smooth mapping $f:M\longrightarrow N$ of a smooth manifold M on a smooth manifold

N is called an *isomorphism* if the inverse mapping $f^{-1}:N\longrightarrow M$ is also smooth. Automorphisms, i.e., isomorphisms $f:M\longrightarrow M$, are called *transformations* of a smooth manifold M.

Let $f:M\longrightarrow N$ be a smooth mapping. The *tangent mapping* $f_*:TM\longrightarrow TN$ takes a vector $\zeta\in T_aM$, $a\in M$, to the vector $\eta=f_*\zeta\in T_bN$, where $b=f(a)$, $(f_*\zeta)(\phi)=\zeta(\phi\circ f)$ for all functions $\phi\in\mathscr{C}^\infty(\{b\})$ (note that in this case $\phi\circ f\in\mathscr{C}^\infty(\{a\})$). The *cotangent mapping* $f^*:T^*N\longrightarrow T^*M$ takes a covector $\omega\in T_b^*N$ to the covector $\theta=f^*\omega\in T_a^*M$, where $(f^*\omega)(\zeta)=\omega(f_*\zeta)$ for all vectors $\zeta\in T_aM$. The last rule defines the action of the cotangent mapping (the *pull-back*) f^* on forms, $f^*:\Lambda^q(Q)\longrightarrow\Lambda^q(\mathcal{O})$, where an open set $\mathcal{O}\subset M$, an open set $Q\subset N$, the image $f(\mathcal{O})\subset Q$,

$$(f^*\omega)(\zeta_1,\dots,\zeta_q) = \omega(f_*\zeta_1,\dots,f_*\zeta_q)$$

for all forms $\omega\in\Lambda^q(Q)$ and fields $\zeta_1,\dots,\zeta_q\in\mathcal{T}(\mathcal{O})$, $q\in\mathbb{N}$ (note that $f_*\zeta_1,\dots,f_*\zeta_q$ are not defined as fields on $f(\mathcal{O})$). In particular, $f^*\phi=\phi\circ f$ for any $\phi\in\Lambda^0(\mathcal{O})=\mathscr{C}^\infty(\mathcal{O})$. Moreover,

$$f^*(\omega\wedge\theta) = (f^*\omega)\wedge(f^*\theta)$$

for all forms $\omega\in\Lambda^q(Q)$ and $\theta\in\Lambda^p(Q)$. Using these two properties, one can check that

$$d(f^*\omega) = f^*(d\omega)$$

for all forms $\omega\in\Lambda^q(Q)$.

A *local one-parameter transformation group* Γ on a smooth manifold M is determined by an open set $\mathcal{U}$, $\{0\}\times M\subset\mathcal{U}\subset\mathbb{R}\times M$, and a smooth mapping $\Gamma:\mathcal{U}\longrightarrow M$ with the following properties:

$$\Gamma(0,u) = u \text{ for all } u \in M,$$

$$(-t,\Gamma(t,u)) \in \mathcal{U} \text{ for all } (t,u) \in \mathcal{U},$$

$$\Gamma(s,\Gamma(t,u)) = \Gamma(s+t,u) \text{ if } (t,u),(s,\Gamma(t,u)) \text{ and } (s+t,u) \in \mathcal{U}.$$

Every local one-parameter transformation group Γ on a smooth manifold M defines the vector field ζ on M by the rule

$$\zeta(\phi)(u) = \frac{\partial}{\partial t}\phi(\Gamma(t,u))\Big|_{t=0} \quad \text{for all } u \in M \text{ and } \phi \in \mathcal{C}^{\infty}(\{u\}).$$

If a manifold M is finite-dimensional then the converse is also true, i.e., every vector field ζ on M can be integrated to the local one-parameter transformation group. This follows from the theory of the ordinary differential equations. On the other side, if a manifold M is infinite-dimensional there exist vector fields on M which cannot be integrated (see Example 0.6.1 in Section 0.6 below).

In the same way, a *local transformation Lie group* Γ on a smooth manifold M is determined by a Lie group G, an open set $\mathcal{U}$, $\{e\}\times M\subset\mathcal{U}\subset G\times M$, and a smooth mapping $\Gamma:\mathcal{U}\longrightarrow M$ with the following properties:

$$\Gamma(e,u) = u \text{ for all } u \in M,$$

$$(g^{-1},\Gamma(g,u)) \in \mathcal{U} \text{ for all } (g,u) \in \mathcal{U},$$

$$\Gamma(g,\Gamma(h,u)) = \Gamma(g\cdot h,u) \text{ if } (h,u),(g,\Gamma(h,u)) \text{ and } (g\cdot h,u) \in \mathcal{U},$$

where e is the identity element of G. Every one-parameter subgroup of Γ defines the vector field on M. The set $\mathfrak{g}$ of all such vector fields is a Lie algebra, a subalgebra of the Lie algebra $\mathcal{T}(M)$ of all vector fields on M. By the Lie theorem, when the manifold M is finite-dimensional every finite-dimensional Lie subalgebra of $\mathcal{T}(M)$ gives rise to a local transformation Lie group on M. In the infinite-dimensional case the Lie theorem is no longer true.

A smooth mapping $f: M \longrightarrow N$ is called an *immersion* if for each point $a \in M$ there exist a chart $(U, \varphi, \mathbb{R}^I)$ on M, $a \in U$, with coordinates u^i, $i \in I$, and a chart $(W, \chi, \mathbb{R}^{I+J})$ on N, $f(U) \subset W$, with coordinates u^i, v^j, $i \in I$, $j \in J$, such that $(\chi \circ f \circ \varphi^{-1})^i(u) = u^i$, $i \in I$, $(\chi \circ f \circ \varphi^{-1})^j(u) = 0$, $j \in J$. In other words, $\chi \circ f = \nu_I \circ \varphi$ on U, where ν_I is the natural injection of $\mathbb{R}^I$ into $\mathbb{R}^I \times \mathbb{R}^J$. We shall call these charts *defining charts*.

Let $f: M \longrightarrow N$ be an immersion and $f_*: TM \longrightarrow TN$ be its tangent mapping. For a point $a \in M$ let us consider defining charts $(U, \varphi, \mathbb{R}^I)$ on M and $(W, \chi, \mathbb{R}^{I+J})$ on N, $a \in U$, $b = f(a) \in W$. Then $T_a M \simeq \mathbb{R}^I$, $T_b N \simeq \mathbb{R}^I \times \mathbb{R}^J$ and $(f_*)_a \approx \nu_I$, where $(f_*)_a : T_a M \longrightarrow T_b N$. In other words, the linear mapping $(f_*)_a$ is an injection (i.e., its kernel $\mathrm{Ker}(f_*)_a = \{0\}$ and its image $\mathrm{Im}(f_*)_a$ has a closed complement) for any point $a \in M$.

A smooth mapping $f: M \longrightarrow N$ is called a *submersion* if for each point $a \in M$ there exist a chart $(W, \chi, \mathbb{R}^{I+J})$ on M, $a \in W$, with coordinates u^i, v^j, $i \in I$, $j \in J$, and a chart $(V, \psi, \mathbb{R}^J)$ on N, $f(W) \subset V$, with coordinates v^j, $j \in J$, such that $(\psi \circ f \circ \chi^{-1})^j(u, v) = v^j$, $j \in J$. In other words, $\psi \circ f = \pi_J \circ \chi$ on W, where π_J is the natural projection of $\mathbb{R}^I \times \mathbb{R}^J$ onto $\mathbb{R}^J$. We shall call these charts *defining charts*.

For example, if Γ is a local transformation Lie group on a smooth manifold M then the mapping $\Gamma: \mathcal{U} \longrightarrow M$ is the submersion.

Let $f: M \longrightarrow N$ be a submersion and $f_*: TM \longrightarrow TN$ be its tangent mapping. For a point $a\in M$ let us consider defining charts $(W,\chi,\mathbb{R}^{I+J})$ on M and $(V,\psi,\mathbb{R}^{J})$ on N, $a\in W$, $b=f(a)\in V$. Then $T_aM\simeq\mathbb{R}^I\times\mathbb{R}^J$, $T_bN\simeq\mathbb{R}^J$ and $(f_*)_a\approx\pi_J$, where $(f_*)_a:T_aM\longrightarrow T_bN$. In other words, the linear mapping $(f_*)_a$ is a projection (i.e., its image $\mathrm{Im}(f_*)_a=T_bN$ and its kernel $\mathrm{Ker}(f\ )$ has a closed complement) for any point $a\in M$.

Let M and N be smooth manifolds and $M\times N$ be their direct product. Then the natural projection mappings $\mathrm{pr}_M:M\times N\longrightarrow M$ and $\mathrm{pr}:M\times N\longrightarrow N$ are submersions. Hence, the tangent mappings $(\mathrm{pr}_{M*})_c$ and $(\mathrm{pr}_{N*})_c$ are projections for all points $c\in M\times N$. Further, the natural injective mappings

$$\mathrm{inj}_M\ :\ M\longrightarrow M\times N,\ x\ \mapsto\ (x,b)\ \text{for all}\ x\in M,$$

and

$$\mathrm{inj}_N\ :\ N\longrightarrow M\times N,\ y\ \mapsto\ (a,y)\ \text{for all}\ y\in N,$$

are defined for every point $b\in N$ and for every point $a\in M$. These mappings are immersions. Hence, the tangent mappings $T_aM\longrightarrow T_c(M\times N)$ and $T_bN\longrightarrow T_c(M\times N)$ are injections for any point $a\in M$ any point $b\in N$, where $c=(a,b)$. Moreover, these mappings define the isomorphism $T_aM\times T_bN\longrightarrow T_c(M\times N)$ for each point $c=(a,b)$. For brevity, we shall write $T_c(M\times N)=T_aM+T_bN$ in this case.

0.4. Submanifolds

Let M be a smooth manifold. A subset $S\subset M$ is called a *submanifold* of M if for every point $a\in M$ there exists a chart (we call it a *defining chart*) $(W,\chi,\mathbb{R}^{I+J})$ on M, $a\in W$, with coordinates u^i, v^j, $i\in I$, $j\in J$, such that in this chart

$$\chi(S\cap W) = \left\{(u,v) \in \chi(W) \subset \mathbb{R}^{I+J} \; : \; v^j = 0, \; j \in J\right\}$$

or $S\cap W=\{u,v:v=0\}$, for brevity. In this case there is the natural smooth manifold structure on S induced from M. The defining atlas on S is constructed by restriction on S of defining charts on M. Namely, to every chart $(W,\chi,\mathbb{R}^{I+J})$ on M there corresponds the chart $(U,\varphi,\mathbb{R}^I)$ on S, where $U=S\cap W$, $\varphi=\pi_I\circ\chi$, π_I is the natural projection of $\mathbb{R}^{I+J}$ onto $\mathbb{R}^I$.

Let $f:S\longrightarrow M$ be an immersion. Let $(U,\varphi,\mathbb{R}^I)$ and $(W,\chi,\mathbb{R}^{I+J})$ be defining charts, $\chi\circ f=\nu_I\circ\varphi$ on U. Then the image $f(U)$ will be a submanifold of M. In other words, a local image of S is a submanifold of M. Because of this fact, the manifold S is often called the *immersed submanifold* of M.

Let $f:M\longrightarrow N$ be a submersion. Let us fix a point $b\in N$ and consider the inverse image $f^{-1}(b)\subset M$ of b. For every point $a\in f^{-1}(b)$ there exist defining charts $(W,\chi,\mathbb{R}^{I+J})$ and $(V,\psi,\mathbb{R}^J)$ such that $a\in W$ and $\psi\circ f=\pi_J\circ\chi$ on W. Without loss of generality we may assume $v=\psi(b)=0$. Thus, we have the representation $f^{-1}(b)\cap W=\{u,v:v=0\}$, i.e., the set $f^{-1}(b)$ is the submanifold of M.

Let M be a smooth manifold and $(W,\chi,\mathbb{R}^K)$ be a chart on M with coordinates $w=(u,v)$, $u=(u^i:i\in I)$, $v=(v^j:j\in J)$, $K=I+J$ (i.e., $I,J\subset K$, $I\cup J=K$, $I\cap J=\varnothing$). Every set

$$S = \left\{w \; : \; v = f(u)\right\} \subset W,$$

where $f^i(u)$ are smooth functions, is a submanifold of M.

The standard way of specifying a submanifold S of a manifold M is to represent it as a solution set of a system of equations

$$S = \left\{ w \in M \;:\; F^p(w) = 0,\; p \in P \right\}$$

(or $\{F=0\}$ for brevity), where F^p are smooth functions on M. In a chart $(W, \chi, \mathbb{R}^K)$ on M with coordinates w^k, $k \in K$, these equations take the form

$$S = \left\{ w \in \chi(W) \subset \mathbb{R}^K \;:\; F^p(w) = 0,\; p \in P \right\},$$

where F^p are smooth functions on the image $\chi(W)$. Such set will be a submanifold of M if in a neighborhood of each point $a \in W$ the coordinates w^k can be split into two classes $w=(u,v)$, in a way that the system $\{F=0\}$ will be solvable in this neighborhood with respect to the coordinates v, i.e. if it can be written in the above form $S=\{v=f(u)\}$. If the system is finite, i.e., if the number $\#P$ of the equations is finite, then the condition

$$\mathrm{rank} \left\| \frac{\partial F^p}{\partial w^k} \right\| = \#P \text{ on } S \cap W$$

is sufficient for the local solvability.

On the other hand, a set $S \subset W$, where $(W, \chi, \mathbb{R}^K)$ is a chart on a smooth manifold M with coordinates w will be a finite dimensional submanifold of M iff it has a parameterization

$$S = \left\{ w = W(t),\; t \in O \right\},$$

where $W^k(t)$ are smooth functions on an open set $O \subset \mathbb{R}^m$, m is a dimension of the submanifold S, and

$$\operatorname{rank}\left\|\frac{\partial W^k}{\partial t^\mu}\right\| = m \quad \text{on } S.$$

Let S be a submanifold of a smooth manifold M. Let us denote by $i:S\longrightarrow M$ the natural imbedding of S into M. For every open subset $Q\subset M$ the restriction mapping $i^*:\mathscr{C}^\infty(Q)\longrightarrow\mathscr{C}^\infty(O)$ is defined, where $O=S\cap Q$ is the open subset of S. We denote by $I\mathscr{C}^\infty(Q)$ the kernel of this mapping, i.e., $I\mathscr{C}^\infty(Q)=\operatorname{Ker} i^*$. The set $I\mathscr{C}^\infty(Q)$ is an ideal of the algebra $\mathscr{C}^\infty(Q)$, so the quotient algebra $\mathscr{C}^\infty(O)_{ext}=\mathscr{C}^\infty(Q)\big/I\mathscr{C}^\infty(Q)$ is defined. It is clear that the sequence of the linear spaces

$$0 \longrightarrow I\mathscr{C}^\infty(Q) \longrightarrow \mathscr{C}^\infty(Q) \xrightarrow{\quad i^* \quad} \mathscr{C}^\infty(O) \longrightarrow 0$$

is locally exact (for example, in every defining chart on M). Hence, we can identify $\mathscr{C}^\infty(O)=\mathscr{C}^\infty(O)_{ext}$ at least locally.

Let $(W,\chi,\mathbb{R}^{I+J})$ be a defining chart on M and a function $F(u,v)\in\mathscr{C}^\infty(W)$. Then the restriction $f(u)=(i^*F)(u)=F(u,0)\in\mathscr{C}^\infty(U)$ (remind that $U=S\cap W$).

Let S be a submanifold of a smooth manifold M. Denote by T_SM the restriction of the vector bundle TS on S, i.e.,

$$T_SM = \bigcup_{a\in S} T_aM.$$

The tangent mapping $i_*:TS\longrightarrow TM$ is injective, and the image $i_*(TS)\subset T_SM$, so we have the following exact sequence of vector bundles on S:

$$0 \longrightarrow TS \xrightarrow{\quad j \quad} T_S M \xrightarrow{\quad \nu \quad} NS \longrightarrow 0,$$

where $NS=T_S M \big/ j(TS)$ is the quotient bundle, the bundle monomorphism j is defined by the decomposition $i_*=\rho\circ j$, $\rho: T_S M \longrightarrow TM$ is the natural imbedding, the canonical bundle morphism ν takes a vector $\zeta\in T_a M$, $a\in S$, to its equivalence class $\zeta+T_a S\in N_a S$. The mapping ν we shall call the *normal projection.* Local smooth sections of the bundles TS, $T_S M$ and NS on S we shall call *tangent fields, vector fields* and *normal fields,* respectively. We denote by $\mathcal{T}(\mathcal{O})$, $\mathcal{V}(\mathcal{O})$ and $\mathcal{N}(\mathcal{O})$ the corresponding linear spaces, where $\mathcal{O}$ is an open subset of the submanifold S. The space $\mathcal{T}(\mathcal{O})$ has the structure of the Lie algebra (see Section 0.2) but there are no such structures on the spaces $\mathcal{V}(\mathcal{O})$ and $\mathcal{N}(\mathcal{O})$.

In a defining chart $(W,\chi,\mathbb{R}^{I+J})$ on M a tangent field $\xi\in\mathcal{T}(U)$ has the representation

$$\xi = \sum_{i\in I} \xi^i(u)\, \frac{\partial}{\partial u^i},$$

a vector field $\zeta\in\mathcal{V}(U)$ has the representation

$$\zeta = \sum_{i\in I} \xi^i(u)\, \frac{\partial}{\partial u^i} + \sum_{j\in J} \eta^j(u)\, \frac{\partial}{\partial v^j}$$

and a normal field $\eta\in\mathcal{N}(U)$ has the representation

$$\eta = \sum_{j\in J} \eta^j(u)\, \frac{\partial}{\partial v^j},$$

where ξ^i and η^j are smooth functions on U.

Let Q be an open subset of M and $\mathcal{O}=S\cap Q$. Consider the restriction mapping $\rho^*:\mathcal{T}(Q)\longrightarrow\mathcal{V}(\mathcal{O})$. Its kernel $R\mathcal{T}(Q)=\operatorname{Ker}\rho^*$ is a Lie subalgebra but not an ideal of $\mathcal{T}(Q)$, so $\mathcal{V}(\mathcal{O})_{ext}=\mathcal{T}(Q)\Big/R\mathcal{T}(Q)$ is a linear space but not a Lie algebra. It is clear that the sequence of the linear spaces

$$0 \longleftarrow \mathcal{V}(\mathcal{O}) \xleftarrow{\ \rho^*\ } \mathcal{T}(Q) \longleftarrow R\mathcal{T}(Q) \longleftarrow 0$$

is locally exact (for example, in every defining chart on M). Hence, we can identify $\mathcal{V}(\mathcal{O})=\mathcal{V}(\mathcal{O})_{ext}$ at least locally. Further, let a vector field $Z\in\mathcal{T}(Q)$ then the restriction $\zeta=\rho^*Z\in\mathcal{T}(\mathcal{O})\subset\mathcal{V}(\mathcal{O})$ iff $Z:I\mathcal{C}^\infty(Q)\longrightarrow I\mathcal{C}^\infty(Q)$, i.e., if $Z(F)=0$ on $\mathcal{O}$ for all $F\in\mathcal{C}^\infty(Q)$, $F=0$ on $\mathcal{O}$.

Let $(W,\chi,\mathbb{R}^{I+J})$ be a defining chart on M and a vector field

$$Z = \sum_{i\in I} \Xi^i(u,v)\,\frac{\partial}{\partial u^i} + \sum_{j\in J} \mathrm{H}^j(u,v)\,\frac{\partial}{\partial v^j} \in \mathcal{T}(W).$$

Then the restriction

$$\zeta = \rho^*Z = \sum_{i\in I} \xi^i(u)\,\frac{\partial}{\partial u^i} + \sum_{j\in J} \eta^j(u)\,\frac{\partial}{\partial v^j} \in \mathcal{V}(U),$$

where $\xi^i(u)=\Xi^i(u,0)$ and $\eta^j(u)=\mathrm{H}^j(u,0)$.

Let T^*S, T^*_SM and N^*S be the dual bundles on S, where

$$T^*_SM = \bigcup_{a\in S} T^*_aM, \qquad N^*S = \bigcup_{a\in S} N^*_aS,$$

dual spaces

$$N^*_aS = \left\{\omega \in T^*_aM \ :\ \omega(\zeta) = 0 \text{ for all } \zeta \in T_aS\right\}$$

are polar to linear subspaces T_aS in T_aM. The dual sequence

$$o \longleftarrow T^*S \xleftarrow{\;\;j^*\;\;} T^*_SM \xleftarrow{\;\;\nu^*\;\;} N^*S \longleftarrow 0$$

is also exact, where j^* is the restriction and ν^* is the imbedding bundle morphisms on S. Local smooth sections of the bundles T^*S, T^*_SM and N^*S we shall call *tangent 1-forms*, *1-forms* and *normal 1-forms*, respectively, and we denote by $\Lambda(\mathcal{O})$, $\Omega(\mathcal{O})$ and $N\Omega(\mathcal{O})$ the corresponding linear spaces, where $\mathcal{O}$ is an open subset of the submanifold S.

In a defining chart $(W,\chi,\mathbb{R}^{I+J})$ on M a tangent 1-form $\varphi\in\Lambda(U)$ has the representation

$$\varphi = \sum_{i\in I} \varphi_i(u)\ du^i,$$

a 1-form $\omega \in \Omega(U)$ has the representation

$$\omega = \sum_{i\in I} \varphi_i(u)\ du^i + \sum_{j\in J} \psi_j(u)\ dv^j$$

and a normal 1-form $\psi \in N\Omega(U)$ has the representation

$$\psi = \sum_{j\in J} \psi_j(u)\ dv^j,$$

where φ_i and ψ_j are smooth functions on U.

Let Q be an open subset of M and $\mathcal{O}=S\cap Q$. Consider the restriction mapping $\rho^*:\Lambda(Q)\longrightarrow\Omega(\mathcal{O})$. Its kernel $R\Lambda(Q)=\mathrm{Ker}\rho^*$ is the linear subspace of $\Lambda(Q)$ and we denote by $\Omega(\mathcal{O}) \quad =\Lambda(Q)\big/R\Lambda(Q)$ the quotient space. It is clear that the sequence of the linear spaces

$$0 \longleftarrow \Omega(\mathcal{O}) \xleftarrow{\;\;\rho^*\;\;} \Lambda(Q) \longleftarrow R\Lambda(Q) \longleftarrow 0$$

is locally exact (for example, in every defining chart on M). Hence, we can identify $\Omega(\mathcal{O})=\Omega(\mathcal{O})_{ext}$ at least locally. In a defining chart $(W,\chi,\mathbb{R}^{I+J})$ on M a 1-form

$$\Omega = \sum_{i\in I} \Phi_i(u,v)\ du^i + \sum_{j\in J} \Psi_j(u,v)\ dv^j \in \Lambda(W)$$

has the restriction

$$\omega = \rho^*\Omega = \sum_{i\in I} \phi_i(u)\ du^i + \sum_{j\in J} \psi_j(u)\ dv^j \in \Omega(U),$$

where $\phi_i(u)=\Phi_i(u,0)$ and $\psi_j(u)=\Psi_j(u,0)$.

Let us denote by $\Lambda^q_S M$ the restriction of the bundle $\Lambda^q M$ on S, i.e.,

$$\Lambda^q_S M = \underset{a\in S}{\cup} \Lambda^q_a M,$$

$q\in\mathbb{N}$. Further, for a point $a\in S$ and $p,q\in\mathbb{N}$ let us denote by $N^p\Lambda^q_a M$ the linear space of all *p-normal q-covectors*, i.e., q-covectors $\omega\in\Lambda^q_a M$ such that $\omega(\zeta_1,\ldots,\zeta_q)=0$ if among vectors $\zeta_1,\ldots,\zeta_q\in T_aM$ there are at least $q+1-p$ tangent, i.e., belong to T_aS. By the definition, $N^0\Lambda^q_a M=\Lambda^q_a M$, $N^p\Lambda^q_a M\supset N^{p+1}\Lambda^q_a M$ if $p=1,\ldots,q$, and $N^p\Lambda^q_a M=0$ if $p>q$. Hence, we have the filtration

$$\Lambda^q_a M \supset N^1\Lambda^q_a M \supset \ldots \supset N^q\Lambda^q_a M \supset 0$$

for every $a\in S$ and $q\in\mathbb{N}$. Note that elements of the space $N^q\Lambda^q_a M$ can be considered as q-covectors on $N\ S$. The linear spaces $N^p\Lambda^q M$ give rise to the bundles

$$N^p\Lambda^q_S M = \underset{a\in S}{\cup} N^p\Lambda^q_a M$$

on S. Clear that the sequence of bundles on S

$$0 \longleftarrow \Lambda^q S \xleftarrow{\;j^*\;} \Lambda^q_S M \xleftarrow{\;\nu^*\;} N\Lambda^q_S M \longleftarrow 0$$

is exact, where $\Lambda^q_S M = N^0\Lambda^q_S M$, $N\Lambda^q_S M = N^1\Lambda^q_S M$, $q\in\mathbb{N}$. Local smooth sections of the bundles $\Lambda^q S$, $\Lambda^q_S M$ and $N^p\Lambda^q_S M$ we shall call *tangent q-forms*, *q-forms* and *p-normal q-forms* on S, respectively, and we denote by $\Lambda^q(\mathcal{O})$, $\Omega^q(\mathcal{O})$ and $N^p\Omega^q(\mathcal{O})$ the corresponding linear spaces, where $\mathcal{O}$ is an open subset of the submanifold S, $p,q\in\mathbb{N}$. Clear that there is the filtration

$$\Omega^q(\mathcal{O}) \supset N^1\Omega^q(\mathcal{O}) \;\ldots\; \supset N^q\Omega^q(\mathcal{O}) \supset 0$$

for every open $\mathcal{O}\subset S$, $q\in\mathbb{N}$.

For an open set $\mathcal{O}\subset M$ and $p\in\mathbb{N}$ let us denote by $N^p\Omega^*(\mathcal{O})$ the direct sum of all linear spaces $N^p\Omega^q(\mathcal{O})$, $q=0,1,2,\ldots$, i.e.

$$N^p\Omega^*(\mathcal{O}) = \sum_{q\ge p} N^p\Omega^q(\mathcal{O}).$$

In particular, $\Omega^*(\mathcal{O})=N^0\Omega^*(\mathcal{O})$ and $N\Omega^*(\mathcal{O})=N^1\Omega^*(\mathcal{O})$. The wedge product $\omega\wedge\theta\in N^{p+r}\Omega^{q+s}(\mathcal{O})$ of forms $\omega\in N^p\Omega^q(\mathcal{O})$ and $\theta\in N^r\Omega^s(\mathcal{O})$ is defined by the rule

$$\omega\wedge\theta(\zeta_1,\ldots,\zeta_{q+p}) = \mathrm{A.P.}\Big(\omega(\zeta_1,\ldots,\zeta_q)\theta(\zeta_{q+1},\ldots,\zeta_{q+p})\Big)$$

for all fields $\zeta_1,\ldots,\zeta_{q+s}\in V(\mathcal{O})$ (see Section 0.2). Hence, $\Omega^*(\mathcal{O})$ is the algebra, and $N^p\Omega^*(\mathcal{O})$ are its proper two-sided ideals for $p=1,2,\ldots$. The interior product $\zeta\rfloor\omega\in\Omega^{q-1}(\mathcal{O})$ of a vector field $\zeta\in V\mathcal{O})$ and a form $\omega\in\Omega^q(\mathcal{O})$ is defined by the rule

$$(\zeta \rfloor \omega)(\zeta_1,\dots,\zeta_{q-1}) = \omega(\zeta,\zeta_1,\dots,\zeta_{q-1})$$

for all fields $\zeta_1,\dots,\zeta_{q-1}\in \mathcal{V}(\mathcal{O})$, $q=1,2,\dots$ (see Section 0.2). If $\omega\in N^p\Omega^q(\mathcal{O})$ then

$$\zeta \rfloor \omega \in N^{p-1}\Omega^{q-1}(\mathcal{O}) \text{ when } \zeta \in \mathcal{V}(\mathcal{O})$$

and

$$\zeta \rfloor \omega \in N^{p}\Omega^{q-1}(\mathcal{O}) \text{ when } \zeta \in \mathcal{T}(\mathcal{O}),$$

where $p,q\in\{1,2,\dots\}$. The interior product acts as an anti-differentiation on the algebra $\Omega^*(\mathcal{O})$, i.e.,

$$\zeta \rfloor \left(\omega\wedge\theta\right) = \left(\zeta \rfloor \omega\right)\wedge\theta + (-1)^q\omega\wedge\left(\zeta \rfloor \theta\right)$$

for all forms $\omega\in\Omega^q(\mathcal{O})$ and $\theta\in\Omega^s(\mathcal{O})$. As we said above, the Lie bracket is not defined on the space $\mathcal{V}(\mathcal{O})$, hence, the exterior derivative and the Lie derivative cannot be defined on $\Omega^*(\mathcal{O})$ in any natural way.

Let Q be an open subset of M and $\mathcal{O}=S\cap Q$. Consider the restriction mapping $\rho^*:\Lambda^q(Q)\longrightarrow\Omega^q(\mathcal{O})$, $q\in\mathbb{N}$. Its kernel $R\Lambda^q(Q)=\mathrm{Ker}\rho^*$ is the linear subspace of $\Lambda^q(Q)$ and we denote by $\Omega^q(\mathcal{O})_{ext}=\Lambda^q(Q)\big/R\Lambda^q(Q)$ the quotient space. It is clear that the sequence of the linear spaces

$$0 \longleftarrow \Omega^q(\mathcal{O}) \xleftarrow{\ \rho^*\ } \Lambda^q(Q) \longleftarrow R\Lambda^q(Q) \longleftarrow 0$$

is locally exact (for example, in every defining chart on M). Hence, we can identify $\Omega(\mathcal{O})=\Omega(\mathcal{O})_{ext}$, at least locally.

0.5. Involutive Distributions

Let M be a smooth manifold. A distribution DTM on M is a smooth subbundle of a fixed finite rank of the tangent bundle TM. In other words, DTM is a correspondence

$$M \ni u \mapsto DT_uM \subset T_uM$$

with the following properties:

DT_uM is a subspace of the tangent space T_uM, $\dim DT_uM=m$, where a finite number m does not depend on a point $u \in M$;

for every point $a \in M$ there exist an open set $\mathcal{O}$, $a \in \mathcal{O} \subset M$, and vector fields $\zeta_1,\ldots,\zeta_m \in \mathcal{T}(\mathcal{O})$ such that vectors $\zeta_1(u),\ldots,\zeta_m(u)$ are basis vectors for the space DT_uM for every $u \in \mathcal{O}$.

In this case the number m is called the *dimension* of DTM and the vector fields are called the *basis fields* of DTM in $\mathcal{O}$.

Let DTM be a m-dimensional distribution on a smooth manifold M. A connected submanifold S of M is called an *integral submanifold* of DTM if $T_uS=DT_uM$ for every point $u \in S$. Clear that in this case $\dim S=m$. If the distribution DTM is defined by basis fields $\zeta_1,\ldots,\zeta_m$ and the submanifold S is defined by equations

$$S = \left\{ u \,:\, F^p(u) = 0,\ p \in P \right\}$$

then S is integral iff

$$\zeta_\mu(F^p) = 0 \text{ on } S \text{ for all } \mu = 1,\ldots,m \text{ and } p \in P.$$

For an open set $O \subset M$ let us denote by $D\mathcal{T}(O)$ the set of all local smooth sections of the subbundle DTM, i.e., a vector field $\zeta \in \mathcal{T}(O)$ belongs to $D\mathcal{T}(O)$ iff $\zeta(u) \in DT_u M$ for all $u \in O$. The distribution DTM is called *involutive* if for every open $O \subset M$ the linear subspace $D\mathcal{T}(O)$ is a Lie subalgebra of $\mathcal{T}(O)$, i.e., if $[\zeta,\eta] \in D\mathcal{T}(O)$ for all $\zeta,\eta \in D\mathcal{T}(O)$.

The standard way to specify a m-dimensional distribution DTM on a smooth manifold M is to define it as the *span* (i.e., as the linear envelope) of some linear-independent system of vector fields $\zeta_1,\dots,\zeta_m$ on M. Here,

$$DT_u M = \left\{ \zeta = \sum_{\mu=1}^{m} C^{\mu}(u)\ \zeta_{\mu}(u) \right\}, \quad u \in M,$$

where C^{μ} are arbitrary smooth functions on M. Thus defined distribution DTM will be involutive iff the fields $\zeta_1,\dots,\zeta_m$ *generate* a Lie algebra, i.e., iff

$$\left[\zeta_{\mu},\zeta_{\nu}\right] = \sum_{\lambda=1}^{m} C^{\lambda}_{\mu\nu}\ \zeta_{\lambda}, \quad \mu,\nu = 1,\dots,m,$$

where $C^{\lambda}_{\mu\nu}$ are smooth functions on M.

The dual way to specify a m-dimensional distribution DTM on a smooth manifold M is to define it as the common *null subbundle* of some system of 1-forms ω^k, $k \in K$, on M. In more detail,

$$DT_u M = \left\{ \zeta \in T_u M \ :\ \omega^k(\zeta) = 0,\ k \in K \right\}, \quad u \in M.$$

Here, one should assure that $\dim DT_u M = m$ for all $u \in M$. Thus defined distribution DTM will be involutive iff

$$d\omega^k(\zeta, \eta) = 0 \quad \text{for all } k \in K \text{ and } \zeta, \eta \in D\mathcal{T}(M).$$

The Frobenius' theorem states that if a smooth manifold M is finite-dimensional and a distribution DTM on M is involutive then through every point $a \in M$ there passes one and only one integral submanifold. Moreover, in this case for each point $a \in M$ there exists a *foliating chart* $(W, \varphi, \mathbb{R}^J)$, $a \in W$, with coordinates $w=(x,u)$, $x=(x^1,\ldots,x^m)$, $u=(u^i)$, $i \in I$, $J=\{1,\ldots,m\}+I$, such that the fields

$$\zeta_1 = \frac{\partial}{\partial x^1}, \ldots, \zeta_m = \frac{\partial}{\partial x^m}$$

are basis fields of DTM in W, where m is the dimension of DTM. In particular, a set $S \subset W$ is an integral submanifold iff

$$S = \left\{ x,u \ : \ u^i = c^i, \ i \in I \right\},$$

where c^i are some constants. In other words, integral submanifolds are *folios* in this chart. If the manifold M is infinite-dimensional or/and the distribution DTM is noninvolutive then the Frobenius' theorem is no longer true (see Section 0.6).

0.6. Discussion

We assume that the reader is acquainted with the foundations of the differential geometry in the finite-dimensional case (see, for example, Kobayashi and Nomizu [1], Sternberg [1]) and in the Banach case (see, for example, Bourbaki [1], Lang [1]). In the short review above we paid attention to notions and constructions which are essential to our further exposition or have different infinite-dimensional formulations. In the second case we prefer formulations convenient for our purposes.

Let us discuss some consequences of an infinite number of dimensions. Let an index set I be countable. The basic property of the space $\mathbb{R}^I$ in this case is that every compact subset of it has the empty interior. Hence, the only smooth function with compact support in $\mathbb{R}^I$ is zero. In particular, there is no partition of unity in the usual sense. Further, the space $\mathbb{R}^I$ is nonparacompact and this does not allow us to construct some substitution of a partition of unity, suitable for smooth functions of a finite number of variables. All this essentially restricts our considerations to a local situation covered by one chart.

Now, let us consider some model situations.

Example 0.6.1. The simplest example of an infinite-dimensional smooth manifold is the space $\mathbb{R}_{\mathbb{N}}$. Elements of this space are real sequences $\mathbf{u}=\{u\ ,u\ ,\dots\}=(u\ ;i\in\mathbb{N})$. The defining atlas consists of one chart with global coordinates u , $i\in\mathbb{N}$. The natural topology of $\mathbb{R}_{\mathbb{N}}$ is the Fréchet topology of the projective limit of the Euclidean spaces $\mathbb{R}_n$, $n\to\infty$.

For further convenience we slightly change the notation and denote the previous space by $\mathbb{R}\times\mathbb{R}_{\mathbb{N}}$. Its elements we denote by $\mathbf{w}=(x,\mathbf{u})$, $x\in\mathbb{R}$, $\mathbf{u}=(u_i)\in\mathbb{R}_{\mathbb{N}}$. Let us consider the vector field

$$D_x = \frac{\partial}{\partial x} + \sum_{i\in\mathbb{N}} u_{i+1} \frac{\partial}{\partial u_i}$$

on $\mathbb{R}^{1+\mathbb{N}}$. The linear mapping $D_x : \mathcal{C}^\infty(\mathbb{R}\times\mathbb{R}_{\mathbb{N}})\to\mathcal{C}^\infty(\mathbb{R}\times\mathbb{R}_{\mathbb{N}})$ has the kernel

$$\text{Ker } D_x = \left\{\varphi \in \mathcal{C}^\infty(\mathbb{R}\times\mathbb{R}_{\mathbb{N}}) \; : \; D_x\varphi = 0\right\} = \mathbb{R},$$

i.e., *the equation* $D_x\varphi=0$ *has only constant solutions.* Indeed, let $\varphi(\mathbf{w})=\varphi(x,\mathbf{u})=\varphi(x,u_0,\dots,u_p)$ be a smooth function on $\mathbb{R}\times\mathbb{R}_{\mathbb{N}}$ then

$$D_x\varphi = \varphi_x + \varphi_{u_0}u_1 + \varphi_{u_1}u_2 + \dots + \varphi_{u_p}u_{p+1} = \varphi_{u_p}u_{p+1} + \dots,$$

where omitted terms do not depend on u_{p+1}. Hence, the equality $D_x\varphi=0$ implies $\varphi_{u_p}=0$, so $\varphi=\varphi(x,u_0,\dots,u_{p-1})$. By induction, $\varphi=\varphi(x)$ and $D_x\varphi=\varphi_x=0$, i.e., $\varphi=const$.

Let us compare this result with the finite-dimensional case. Here, a first order partial differential equation (PDE)

$$A_1(x)\frac{\partial\varphi(x)}{\partial x^1} + \dots + A_n(x)\frac{\partial\varphi(x)}{\partial x^n} = 0,$$

where $A_1,\dots,A_n$ are smooth coefficients, $A_1^2(x)+\dots+A_n^2(x)>0$ for each $x\in\mathbb{R}^n$, locally has exactly $n-1$ functionally independent partial solutions and the general solution is an arbitrary function of these partial solutions (see, for example, Cartan [1]).

Now, let us try to integrate the vector field D_x, i.e., let us look for a local transformation group

$$x(t) = X(x,\mathbf{u},t), \qquad x(0) = x,$$
$$u_i(t) = U_i(x,\mathbf{u},t), \quad u_i(0) = u_i, \quad i \in \mathbb{N},$$

where X, U_i are smooth functions in an open subset of $\mathbb{R}\times\mathbb{R}_{\mathbb{N}}\times\mathbb{R}$. Such group generates the vector field D_x iff functions X, U_i satisfy to the infinite system of the ordinary differential equations:

$$\frac{dX}{dt} = 1, \qquad X(x,\mathbf{u},0) = x,$$
$$\frac{dU_i}{dt} = U_{i+1}, \quad U_i(x,\mathbf{u},0) = u_i, \quad i \in \mathbb{N}.$$

The first equation has the solution $x(t)=x+t$. The other imply $U_i=\varphi^{(i)}$, $i\in\mathbb{N}$, where $\varphi(x,\mathbf{u},t)$ is a smooth function in an open subset of $\mathbb{R}\times\mathbb{R}_{\mathbb{N}}\times\mathbb{R}$, $\varphi^{(i)}=\partial^i\varphi\big/\partial t^i$. The initial data take the form

$$U_i(x,\mathbf{u},0) = \varphi^{(i)}(x,\mathbf{u},0) = u_i, \quad i = 0,1,2,\ldots.$$

This clearly contradicts to the basic assumption that a smooth function depends only on a finite number of variables. Thus, *the vector field D_x cannot be integrated to a local transformation group on $\mathbb{R}\times\mathbb{R}_{\mathbb{N}}$.*

Let us consider further the 1-dimensional distribution on $\mathbb{R}\times\mathbb{R}_{\mathbb{N}}$ defined by the vector field D_x. A subset

$$S = \left\{x = X(t),\ u_i = U_i(t)\ ;\ t \in \mathbb{R},\ i \in \mathbb{N}\right\} \subset \mathbb{R}\times\mathbb{R}_{\mathbb{N}},$$

where X, U_i are smooth functions, will be a submanifold of $\mathbb{R}\times\mathbb{R}_{\mathbb{N}}$ if for each $t\in\mathbb{R}$ at least one of derivatives $X'(t)$, $U_i'(t)$ differs from zero. Suppose that this is the case. The submanifold S will be integral if the vectors $\zeta=(X'(t),\mathbf{U}'(t))$ and $D_x(X(t),\mathbf{U}(t))$ are collinear for all $t\in\mathbb{R}$, i.e., if

$$X'(t) = \lambda(t),$$

$$U_i'(t) = \lambda(t)U_{i+1}(t), \quad i \in \mathbb{N},$$

where a smooth function $\lambda(t)\neq 0$ for any $t\in\mathbb{R}$. Hence, without loss of generality we may assume $x=t$. Then we shall have

$$S = \left\{x,\mathbf{u} \;:\; u_i = U_i(x),\; i \in \mathbb{N}\right\},$$

and the vectors $\zeta=(1,\mathbf{U}'(x))$ and $D_x(x,\mathbf{U}(x))$ will be collinear if

$$U_i'(x) = U_{i+1}(x), \quad i \in \mathbb{N}.$$

This infinite system of the ordinary differential equations has the general solution $U_i(x)=\varphi^{(i)}(x)$, where φ is an arbitrary smooth function, $\varphi^{(i)}=d^i\varphi\big/dx^i$. Now, an integral submanifold

$$S = \left\{x,\mathbf{u} \;:\; u_i = \varphi^{(i)}(x),\; i \in \mathbb{N}\right\},$$

will pass through a point $\mathbf{c}=(a,\mathbf{b})\in\mathbb{R}\times\mathbb{R}_{\mathbb{N}}$ if

$$\varphi^{(i)}(a) = b_i, \quad i = 0,1,2,\dots .$$

One of the well-known results of the function theory is that there are infinitely many functions satisfying above conditions for any sequence (b_i). Thus, *through each point of* $\mathbb{R}\times\mathbb{R}_{\mathbb{N}}$ *there passes infinitely many submanifolds tangent to the vector field* D .

At last, let us try to find a chart on $\mathbb{R}\times\mathbb{R}_{\mathbb{N}}$ foliating the vector field D_x. Namely, let us look for a chart $(W,\chi,\mathbb{R}\times\mathbb{R}_{\mathbb{N}})$ with local coordinates $y\in\mathbb{R}$, $\mathbf{v}=(v_i)\in\mathbb{R}_{\mathbb{N}}$ such that

$$D_x = \frac{\partial}{\partial y}$$

in these coordinates. In this case the transition functions

$$y = Y(x,\mathbf{u}), \quad v_i = V_i(x,\mathbf{u}), \quad i \in \mathbb{N},$$

are smooth functions in an open subset of $\mathbb{R}\times\mathbb{R}_{\mathbb{N}}$ satisfying to the system of PDE:

$$D_x(Y) = 1,$$

$$D_x(V_i) = 0, \quad i \in \mathbb{N}.$$

The general solution of this system is

$$Y = x + a, \quad V_i = b_i, \quad i \in \mathbb{N},$$

where a, b_i are arbitrary constants. It is obvious that transition functions cannot be constant. Thus, *the foliation chart does not exist*.

It is easy to see that the origin of the difference between our case and the finite-dimensional case lies in the fact that by the definition a smooth function has a *finite order*, i.e., it depends on a finite number of variables. What are the reasons of such choice? The first reason is that there is no acceptable theory of functions of infinitely many variables (see, for example, Zharinov [1]). The second reason is that in the geometrical approach to PDE functions of a finite number of variables correspond to differential operators of a finite order and this is quite enough for traditional applications. Certainly, one needs functions of infinitely many variables for a geometrical study of *pseudodifferential* equations (differential equations of an infinite order) but at my knowledge nothing essential have been done in this direction yet. Thus, there is a wide and promising field of study for an interested reader.

In the frames of our definition of a smooth function it is convenient to formalize a notion of the order we used above. Namely, let an index set I be filtrated by its finite subsets I_n,

$$I_0 \subset I_1 \subset I_2 \subset \dots \subset I = \bigcup_{n=0}^{\infty} I_n,$$

then

$$\mathbb{R}^I = \lim_{n\to\infty} \operatorname{proj} \mathbb{R}^{I_n},$$

where the projective limit is taken with respect to projections

$$\mathbb{R}^{I_n} \ni u=(u^i; i\in I_n) \mapsto \pi^n_{n-1}(u)=(u^i; i\in I_{n-1}) \in \mathbb{R}^{I_{n-1}}.$$

By the definition, projections

$$\mathbb{R}^I \ni u=(u^i; i\in I) \mapsto \pi_n(u)=(u^i; i\in I_n) \in \mathbb{R}^{I_n}$$

are continuous for any $n\in\mathbb{N}$. Moreover, they are *open*, i.e., the image $\pi_n U$ is open in $\mathbb{R}^{I_n}$ for any open subset $U\subset\mathbb{R}^I$ and any $n\in\mathbb{N}$.

Let U be an open subset of $\mathbb{R}^I$. The algebra $\mathscr{C}^\infty(U)$ of all smooth functions in an open subset $U\subset\mathbb{R}^I$ (see Section 0.1) has the representation

$$\mathcal{C}^\infty(U) = \lim_{n\to\infty} \text{ind}\, \mathcal{C}^\infty(\pi_n U),$$

where the inductive limit is taken with respect to injections

$$\mathcal{C}^\infty(\pi_n U) \ni f(u) \mapsto (\pi_n^{n+1})^* f(u)=f(\pi_n^{n+1} u) \in \mathcal{C}^\infty(\pi_{n+1} U).$$

In particular, $f\in\mathcal{C}^\infty(U)$ iff there exist a number $n\in\mathbb{N}$ and a function $\phi\in\mathcal{C}^\infty(\pi_n U)$ such that $f(u)=\pi_n\phi(u)=\phi(\pi_n u)$. In the last case we identify the functions f and ϕ. In other words, the algebra $\mathcal{C}^\infty(U)$ is filtrated by subalgebras $\mathcal{C}^\infty(\pi_n U)$, i.e.,

$$\mathcal{C}^\infty(\pi_0 U) \subset \mathcal{C}^\infty(\pi_1 U) \subset \mathcal{C}^\infty(\pi_2 U) \subset \dots \subset \mathcal{C}^\infty(U) = \bigcup_{n=0}^{\infty} \mathcal{C}^\infty(\pi_n U).$$

The number $p\in\mathbb{N}$ is called the *order* of a function $f\in\mathcal{C}^\infty(U)$ if

$$f \in \mathcal{C}^\infty(\pi_p U)\backslash\mathcal{C}^\infty(\pi_{p-1} U).$$

An operator $F:\mathcal{C}^\infty(U)\longrightarrow\mathcal{C}^\infty(U)$ is said to be *subordinated* to the filtration of $\mathcal{C}^\infty(U)$ if the image $F(\mathcal{C}^\infty(\pi_n U))\subset\mathcal{C}^\infty(\pi_n U)$ for all $n\geq p$, $p=p(F)$ is some natural number. In this case the restrictions $F_n:\mathcal{C}^\infty(\pi_n U)\longrightarrow\mathcal{C}^\infty(\pi_n U)$, $n\geq p$, are defined. These restrictions completely determine F, hence the situation is in fact finite. Unfortunately, most operators arising in applications are nonsubordinated, they increase the order of a function they act on. For instance, the vector field D_x is nonsubordinated, it increases the order by one. Properties of nonsubordinated operators are rather specific and they are still waiting for a serious study.

Note that infinite-dimensional effects can appear in the subordinate case too.

Example 0.6.2. Let us consider the *1*-dimensional distribution on $\mathbb{R}\times\mathbb{R}_{\mathbb{N}}$ defined by the basis vector field

$$\zeta = \frac{\partial}{\partial x} + \sum_{i\in\mathbb{N}} 2xu_i^2 \frac{\partial}{\partial u_i}.$$

As above, one can check that every integral submanifold S passing through a point $\mathbf{c}=(a,\mathbf{b})$ has the form

$$S = \left\{x,\mathbf{u} \ : \ u_i = U_i(x), \ i \in \mathbb{N}\right\},$$

where $U_i(x)$ are smooth functions in a common interval $-\varepsilon<x-a<\varepsilon$, $\varepsilon>0$, satisfying the infinite Cauchy problem

$$U'_i(x) = 2xU_i^2(x), \ U_i(a) = b_i, \ i\in\mathbb{N}.$$

This Cauchy problem has the unique solution

$$U_i(x) = \frac{b_i}{1+(a^2-x^2)b_i}, \ i\in\mathbb{N}.$$

It is easy to see that the functions $U_i(x)$ have no common interval of definition if $\lim_{i\to\infty} b_i=+\infty$. Hence, in the last case there is no integral submanifold passing through the point $\mathbf{c}=(a,\mathbf{b})$.

To finish the discussion we present an infinite-dimensional manifold with a nontrivial atlas.

Example 0.6.3. Consider a real plane $\mathbb{V}^2$ and denote by Γ the set of all smooth curves (i.e., *1*-dimensional submanifolds) in $\mathbb{V}^2$. Two smooth curves $\gamma_1,\gamma_2\in\Gamma$ are said to have *contact of n-th order* in a point $z\in\gamma_1\cap\gamma_2$ if there are coordinates (ξ,ϑ) on $\mathbb{V}^2$ such that

$$z=(x,y), \ \gamma_1=\{\vartheta=f_1(\xi)\}, \ \gamma_2=\{\vartheta=f_2(\xi)\}$$

and

$$f_1^{(i)}(x) = f_2^{(i)}(x), \quad i=0,1,\dots,n,$$

where (i) denotes the i-th derivative. We say that curves γ_1 and γ_2 have *contact of ∞-th order* in a point z if they have contacts of all orders in this point. Let us denote by $\mathcal{P}$ the set of all pairs (z,γ), where $z\in\gamma\in\Gamma$. Two pairs (z_1,γ_1) and (z_2,γ_2) from $\mathcal{P}$ are called *equivalent* if $z_1=z_2=z$ and the curves γ_1 and γ_2 have contact of ∞-th order in the point z. This defines the equivalence relation in $\mathcal{P}$. The equivalence class $\mathbf{j}=\mathbf{j}(z,\gamma)$ of a pair $(z,\gamma)\in\mathcal{P}$ is called the *∞-jet* of the curve γ in the point z. We denote by $\mathbb{J}=\mathbb{J}^{\infty}(1,\mathbb{V}^2)$ the collection of all ∞-jets corresponding to all pairs $(z,\gamma)\in\mathcal{P}$. Now, let (ξ,ϑ) be some fixed coordinates on $\mathbb{V}^2$. Let us denote by $\mathbb{J}_\xi$ the set of all ∞-jets $\mathbf{j}(z,\gamma)$ for which $\gamma=\{\vartheta=f(\xi)\}$. Every ∞-jet $\mathbf{j}=\mathbf{j}(z,\gamma)\in\mathbb{J}_\xi$ has the unique representation $\mathbf{j}(x,\mathbf{u})$, $x\in\mathbb{R}$, $\mathbf{u}\in\mathbb{R}_{\mathbb{N}}$, where $z=(x,y)$, $u_i=f^{(i)}(x)$, $i\in\mathbb{N}$. Thus, we have the chart $(\mathbb{J}_\xi,\varphi,\mathbb{R}\times\mathbb{R}_{\mathbb{N}})$ with coordinates $(x,\mathbf{u})$, where the mapping φ is defined by the above procedure $\mathbf{j}\mapsto(x,\mathbf{u})$. The image of this mapping is the whole space $\mathbb{R}\times\mathbb{R}_{\mathbb{N}}$, because, as we said above, there exist functions with any prescribed derivatives of all orders in a point. In the same way, the chart $(\mathbb{J}_\vartheta,\psi,\mathbb{R}\times\mathbb{R}_{\mathbb{N}})$ with coordinates $(y,\mathbf{v})$ is defined. These two charts cover the whole space $\mathbb{J}$. The intersection $\mathbb{J}_\xi\cap\mathbb{J}_\vartheta$ consists of all ∞-jets $\mathbf{j}(z,\gamma)$ for which $\gamma=\{\vartheta=f(\xi)\}=\{\xi=g(\vartheta)\}$, where $f\circ g=id_\vartheta$ and $g\circ f=id_\xi$. The transition functions

$$v_j = (\psi\circ\varphi^{-1})_i(x,\mathbf{u}) \text{ and } u_j = (\varphi\circ\psi^{-1})_i(y,\mathbf{v})$$

are defined by the usual rule of the differentiation of inverse functions. For example,

$$y = f(x) = u_0,$$

$$v_0 = g(y) = x,$$

$$v_1 = \left.\frac{dg}{d\vartheta}\right|_{\vartheta=y} = 1\Big/\left.\frac{df}{d\xi}\right|_{\xi=x} = 1\Big/u_1,$$

$$v_2 = \left.\frac{d^2g}{d\vartheta^2}\right|_{\vartheta=y} = -\left.\left(\frac{d^2f}{d\xi^2}\Big/\left(\frac{df}{d\xi}\right)^3\right)\right|_{\xi=x} = -u_2\Big/u_1^3$$

and so on. Clear that

$$v_j = v_j(x,u_0,u_1,\ldots,u_j) \text{ and } u_i = u_i(y,v_0,v_1,\ldots,v_i)$$

are smooth functions on $\mathbb{J}_\xi \cap \mathbb{J}_\vartheta$. Hence, the space $\mathbb{J}$ is endowed with a structure of a smooth manifold with a defining atlas consisting of two charts $(\mathbb{J}_\xi, \varphi, \mathbb{R}\times\mathbb{R}_{\mathbb{N}})$ and $(\mathbb{J}_\vartheta, \psi, \mathbb{R}\times\mathbb{R}_{\mathbb{N}})$. The rule $\mathbb{J}\ni \mathbf{j}(z,\gamma)\mapsto z\in\mathbb{V}^2$ defines the projection $\pi: \mathbb{J}\longrightarrow\mathbb{V}^2$.

CHAPTER I. INTERNAL GEOMETRY OF PDE

1. Differential Manifolds

A *differential manifold* is defined to be a pair $\mathcal{M}=(M,CTM)$, where M is a smooth manifold and CTM is an involutive distribution on M. In this case the distribution CTM is called the *Cartan distribution*, the dimension of CTM is called the *Cartan dimension* of the differential manifold $\mathcal{M}$. Local smooth sections of the subbundle CTM are called *Cartan fields*. Note that the Cartan dimension of $\mathcal{M}$ is always finite and the dimension of M is in the general case infinite. If M is finite-dimensional then the differential manifold $\mathcal{M}$ is called *maximally overdetermined* or *holonomic* (cf. Tsujishita [1]). In our context finite-dimensional manifolds arise in the geometrical study of systems of ordinary differential equations and of a narrow class of systems of partial differential equations (PDE). The Cartan dimension coincides with the number of independent variables of the background system of differential equations.

Let us consider several examples.

Example 1.1. Let

$$u \quad = f(t,u)$$

be an ordinary differential equation ($u\ =f$, for brevity), where t is an independent variable, u is a dependent variable, $u_t=du\big/dt$ and f is a smooth function in some open set $\mathcal{O}\subset\mathbb{R}^2$. Remind that a smooth function $s(t)$, defined in an interval of $\mathbb{R}^1$, is called a *local solution* of the equation $u_t=f$ if the substitution

$$u = s(t),\ u_t = \frac{ds(t)}{dt}$$

turns it into an identity. Now, let $M=\mathcal{O}$ be an open set with the global coordinates t,u, and let us define the Cartan distribution CTM by the basis vector field

$$\partial_t = \frac{\partial}{\partial t} + f(t,u)\,\frac{\partial}{\partial u}$$

(here the Cartan dimension $\dim CTM=1$). A curve

$$\gamma = \left\{u = s(t)\right\} \subset \mathcal{O},$$

where $s(t)$ is a smooth function, will be an integral curve of thus obtained differential manifold $\mathcal{M}$ iff

$$\partial_t\left(u - s(t)\right) = 0 \text{ on } \gamma,$$

i.e., iff

$$f(t,s(t)) = \frac{ds(t)}{dt}$$

for all t in a common domain, i.e., iff $s(t)$ is a local solution of the equation $u_t=f$. This property establishes the connection between the original differential equation and the corresponding differential manifold.

Example 1.2. In the same way, let

$$u_t^\alpha = f^\alpha(t,u^1,\dots,u^a), \quad \alpha = 1,\dots,a,$$

be a system of ordinary differential equations ($u_t=f$), where f^α are smooth functions in some open set $\mathcal{O}\subset\mathbb{R}^{1+a}$. On the manifold $M=\mathcal{O}$ with the global coordinates $t,u^1,\dots,u^a$ consider the Cartan distribution CTM defined by the basis vector field

$$\partial_t = \frac{\partial}{\partial t} + \sum_{\alpha=1}^{a} f^\alpha \frac{\partial}{\partial u^\alpha}$$

(here $\dim M=a+1$ and $\dim CTM=1$). As before, a curve

$$\gamma = \left\{u = s(t)\right\} \subset \mathcal{O},$$

where $s(t)=(s^\alpha(t):\alpha=1,\dots,a)$ is a smooth function, will be an integral curve of the differential manifold $\mathcal{M}=(M,CTM)$ iff $s(t)$ is a local solution of the equation $u_t=f$.

Example 1.3. Let

$$u_{t^\mu} = f_\mu(t,u), \quad \mu = 1,\dots,m,$$

be a maximally overdetermined system of partial differential equations, where u is a dependent variable, $t=(t^1,\dots,t^m)$ are independent variables, $u_{t^\mu}=\partial u\big/\partial t^\mu$, f_μ are smooth functions in an open set $\mathcal{O}\subset\mathbb{R}^{m+1}$. On the manifold $M=\mathcal{O}$ with the global coordinates t,u let us define the Cartan distribution CTM by the basis vector fields

$$\partial_{t^\mu} = \frac{\partial}{\partial t^\mu} + f_\mu(t,u)\frac{\partial}{\partial u}, \quad \mu = 1,\ldots,m$$

(here $\dim M=m+1$ and $\dim CTM=m$). The distribution CTM will be involutive iff

$$\left[\partial_{t^\mu}, \partial_{t^\nu}\right] = 0, \quad \mu,\nu = 1,\ldots,m,$$

i.e., iff

$$\partial_{t^\mu} f_\nu = \partial_{t^\nu} f_\mu, \quad \mu,\nu = 1,\ldots,m,$$

i.e., iff

$$(f_\nu)_{t^\mu} + (f_\nu)_u f_\mu = (f_\mu)_{t^\nu} + (f_\mu)_u f_\nu, \quad \mu,\nu = 1,\ldots,m.$$

The last condition is the condition of the *formal integrability* of our system (cf. Pommaret [1]). We assume that functions f_μ satisfy this condition. Again, a submanifold

$$S = \left\{u = s(t)\right\} \subset \mathcal{O},$$

where $s(t^1,\ldots,t^m)$ is a smooth function in some open subset of $\mathbb{R}^m$, will be an integral submanifold of $\mathcal{M}=(M,CTM)$ iff

$$\partial_{t^\mu}\left(u - s(t)\right) = 0 \text{ on } S \text{ for } \mu = 1,\ldots,m,$$

i.e., iff

$$f_\mu(t,s(t)) = \frac{\partial s(t)}{\partial t^\mu}, \quad \mu = 1,\ldots,m,$$

for all t in a common domain, i.e., iff $s(t)$ is a local solution of the system $u_{t^\mu}=f_\mu$.

Example 1.4. Let

$$u_t = f(t,x,u,u_x,\dots,u_{x\dots x})$$

be an evolution equation ($u_t=f$), where u is a dependent variable, t, x are independent variables,

$$u_t = \frac{\partial u}{\partial t}, \quad u_x = \frac{\partial u}{\partial x}, \quad \dots, \quad u_{x\dots x} = \frac{\partial^p u}{\partial x^p},$$

p is the order of the equation, f is a smooth function in some open set $\mathcal{O}\subset\mathbb{R}^2\times\mathbb{R}_{1+p}$. Consider the space $\mathbb{R}^2\times\mathbb{R}_{\mathbb{N}}$ with elements $t,x,\mathbf{u}$, where $t\in\mathbb{R}$, $x\in\mathbb{R}$, $\mathbf{u}=(u_0,u_1,\dots)\in\mathbb{R}_{\mathbb{N}}$ (cf. the space $\mathbb{R}\times\mathbb{R}_{\mathbb{N}}$ in Section 0.6), and let $\pi_n:\mathbb{R}^2\times\mathbb{R}_{\mathbb{N}}\to\mathbb{R}^2\times\mathbb{R}_{1+n}$ be the projection

$$\mathbb{R}^2\times\mathbb{R}_{\mathbb{N}} \ni (t,x,u_0,u_1,\dots) \mapsto (t,x,u_0,\dots,u_n) \in \mathbb{R}^2\times\mathbb{R}_{1+n}$$

for $n\in\mathbb{N}$ (see Section 0.6). Let us denote by M the inverse image $\pi_p^{-1}(\mathcal{O})\subset\mathbb{R}^2\times\mathbb{R}_{\mathbb{N}}$, so

$$M = \left\{(t,x,u_0,u_1,\dots) : (t,x,u_0,\dots,u_p) \in \mathcal{O}\right\}$$

is the infinite-dimensional manifold with the global coordinates $t,x,\mathbf{u}$. The Cartan distribution CTM on M is defined by two basis vector fields

$$\partial_x = \frac{\partial}{\partial x} + \sum_{i=0}^{\infty} u_{i+1} \frac{\partial}{\partial u_i},$$

$$\partial_t = \frac{\partial}{\partial t} + \sum_{i=0}^{\infty} f_i \frac{\partial}{\partial u_i},$$

where $f_0=f(t,x,u_0,u_1,\dots,u_p)$ is the smooth function on M defined by our evolution equation $u_t=f$ and $f_i=\partial_x^i f$ for $i\in\mathbb{N}$. It is easy to check that $[\partial_x,\partial_t]=0$, hence, the distribution CTM is involutive. Now, a 2-dimensional submanifold

$$S = \left\{\mathbf{u} = \mathbf{s}(t,x)\right\} \subset M$$

(i.e., $u_i=s_i(t,x)$, $i\in\mathbb{N}$, where $s_i(t,x)$ are smooth functions) will be an integral submanifold of $\mathcal{M}=(M,CTM)$ iff

$$\partial_x\left(u_i - s_i(t,x)\right) = 0 \quad \text{and} \quad \partial_t\left(u_i - s_i(t,x)\right) = 0$$

on S for all $i\in\mathbb{N}$. The first set of equations yields

$$s_{i+1} = \frac{\partial s_i}{\partial x}, \text{ hence } s_i = \frac{\partial^i s}{\partial x^i}, \text{ for all } i \in \mathbb{N},$$

where $s(t,x)$ is a smooth function. Let us assume that this is the case. Then

$$\left(\partial_x f(t,x,\mathbf{u})\right)\Big|_{\mathbf{u}=\mathbf{s}(t,x)} = \frac{\partial}{\partial x}\left(f(t,x,\mathbf{u})\Big|_{\mathbf{u}=\mathbf{s}(t,x)}\right)$$

and the second set of equations gives

$$\left(\partial_x^i f\right)\Big|_{\mathbf{u}=\mathbf{s}(t,x)} - \frac{\partial}{\partial t}\left(\frac{\partial^i s(t,x)}{\partial x^i}\right) = 0$$

i.e.,

$$\left(\frac{\partial}{\partial x}\right)^i\left(f(t,x,\mathbf{s}(t,x)) - \frac{\partial s(t,x)}{\partial t}\right) = 0$$

for all $i\in\mathbb{N}$. The last system is equivalent to the original evolution equation

$$\frac{\partial s(t,x)}{\partial t} = f(t,x,\mathbf{s}(t,x)).$$

Thus, the submanifold S is integral iff $s_i=\partial^i s\big/\partial x^i$, where $s(t,x)$ is a local solution of the equation $u_t=f$.

Example 1.5. Let $M=\mathbb{R}\times\mathbb{R}_{\mathbb{N}}$ be the manifold with global coordinates x, $\mathbf{u}=(u_i;i\in\mathbb{N})$ and let us define the Cartan distribution $DTM=D_x$ by the vector field

$$D_x = \frac{\partial}{\partial x} + \sum_{i\in\mathbb{N}} u_{i+1}\frac{\partial}{\partial u_i}.$$

The differential manifold $\mathcal{M}=(\mathbb{R}\times\mathbb{R}_{\mathbb{N}},D_x)$ has integral curves

$$\gamma = \left\{x,\mathbf{u}\ ;\ u_i = \varphi^{(i)},\ i\in\mathbb{N}\right\},$$

where φ is an arbitrary smooth function (see Example 0.6.1). In a more general way, integral curves have the parameterization

$$\gamma = \left\{x = X(t),\ u_i = \left(\frac{1}{X'}\frac{d}{dt}\right)^i U,\ i\in\mathbb{N}\right\},$$

where $X,U(t)$ are arbitrary smooth functions of a parameter t, the derivative $X'\neq 0$.

Example 1.6. Let $\mathbb{J}=\mathbb{J}^\infty(1,\mathbb{V}^2)$ be the jet manifold from Example 0.6.3. For a jet $\mathbf{j}(z,\gamma)\in\mathbb{J}$, where γ is a smooth curve in $\mathbb{V}^2$, $z\in\gamma$, let us define the lifting mapping

$$\gamma \ni s \mapsto \mathbf{j}(s,\gamma) \in \mathbb{J}.$$

The lift $\mathbf{g}=\{\mathbf{j}(s,\gamma):s\in\gamma\}$ is the smooth curve in $\mathbb{J}$. The original jet $\mathbf{j}$ belongs to this curve and we put

$$CT_{\mathbf{j}}\mathbb{J} = T_{\mathbf{j}}\mathbf{g} \subset T_{\mathbf{j}}\mathbb{J}.$$

A parameterization $\gamma=\gamma(t)$ induces the parameterization $\mathbf{g}=\mathbf{g}(t)$. In this case the differentiation

$$\mathscr{C}^{\infty}(\{\mathbf{j}\}) \ni \Phi \mapsto D(\Phi) = \left[\frac{d}{dt}\Phi(\mathbf{g}(t))\right]\bigg|_{t=0} \in \mathbb{R}$$

will be a basis vector of $CT_{\mathbf{j}}\mathbb{J}$. In the chart $(\mathbb{J}_{\xi}, \varphi, \mathbb{R}\times\mathbb{R}_{\mathbb{N}})$ the jet $\mathbf{j}$ has coordinates $(x, \mathbf{u})$ and the curve γ can be parameterized

$$\gamma(t) = \left\{\xi = x+t, \vartheta = f(x+t)\right\}$$

then

$$\mathbf{g}(t) = \left\{x(t) = x+t,\ u_i(t) = f^{(i)}(x+t),\ i \in \mathbb{N}\right\}$$

$(x=x(0)$ and $\mathbf{u}=\mathbf{u}(0))$. Now, let $\Phi(x,\mathbf{u})\in\mathscr{C}^{\infty}(\mathcal{O})$, where $\mathcal{O}$ is an open subset of $\mathbb{J}_{\xi}$, then

$$\begin{aligned} D(\Phi)(x,\mathbf{u}) &= \left[\frac{d}{dt}\Phi(x(t),\mathbf{u}(t))\right]\bigg|_{t=0} \\ &= \frac{\partial\Phi(x,\mathbf{u})}{\partial x} + \sum_{i=0}^{\infty} u_{i+1}\frac{\partial\Phi(x,\mathbf{u})}{\partial u_i}. \end{aligned}$$

Hence, in the chart $(\mathbb{J}_{\xi}, \varphi, \mathbb{R}\times\mathbb{R}_{\mathbb{N}})$ the distribution $CT\mathbb{J}$ is defined by the basis field D_x from Example 0.6.1. In the chart $(\mathbb{J}_{\vartheta}, \varphi, \mathbb{R}\times\mathbb{R}_{\mathbb{N}})$ $CT\mathbb{J}$ is defined by the analogous field D_y. In the intersection $\mathbb{J}_{\xi}\cap\mathbb{J}_{\vartheta}$ there is the transition formula

$$D_x = \left(D_x y\right)\frac{\partial}{\partial y} + \left(D_x v_0\right)\frac{\partial}{\partial v_0} + \left(D_x v_1\right)\frac{\partial}{\partial v_1} + \ldots = \frac{1}{v_1} D_y,$$

because

$$D_x y = D_x u_0 = u_1 = \frac{1}{v_1}, \quad D_x v_0 = D_x x = 1 = \frac{v_1}{v_1},$$

$$D_x v_1 = D_x \frac{1}{u_1} = -\frac{u_2}{u_1^2} = \frac{v_2}{v_1}, \ldots .$$

According to Example 0.6.1, a curve $\mathbf{g}\subset\mathbb{J}_\xi$ will be integral iff it has the form

$$\mathbf{g} = \left\{x,\mathbf{u} \; : \; u_i = f^{(i)}(x), \;\; i \in \mathbb{N}\right\},$$

where f is an arbitrary smooth function. In other words, $\mathbf{g}$ is the lift of a curve $\gamma=\{\vartheta=f(\xi)\}\subset\mathbb{V}^2$. Thus, locally, integral curves of the differential manifold $(\mathbb{J},CT\mathbb{J})$ are just lifts of smooth curves in $\mathbb{V}^2$. Globally, the projection $\pi(\mathbf{g})$ of an integral curve $\mathbf{g}\subset\mathbb{J}$ into $\mathbb{V}^2$ is a smooth curve with possible intersections.

Let us derive some elementary properties of differential manifolds.

Let $\mathcal{M}=(M,CTM)$ be a differential manifold. Let $(U,\varphi,\mathbb{R}^I)$ be a chart on M and $\zeta=(\zeta_1,\ldots,\zeta_m)$ be basis vector fields of CTM on U, where m is the Cartan dimension of $\mathcal{M}$, then the four $(U,\varphi,\mathbb{R}^I,\zeta)$ is called a *chart on* $\mathcal{M}$. A chart $(U,\varphi,\mathbb{R}^{m+I},\partial)$ on $\mathcal{M}$ with coordinates x^μ, u^i, $\mu=1,\ldots,m$, $i\in I$, is called *standard* if basis vector fields ∂_μ have the form

$$\partial_\mu = \frac{\partial}{\partial x^\mu} + \sum_{i\in I} A^i_\mu(x,u)\,\frac{\partial}{\partial u^i}, \qquad \mu = 1,\ldots,m.$$

In this case the distribution CTM is involutive iff

$$\left[\partial_\mu,\partial_\nu\right] = 0, \qquad \mu,\nu = 1,\ldots,m,$$

i.e., iff

$$\partial_\mu A^i_\nu = \partial_\nu A^i_\mu, \qquad \mu,\nu = 1,\ldots,m, \;\; i \in I.$$

The basis vector fields ∂_μ are called *standard basis fields*.

Proposition 1.7. *Every differential manifold has a standard atlas* (i.e., an atlas consisting of standard charts).

Proof. Let $\mathcal{M}=(M,CTM)$ be a differential manifold and a point $c\in M$. Let $(U,\varphi,\mathbb{R}^J,\zeta)$ be a chart on $\mathcal{M}$, $c\in U$, with coordinates u^i, $i\in J$, and basis vector fields

$$\zeta_\mu = \sum_{i\in J} \zeta_\mu^i(u) \frac{\partial}{\partial u^i}, \quad \mu = 1,\dots,m,$$

where m is the Cartan dimension of $\mathcal{M}$. By the definition,

$$\operatorname{rank} \left\| \zeta_\mu^i(u) \right\| = m \text{ on } U.$$

Hence, there are indices $i_1,\dots,i_m\in J$ such that

$$\det \left\| \zeta_\mu^{i_\nu}(u) \right\| \neq 0 \text{ on } V,$$

where V is an open set, $c\in V\subset U$. Now, let $x^1=u^{i_1},\dots,x^m=u^{i_m}$, $I=J\setminus\{i_1,\dots,i_m\}$ and

$$\partial_\mu = \sum_{\nu=1}^{m} S_\mu^\nu \zeta_\nu = \frac{\partial}{\partial x^\mu} + \sum_{i\in I} A_\mu^i(x,u) \frac{\partial}{\partial u^i}, \quad \mu = 1,\dots,m,$$

where S_μ^ν is the inverse matrix to $\zeta_\nu^{i_\lambda}$, so

$$\sum_{\nu=1}^{m} S_\mu^\nu(u)\zeta_\nu^{i_\lambda}(u) = \delta_\mu^\lambda \text{ and } A_\mu^i(x,u) = \sum_{\nu=1}^{m} S_\mu^\nu(u)\zeta_\nu^i(u),$$

$\mu,\lambda=1,\dots,m$, $i\in I$. Thus, for an arbitrary point $c\in M$ there exists a standard chart, containing this point and compatible with all charts from the defining atlas of the smooth manifold M. This proves the assertion.

Let $\mathcal{M}=(M,CTM)$ be a differential manifold and let $(U,\varphi,\mathbb{R}^{m+I},\partial)$ be a standard chart on $\mathcal{M}$.

Proposition 1.8. *A m-dimensional smooth submanifold $S\subset U$ is integral iff it has a local parameterization*

$$S = \left\{x,u \;:\; u^i = s^i(x),\ i \in I\right\},$$

where smooth functions $s^i(x)$ satisfy the contact equations

$$\frac{\partial s^i}{\partial x^\mu} = A^i_\mu(x,s), \quad \mu = 1,\dots,m,\ i \in I.$$

Proof. Let S be a m-dimensional submanifold of M, $S\subset U$, and a point $c\in S$. Then in some open set V, $c\in V\subset U$, S has a parameterization

$$S = \left\{x^\mu = X^\mu(t),\ u^i = U^i(t)\right\},$$

where $X^\mu(t)$, $U^i(t)$ are smooth functions, $t\in O\subset\mathbb{R}^m$, $\mu=1,\dots,m$, $i\in I$, such that

$$\operatorname{rank}\left\|\frac{\partial X^\mu}{\partial t^\nu},\frac{\partial U^i}{\partial t^\nu}\right\| = m \quad \text{on } O.$$

This submanifold will be integral iff

$$\left(\frac{\partial}{\partial t^\nu}\right)_* = \sum_{\mu=1}^m \frac{\partial X^\mu}{\partial t^\nu}\frac{\partial}{\partial x^\mu} + \sum_{i\in I}\frac{\partial U^i}{\partial t^\nu}\frac{\partial}{\partial u^i} = \sum_{\mu=1}^m R^\mu_\nu\, \partial_\mu \quad \text{on } S,$$

where $\nu=1,\dots,m$ and $\det\left\|R^\mu_\nu\right\|\neq 0$. In our case,

$$\partial_\mu = \frac{\partial}{\partial x^\mu} + \sum_{i\in I} A^i_\mu(x,u)\,\frac{\partial}{\partial u^i}, \quad \mu = 1,\dots,m,$$

hence,

$$\frac{\partial x^\mu}{\partial t^\nu} = R^\mu_\nu, \quad \mu,\nu = 1,\dots,m,$$

and we can choose $t^\mu = x^\mu$, $\mu = 1,\dots,m$. Then S will take the form

$$S = \left\{x,u \;:\; u^i = s^i(x),\ i \in I\right\}.$$

It will be integral iff

$$\partial_\mu\left(u^i - s^i(x)\right) = 0 \quad \text{on } S,\ \mu = 1,\dots,m,\ i \in I,$$

i.e., iff

$$A^i_\mu(x,u)\Big|_{u=s(x)} = \frac{\partial s^i(x)}{\partial x^\mu}, \quad \mu = 1,\dots,m,\ i \in I,$$

i.e., iff smooth functions $s^i(x)$ satisfy the contact equations.

Proposition 1.9. *A m-dimensional submanifold* $S=\{u=s(x)\}\subset U$ (see Proposition 1.8) *is integral iff*

$$\frac{\partial}{\partial x^\mu}\left(\Phi(x,u)\Big|_{u=s(x)}\right) = \left(\partial_\mu \Phi(x,u)\right)\Big|_{u=s(x)}, \quad \mu = 1,\dots,m,$$

for any smooth function Φ *on* U.

This characteristic property allows us to call the basis fields ∂_μ the *total differentiations*.

Proof. First, let S be the integral submanifold and Φ be a smooth function on U. Then, by the chain rule,

$$\frac{\partial}{\partial x^\mu}\left(\Phi(x,u)\Big|_{u=s(x)}\right) = \left(\frac{\partial\Phi}{\partial x^\mu} + \sum_{i\in I} \frac{\partial\Phi}{\partial u^i}\frac{\partial s^i}{\partial x^\mu}\right)\Bigg|_{u=s(x)}$$

$$= \left(\frac{\partial\Phi}{\partial x^\mu} + \sum_{i\in I} \frac{\partial\Phi}{\partial u^i} A^i_\mu\right)\Bigg|_{u=s(x)} = \left(\partial_\mu\Phi(x,u)\right)\Big|_{u=s(x)},$$

by Proposition 1.8. On the other side, let the submanifold S satisfy the above substitution formula for every smooth function Φ on U. Let us fix an index $i\in I$ and take $\Phi=u^i$. Then

$$\Phi(x,u)\Big|_{u=s(x)} = s^i(x), \quad \partial_\mu\Phi(x,u) = A^i_\mu(x,u)$$

and the substitution formula turns into the contact equations for $s^i(x)$. Hence, by Proposition 1.8, the submanifold S is integral.

Proposition 1.10. *Every vector field $\zeta\in\mathcal{T}(U)$ has the unique representation*

$$\zeta = \sum_{\mu=1}^{m} \xi^\mu(x,u)\,\partial_\mu + \sum_{i\in I} \eta^i(x,u)\frac{\partial}{\partial u^i},$$

where ξ^μ,η^i are smooth functions on U.

Proof. Let ζ be a vector field on U, then

$$\zeta = \sum_{\mu=1}^{m} \zeta^\mu \frac{\partial}{\partial x^\mu} + \sum_{i\in I} \zeta^i \frac{\partial}{\partial u^i}$$

$$= \sum_{\mu=1}^{m} \zeta^\mu\left(\frac{\partial}{\partial x^\mu} + \sum_{i\in I} A^i_\mu \frac{\partial}{\partial u^i}\right) + \sum_{i\in I}\left(\zeta^i - \sum_{\mu=1}^{m}\zeta^\mu A^i_\mu\right)\frac{\partial}{\partial u^i}$$

and we get the desired unique representation with

$$\xi^\mu = \zeta^\mu, \quad \eta^i = \zeta^i - \sum_{\mu=1}^{m}\zeta^\mu A^i_\mu.$$

Consider *1*-forms on U

$$\delta u^i = du^i - \sum_{\mu=1}^{m} A^i_\mu(x,u)\ dx^\mu, \quad i \in I,$$

where $A^i_\mu(x,u)$ are components of the basis fields ∂_μ.

Proposition 1.11. *Every 1-form $\omega\in\Lambda(U)$ has the unique representation*

$$\omega = \sum_{\mu=1}^{m} \varphi_\mu(x,u)\ dx^m + \sum_{i\in I} \chi_i(x,u)\ \delta u^i,$$

where φ_μ, χ_i are smooth functions on U.

Proof. Let ω be a *1*-form on U, then

$$\begin{aligned}\omega &= \sum_{\mu=1}^{m} \omega_\mu\ dx^\mu + \sum_{i\in I} \omega_i\ du^i \\ &= \sum_{\mu=1}^{m} \left(\omega_\mu + \sum_{i\in I} \omega_i\ A^i_\mu\right)\ dx^\mu + \sum_{i\in I} \omega_i\ \left(du^i - \sum_{\mu=1}^{m} A^i_\mu\ dx^\mu\right)\end{aligned}$$

and we get the desired unique representation with

$$\varphi_\mu = \omega_\mu + \sum_{i\in I} \omega_i\ A^i_\mu, \quad \chi_i = \omega_i.$$

Proposition 1.12. *The common null subbundle of the 1-forms δu^i, $i\in I$, is the restriction of the distribution DTM on U.*

Proof. Let a point $u\in U$ and a vector $\zeta\in T_uM$. Then

$$\zeta = \sum_{\mu=1}^{m} \xi^\mu\ \partial_\mu + \sum_{i\in I} \eta^i\ \frac{\partial}{\partial u^i}$$

(see Proposition 1.10) and $\delta u^i(\zeta) = \eta^i$, because $\delta^i(\partial_\mu) = 0$ for all $i \in I$, $\mu = 1,\dots,m$. Hence, $\delta^i(\zeta)=0$ for all $i\in I$ iff

$$\zeta = \sum_{\mu=1}^{m} \xi^{\mu} \partial_{\mu},$$

which is what was required.

Proposition 1.13. *Let Φ be a smooth function on U. Then*

$$d\Phi = \sum_{\mu=1}^{m} \partial_{\mu}\Phi \, dx^{\mu} + \sum_{i\in I} \frac{\partial\Phi}{\partial u^{i}} \delta u^{i}.$$

Proof. We have,

$$d\Phi = \sum_{\mu=1}^{m} \frac{\partial\Phi}{\partial x^{\mu}} dx^{\mu} + \sum_{i\in I} \frac{\partial\Phi}{\partial u^{i}} du^{i}$$

$$= \sum_{\mu=1}^{m} \left(\frac{\partial\Phi}{\partial x^{\mu}} + \sum_{i\in I} A^{i}_{\mu} \frac{\partial\Phi}{\partial u^{i}}\right) dx^{\mu} + \sum_{i\in I} \frac{\partial\Phi}{\partial u^{i}} \left(du^{i} - \sum_{\mu=1}^{m} A^{i}_{\mu} dx^{\mu}\right)$$

$$= \sum_{\mu=1}^{m} \partial_{\mu}\Phi \, dx^{\mu} + \sum_{i\in I} \frac{\partial\Phi}{\partial u^{i}} \delta u^{i}.$$

Proposition 1.14. *The exterior derivatives*

$$d\delta u^{i} = - \sum_{\nu=1}^{m} \sum_{j\in I} \frac{\partial A^{i}_{\nu}}{\partial u^{j}} \delta u^{j} \wedge dx^{\nu}, \quad i \in I.$$

In particular, $d\delta u^{i}(\zeta,\eta)=0$ for all $i\in I$ and $\zeta,\eta\in C\mathcal{T}(U)$.

Remind that $C\mathcal{T}(O)$ is the Lie algebra of Cartan fields on an open set $O\subset M$ (cf. Section 0.5).

Proof. Indeed, by Proposition 1.13,

$$d\delta u^i = -\sum_{\nu=1}^{m} dA^i_\nu \wedge dx^\nu$$

$$= -\sum_{\nu=1}^{m}\left(\sum_{\mu=1}^{m}\partial_\mu A^i_\nu\, dx^\mu + \sum_{j\in I}\frac{\partial A^i_\nu}{\partial u^j}\,\delta u^j\right)\wedge dx^\nu$$

$$= -\sum_{\nu=1}^{m}\sum_{j\in I}\frac{\partial A^i_\nu}{\partial u^j}\,\delta u^j\wedge dx^\nu - \frac{1}{2}\sum_{\mu,\nu=1}^{m}\left(\partial_\mu A^i_\nu - \partial_\nu A^i_\mu\right)\, dx^\mu\wedge dx^\nu$$

$$= -\sum_{\nu=1}^{m}\sum_{j\in I}\frac{\partial A^i_\nu}{\partial u^j}\,\delta u^j\wedge dx^\nu,$$

because $\partial_\mu A^i_\nu - \partial_\nu A^i_\mu = 0$ by the involutivity (see the definition of standard charts above).

If a differential manifold $\mathcal{M}=(M, CTM)$ is finite-dimensional then by the Frobenius' theorem (see Section 0.5) it has a *foliating atlas*, i.e., an atlas consisting of foliating charts $(U,\varphi,\mathbb{R}^{m+I},\partial)$, $\#I$ is finite, with coordinates $x=(x^\mu;\mu=1,\dots,m)$, $u=(u^i;i\in I)$ and basis Cartan fields $\partial_\mu=\partial/\partial x^\mu$, $\mu=1,\dots,m$. In this case integral submanifolds locally are folios, i.e., $S\cap U=\{u=const\}$ for any integral submanifold S.

Example 1.1(a). (The continuation of Example 1.1) Let us fix a point $c=(t_0,u_0)\in\mathcal{O}$ and let us pass to new local coordinates $t=t$, $v=V(t,u)$, where V is a smooth function in an open set $U\subset\mathcal{O}$, $c\in U$, $V_{\cdot}\neq 0$ in U. Now, choose V to be the solution of the Cauchy problem

$$V_t + f\,V_u = 0,\ V(t_0,u) = u,$$

then the basis Cartan field will take the form

$$\partial_t = \partial_t(t)\,\frac{\partial}{\partial t} + \partial_t(V)\,\frac{\partial}{\partial v} = \frac{\partial}{\partial t}.$$

Hence, thus obtained chart will be foliating. Moreover, let $u=s(t)$ be a solution of a Cauchy problem

$$u_t = f(t,u), \quad u(t_0) = s_0.$$

In the coordinates t,v it will take the form $v=\varphi(t)=V(t,s(t))$, where $\varphi(t_0)=V(t_0,s_0)=s_0$ and $\varphi_t=\partial_t V\big|_{u=s(t)}=0$, i.e., $v\equiv s_0$.

Thus, finite-dimensional differential manifolds locally look like $\mathbb{R}^m\times\mathbb{R}^I$. In the infinite-dimensional case it is no longer true: here a differential manifold may have a nontrivial local geometry.

2. Lie-Backlund Mappings

A smooth mapping $f:M\longrightarrow N$ is called a *Lie-Backlund mapping* of a differential manifold $\mathcal{M}=(M,CTM)$ into a differential manifold $\mathcal{N}=(N,CTN)$, written $f:\mathcal{M}\longrightarrow\mathcal{N}$, if it is compatible with the Cartan distributions CTM and CTN, i.e., if $f_*(CTM)\subset CTN$, where $f_*:TM\longrightarrow TN$ is the tangent mapping. Clear that the identity mapping and the composition of Lie-Backlund mappings are Lie-Backlund mappings. Hence, *the category of differential manifolds* is defined, whose objects are differential manifolds and whose morphisms are Lie-Backlund mappings (cf. Vinogradov, Krasil'shchik, Lychagin [1; Ch.VII, §5]). Note that every smooth manifold M can be turned into a differential manifold by equipping it with the zero-dimensional Cartan distribution, i.e., by setting $\mathcal{M}=(M,0)$, where 0 is the zero subbundle of the tangent bundle TM. Thus, the category of smooth manifolds is imbedded into the category of differential manifolds, and this imbedding is clearly full. Every finite-dimensional manifold M can be turned into a differential manifold also in another way by equipping it with the full-dimensional Cartan distribution, i.e., by setting $\mathcal{M}=(M,TM)$. This gives an alternative way of imbedding the category of finite-dimensional smooth manifolds into the category of differential manifolds, a way that is also full (note that the category of finite-dimensional smooth manifolds is the full subcategory of the category of smooth manifolds).

Proposition 2.1. *Let $\mathcal{M}=(M,CTM)$ and $\mathcal{N}=(N,CTN)$ be differential manifolds, and let $(\mathcal{U},\varphi,\mathbb{R}^{m+I},\partial_x)$ and $(\mathcal{V},\psi,\mathbb{R}^{n+J},\partial_y)$ be standard charts on $\mathcal{M}$ and $\mathcal{N}$, respectively, with coordinates x,u and y,v and basis Cartan fields*

$$\partial_{x^\mu} = \frac{\partial}{\partial x^\mu} + \sum_{i\in I} A^i_\mu(x,u)\,\frac{\partial}{\partial u^i},\quad \mu = 1,\dots,m,$$

and

$$\partial_{y^\nu} = \frac{\partial}{\partial y^\nu} + \sum_{j\in J} B^j_\nu(y,v)\,\frac{\partial}{\partial v^j},\quad \nu = 1,\dots,n.$$

Let $f:\mathcal{U}\to\mathcal{V}$ be a smooth mapping given by the formulas

$$y^\nu = Y^\nu(x,u),\quad v^j = V^j(x,u),\quad \nu = 1,\dots,n,\quad j\in J.$$

The mapping f is Lie-Backlund iff smooth functions Y^ν, V^j satisfy the defining equations

$$\partial_{x^\mu}V^j - \sum_{\nu=1}^{n} B^j_\nu(Y,V)\,\partial_{x^\mu}Y^\nu = 0,\quad \mu = 1,\dots,m,\quad j\in J.$$

In this case

$$\partial_{x^\mu}\Big[\varphi\Big|_{y=Y,\,v=V}\Big] = \sum_{\nu=1}^{n}\Big(\partial_{x^\mu}Y^\nu\Big)\Big[\partial_{y^\nu}\varphi\Big]\Big|_{y=Y,\,v=V},\quad \mu = 1,\dots,m,$$

for any smooth function $\varphi(y,v)\in\mathcal{C}^\infty(\mathcal{V})$. In particular,

$$\partial_{x^\mu}\Big[\varphi\Big|_{y=Y,\,v=V}\Big] = \Big[\partial_{y^\mu}\varphi\Big]\Big|_{y=Y,\,v=V},\quad \mu = 1,\dots,m,$$

if $y=x$ (clear, $n=m$ then).

Proof. The mapping f is Lie-Backlund iff

$$f_*\left(\partial_{x^\mu}\right) = \sum_{\nu=1}^{n} R_\mu^\nu \, \partial_{y^\nu}, \quad \mu = 1,\ldots,m,$$

where R_μ^ν are some coefficients. In our case,

$$f_*\left(\partial_{x^\mu}\right) = \sum_{\nu=1}^{n} \left(\partial_{x^\mu} Y^\nu\right) \frac{\partial}{\partial y^\nu} + \sum_{j\in J} \left(\partial_{x^\mu} V^j\right) \frac{\partial}{\partial v^j}$$

$$= \sum_{\nu=1}^{n} \left(\partial_{x^\mu} Y^\nu\right) \partial_{y^\nu} + \sum_{j\in J} \left(\partial_{x^\mu} V^j - \sum_{\nu=1}^{n} B_\nu^j(Y,V)\, \partial_{x^\mu} Y^\nu\right) \frac{\partial}{\partial v^j}$$

(cf. Proposition 1.9). Hence, f is the Lie-Backlund mapping iff the defining equations are valid (note that $R_\mu^\nu = \partial_{x^\mu} Y^\nu$ then).

Example 2.2. Consider ordinary differential equations $u_x = f$, $v_y = g$ and the corresponding differential manifolds $\mathcal{M}$, $\mathcal{N}$ with global coordinates (x,u), (y,v) and basis Cartan fields

$$\partial_x = \frac{\partial}{\partial x} + f(x,u)\, \frac{\partial}{\partial u}, \quad \partial_y = \frac{\partial}{\partial y} + g(y,v)\, \frac{\partial}{\partial v},$$

respectively (see Example 1.1). A smooth mapping $F\colon \mathcal{M} \to \mathcal{N}$ given by the formulas

$$y = Y(x,u), \quad v = V(x,u),$$

will be Lie-Backlund iff smooth functions Y, V satisfy the defining equation

$$\partial_x V - g(Y,V)\, \partial_x Y = 0,$$

i.e., iff

$$(V_x + f(x,u)\, V_u) - g(Y,V)\,(Y_x + f(x,u)\, Y_u) = 0.$$

Here we can choose $Y(x,u)$ to be an arbitrary smooth function and then define $V(x,u)$ as a solution of the quasi-linear first-order PDE. In particular, suppose $f=\varphi(x)/\psi(u)$, $g=\lambda(y)/\rho(v)$, $y=Y(x)$ and $v=V(u)$ then the defining equation takes the form

$$\frac{\varphi(x)}{\psi(u)}\frac{dV}{du} - \frac{\lambda(Y)}{\rho(V)}\frac{dY}{dx} = 0.$$

This equation is easily integrated and we get

$$\int\rho(v)dv = \mu\int\psi(u)du, \quad \int\lambda(y)dy = \mu\int\varphi(x)dx,$$

where μ is an arbitrary constant.

Example 2.3. Consider evolution equations $u_t=f$, $v_s=g$ and the corresponding differential manifolds M, N with global coordinates $(t,x,\mathbf{u})$, $(s,y,\mathbf{v})$ and basis Cartan fields

$$\partial_x = \frac{\partial}{\partial x} + \sum_{i=0}^{\infty} u_{i+1}\frac{\partial}{\partial u_i}, \quad \partial_t = \frac{\partial}{\partial t} + \sum_{i=0}^{\infty} f_i \frac{\partial}{\partial u_i} \quad \text{on } M,$$

$$\partial_y = \frac{\partial}{\partial y} + \sum_{i=0}^{\infty} v_{i+1}\frac{\partial}{\partial v_i}, \quad \partial_s = \frac{\partial}{\partial s} + \sum_{i=0}^{\infty} g_i \frac{\partial}{\partial v_i} \quad \text{on } N,$$

respectively, where $f_i=\partial_x^i f(t,x,\mathbf{u})$, $g_i=\partial_y^i g(s,y,\mathbf{v})$ (see Example 1.4). A smooth mapping $F:M\to N$ given by the formulas

$$s = S(t,x,\mathbf{u}), \quad y = Y(t,x,\mathbf{u}), \quad \mathbf{v} = \mathbf{V}(t,x,\mathbf{u}),$$

will be Lie-Backlund iff smooth functions $S,Y,\mathbf{V}$ satisfy the defining equations

$$\begin{aligned} \partial_x V_i - g_i(S,Y,\mathbf{V})\, \partial_x S - V_{i+1}\, \partial_x Y = 0, \\ \partial_t V_i - g_i(S,Y,\mathbf{V})\, \partial_t S - V_{i+1}\, \partial_t Y = 0, \end{aligned} \qquad i \in \mathbb{N}.$$

In particular, suppose

$$s = t, \quad y = x, \quad \mathbf{v} = \mathbf{V}(t,x,\mathbf{u})$$

then the defining equations take the form

$$\begin{aligned} \partial_x V_i - V_{i+1} = 0, \\ \partial_t V_i - g_i(t,x,\mathbf{V}) = 0, \end{aligned} \qquad i \in \mathbb{N}.$$

The first subsystem easily yields $V_i = \partial_x^i V$, $i \in \mathbb{N}$, i.e., in this case

$$\mathbf{v} = \mathbf{V} = \left(\partial_x^i V\right),$$

where $V(t,x,\mathbf{u})$ is a smooth function. The substitution into the second subsystem gives

$$\partial_t \partial_x^i V - (\partial_y^i g)(t,x,\mathbf{V}) = \partial_x^i \left(\partial_t V - g(t,x,\mathbf{V})\right) = 0, \quad i \in \mathbb{N},$$

where we used Lemma 2.4, below. The last system is clearly equivalent to the single defining equation

$$\partial_t V = g(t,x,\mathbf{V}).$$

Lemma 2.4. *Let*

$$\partial_x = \frac{\partial}{\partial x} + \sum_{i=0}^{\infty} u_{i+1} \frac{\partial}{\partial u_i}, \quad \partial_y = \frac{\partial}{\partial y} + \sum_{i=0}^{\infty} v_{i+1} \frac{\partial}{\partial v_i},$$

then the formula

$$\left[\partial_y f(y,\mathbf{v})\right]\Big|_{y=x,\, \mathbf{v}=\mathbf{V}(x,\mathbf{u})} = \partial_x \left[f(y,\mathbf{v}) \Big|_{y=x,\, \mathbf{v}=\mathbf{V}(x,\mathbf{u})} \right]$$

is valid for all smooth functions $f(y,\mathbf{v})$ and $V(x,\mathbf{u})$, where $\mathbf{V}=(\partial^i V)$.

Proof. Indeed,

$$\left[\partial_y f(y,\mathbf{v})\right]\Big|_{y=x,\,\mathbf{v}=\mathbf{V}(x,\mathbf{u})} = \left[\frac{\partial f}{\partial y} + \sum_{i=0}^{\infty} \frac{\partial f}{\partial v_i} v_{i+1}\right]\Big|_{y=x,\,\mathbf{v}=\mathbf{V}(x,\mathbf{u})}$$

$$= \left[\frac{\partial f(x,\mathbf{v})}{\partial x} + \sum_{i=0}^{\infty} \frac{\partial f(x,\mathbf{v})}{\partial v_i} v_{i+1}\right]\Big|_{\mathbf{v}=\mathbf{V}(x,\mathbf{u})}.$$

On the other side, by the chain rule,

$$\partial_x\left[f(y,\mathbf{v})\Big|_{y=x,\,\mathbf{v}=\mathbf{V}(x,\mathbf{u})}\right] = \partial_x\left[f(x,\mathbf{v})\Big|_{\mathbf{v}=\mathbf{V}(x,\mathbf{u})}\right]$$

$$= \frac{\partial f(x,\mathbf{v})}{\partial x}\Big|_{\mathbf{v}=\mathbf{V}(x,\mathbf{u})} + \sum_{i=0}^{\infty}\left[\frac{\partial f(x,\mathbf{v})}{\partial v_i}\Big|_{\mathbf{v}=\mathbf{V}(x,\mathbf{u})}\right]\partial_x V_i$$

$$= \left[\frac{\partial f(x,\mathbf{v})}{\partial x} + \sum_{i=0}^{\infty} \frac{\partial f(x,\mathbf{v})}{\partial v_i} v_{i+1}\right]\Big|_{\mathbf{v}=\mathbf{V}(x,\mathbf{u})},$$

because $\partial_x V_i = V_{i+1}$, by the definition.

Example 2.3(a). Consider the *Burgers' equation*

$$u_t = u_{xx} + uu_x$$

and the *heat equation*

$$v_s = v_{yy}.$$

In this case $f=u_2+u_1u_0$ and $g=v_2$. The mapping F given by the formulas

$$s = t, \quad y = x, \quad v_i = \partial_x^i V, \quad i \in \mathbb{N},$$

$V=V(t,x,u_0,u_1,\ldots,u_p)$, p is an order of a differential function V, will be Lie-Backlund iff V satisfies the defining equation

$$\partial_t V - \partial_x^2 V = 0.$$

In this case

$$f_i = \partial_x^i\left(u_2 + u_1u_0\right) = u_{i+2} + \sum_{k=0}^{i}\binom{i}{k}u_{k+1}u_{i-k}$$

(note that $\partial_x u_k = u_{k+1}$), hence

$$\partial_t V = V_t + \sum_{i=0}^{p} f_i V_{u_i} = V_t + \sum_{i=0}^{p} u_{i+2}V_{u_i} + \sum_{i=0}^{p}\sum_{k=0}^{i}\binom{i}{k}u_{k+1}u_{i-k}V_{u_i}.$$

Further,

$$\begin{aligned}\partial_x^2 V &= \partial_x\left(V_x + \sum_{i=0}^{p} u_{i+1}V_{u_i}\right)\\ &= V_{xx} + \sum_{i=0}^{p}\left(u_{i+2}V_{u_i} + 2u_{i+1}V_{xu_i}\right) + \sum_{i,k=0}^{p} u_{i+1}u_{k+1}V_{u_iu_k}.\end{aligned}$$

Thus, the defining equation has the form

$$-u_{p+1}^2V_{u_pu_p} + u_{p+1}\left[u_0V_{u_p} - 2\left(V_{xu_p} + \sum_{i=0}^{p-1} u_{i+1}V_{u_iu_p}\right)\right] + \ldots = 0,$$

where omitted terms have orders $\le p$. This implies

$$V_{u_pu_p} = 0 \quad \text{and} \quad u_0V_{u_p} - 2\left(V_{xu_p} + \sum_{i=0}^{p-1} u_{i+1}V_{u_pu_i}\right) = 0.$$

The first equation yields $V=Au_p+B$, where $A, B(t,x,u_0,\ldots,u_{p-1})$, then the second equation takes the form

$$u_0A - 2\partial_x A = -2u_pA_{u_{p-1}} + \ldots = 0$$

and gives $A_{u_{p-1}}=0$, i.e., $A=A(u_0,\ldots,u_{p-2})$. By the induction, we get $A=A(t,x,u_0)$ with

$$u_0 A - 2A_x - 2u_1 A_{u_0} = 0,$$

or $A=0$. Hence, $V=B=V(t,x,u_0,\dots,u_{p-1})$ and, again by the induction, $V=V(t,x,u_0)$ with

$$V_t + u_1 u_0 V_{u_0} - V_{xx} - 2u_1 V_{xu_0} - u_1 u_1 V_{u_0 u_0} = 0,$$

or

$$V_{u_0 u_0}=0, \quad u_0 V_{u_0} - 2V_{xu_0} = 0, \quad V_t - V_{xx} = 0.$$

The simple analysis shows that $V=V(t,x)$ with $V_t - V_{xx}=0$. That means that *Lie-Backlund mappings from the chosen class take solutions of the Burgers' equation into a fixed solution of the heat equation.*

Example 2.3(b). Let us replace the Burgers' equation by the *modified Burgers' equation*

$$u_t = u_{xx} + u_x^2.$$

In this case

$$f = u_2 + u_1^2 \quad \text{and} \quad f_i = u_2 + \sum_{k=0}^{i} \binom{i}{k} u_{k+1} u_{i-k+1}, \qquad i \in \mathbb{N}.$$

Here, the mapping F given by the formulas

$$s = t, \quad y = x, \quad v_i = \partial_x^i V, \quad i \in \mathbb{N},$$

$V=V(t,x,u_0,u_1,\dots,u_p)$, p is an order of a differential function V, will be Lie-Backlund iff V satisfies the defining equation

$$\partial_t V - \partial_x^2 V = V_t - V_{xx} + \sum_{i=0}^{p} \sum_{k=0}^{i} \binom{i}{k} u_{k+1} u_{i-k+1} V_{u_i}$$

$$- \left(2 \sum_{i=0}^{p} u_{i+1} V_{xu_i} + \sum_{i,k=0}^{p} u_{i+1} u_{k+1} V_{u_i u_k} \right)$$

$$= -u_{p+1}^2 V_{u_p u_p} + 2u_{p+1} \left(u_1 V_{u_p} - V_{xu_p} - \sum_{i=0}^{p-1} u_{i+1} V_{u_i u_p} \right) + \ldots = 0.$$

Again, $V = Au_p + B$, where $A,B = A,B(t,x,u_0,\ldots,u_{p-1})$. Note that

$$\left(\partial_t - \partial_x^2 \right) \partial_x^p e^{u_0} = \partial_x^p \left(\partial_t - \partial_x^2 \right) e^{u_0} = \partial_x^p \left(f - u_2 - u_1^2 \right) e^{u_0} = 0$$

and

$$\partial_x^p e^{u_0} = \left(u_p + (p-1) u_{p-1} + \ldots \right) e^{u_0}.$$

Thus, without loss of generality we can take

$$V = A \partial_x^p e^{u_0} + B, \quad A,B(t,x,u_0,\ldots,u_{p-1}).$$

Then $V_{u_p} = Ae^{u_0}$ and a function A should satisfy the equation

$$u_1 Ae^{u_0} - \partial_x \left(Ae^{u_0} \right) = u_1 Ae^{u_0} - \left(\partial_x A \right) e^{u_0} - Au_1 e^{u_0} = e^{u_0} \partial_x A = 0.$$

By Lemma 2.5 below, $A = a_p(t)$, p indicates the order of the function V. Let us define the operator Q acting on smooth functions $\chi(t,x)$ by the formula

$$Q\chi = \frac{1}{2} \int_0^x L\chi dx, \quad \text{where} \quad L\chi = \chi_t - \chi_{xx} \quad \text{is the heat operator,}$$

then the equality

$$L\chi - 2 \left(Q\chi \right)_x = 0$$

will be valid. Further, let us redefine the function V in the following way

$$V = \sum_{i=0}^{p} \left(Q^{p-i} a_p\right) \partial_x^i e^{u_0} + U(t, x, u_0, \dots, u_{p-1}).$$

Then,

$$\left(\partial_t - \partial_x^2\right) \sum_{i=0}^{p} \left(Q^{p-i} a_p\right) \partial_x^i e^{u_0} = \sum_{i=0}^{p} \left(LQ^{p-i} a_p\right) \partial_x^i e^{u_0} - 2 \sum_{i=0}^{p} \left(Q^{p-i} a_p\right)_x \partial_x^i e^{u_0}$$

$$+ \sum_{i=0}^{p} \left(Q^{p-i} a_p\right) \left(\partial_t - \partial_x^2\right) \partial_x^i e^{u_0}$$

$$= -2\left(a_p\right)_x \partial_x^p e^{u_0} + \left(LQ^p a_p\right) e^{u_0},$$

where we used the above equality for Q. In our case $a_p = a_p(t)$, hence,

$$\partial_t V - \partial_x^2 V = \left(LQ^p a_p\right) e^{u_0} + \partial_t U - \partial_x^2 U = -u_p^2 U_{u_{p-1} u_{p-1}}$$

$$+ 2u_p \left(u_1 U_{u_{p-1}} - U_{x u_{p-1}} - \sum_{i=0}^{p-2} u_{i+1} U_{u_i u_{p-1}}\right) + \dots = 0.$$

By the induction,

$$V = \sum_{i=0}^{p} \sum_{k=0}^{i} \left(Q^{i-k} a_i\right) \partial_x^k e^{u_0} + \varphi,$$

where $a_i = a_i(t)$, $i = 0, \dots, p$, $\varphi = \varphi(t, x)$ are smooth functions,

$$\sum_{i=0}^{p} LQ^i a_i = 0 \quad \text{and} \quad L\varphi = 0.$$

This general solution of the defining equation can be made more explicit. Namely, note that

$$LQ^n\chi = \sum_{0\le i\le\frac{n}{2}} c_i \frac{x^{n-2i}}{(n-2i)!} \quad \text{for any } \chi=\chi(t), \quad n\in\mathbb{N},$$

where coefficients $c_i=c_i(t)$, $c_0=2^{-n}\chi^{(n+1)}$, $\chi^{(i)}=d^i\chi\big/dt^i$. Thus,

$$\sum_{i=0}^{p} LQ^i a_i = 2^{-p}a_p^{(p+1)}\frac{x^p}{p!}+\dots = 0,$$

hence, $a_p^{(p+1)}=0$ and $a_p(t)$ is some polynomial of the pth degree. Now, let a_p be an arbitrary polynomial of the pth degree and consider a function

$$\varphi_p = \sum_{0\le k\le\frac{p}{2}} \sum_{i=0}^{p-2k} \left(Q^{p-2k-i}a_{p-2k}\right)\partial_x^i e^{u_0}.$$

Let us show that by choosing functions $a_{p-2}(t)$, $a_{p-4}(t),\dots$ in a proper way we can ensure φ_p to be a solution of the defining equation. Indeed,

$$\partial_t\varphi_p - \partial_x^2\varphi_p = \sum_{0\le k\le\frac{p}{2}} \left(LQ^{p-2k}a_{p-2k}\right)e^{u_0}$$

and we need to satisfy the equality

$$\sum_{0\le k\le\frac{p}{2}} LQ^{p-2k}a_{p-2k} = \sum_{0\le k\le\frac{p}{2}} \sum_{0\le i\le\frac{p}{2}-k} c_{i,p-2k}\frac{x^{p-2k-2i}}{(p-2k-2i)!}$$

$$= \sum_{0\le k\le\frac{p}{2}} \sum_{k\le i\le\frac{p}{2}} c_{i-k,p-2k}\frac{x^{p-2i}}{(p-2i)!}$$

$$= \sum_{0\le i\le \frac{p}{2}} \left(\sum_{k=0}^{i} c_{i-k,\,p-2k} \right) \frac{x^{p-2i}}{(p-2i)!} = 0,$$

or

$$\sum_{k=0}^{i} c_{i-k,\,p-2k} = 0 \quad \text{for } 0\le i\le \frac{p}{2},$$

where coefficients $c_{i,n}$ are defined by the functions $a_i(t)$. In particular, $c_{0,p-2k}=2^{-p+2k}a_{p-2k}^{(p-2k+1)}(t)$. In more detail,

for $i=0$ $c_{0,p} = 2^{-p}a_p^{(p+1)} = 0,$

by the definition of the function φ_p,

for $i = 1,2,\dots,\left[\frac{p}{2}\right]$ the equality $c_{0,p-2i} + \sum_{k=1}^{i} c_{i-k,\,p-2k} = 0$

will be satisfied if

$$a_{p-2i}^{(p-2i+1)} = 2^{p-2i}c_{0,p-2i} = -2^{p-2i}\sum_{k=1}^{i} c_{i-k,\,p-2k},$$

where the last sum does not depend on a_{p-2i}. For example, we can take

$$a_{p-2i}(t) = -2^{p-2i}\left({}_0\!\int^t\right)^{p-2i+1} \sum_{k=1}^{i} c_{i-k,\,p-2k}(t).$$

Thus, *the general* p^{th} *order solution of the defining equation has the form*

$$V = \sum_{q=0}^{p} \varphi_q + \varphi,$$

where

$$\varphi = \varphi(t,x), \quad \varphi_t - \varphi_{xx} = 0,$$

$$\varphi_q = \sum_{0\le k\le \frac{q}{2}} \sum_{i=0}^{q-2k} \left(Q^{q-2k-i} a_{q,q-2k}\right) \partial_x^i e^{u_0},$$

$a_{q,q}=a_q(t)$ are arbitrary polynomials of qth degree, $a_{q,q-2k}(t)$ are compensating polynomials.

Lemma 2.5. *Let*

$$\partial_x = \frac{\partial}{\partial x} + \sum_{i\in\mathbb{N}} u_{i+1} \frac{\partial}{\partial u_i},$$

then a smooth function $\varphi(x,u_0,\dots,u_p)$ *satisfies the equation* $\partial_x\varphi=0$ *iff* $\varphi=const$.

Proof. Indeed,

$$\partial_x\varphi = u_{p+1}\varphi_{u_p} + \dots = 0$$

implies $\varphi_{u_p}=0$, i.e., $\varphi=\varphi(x,u_0,\dots,u_{p-1})$. By the induction, $\varphi=\varphi(x)$ with $\varphi_x=0$, hence, $\varphi=const$.

Let us consider the finite-dimensional case.

Proposition 2.6. *Let* $\mathcal{M}=(M,CTM)$ *and* $\mathcal{N}=(N,CTN)$ *be finite-dimensional differential manifolds and* $f:\mathcal{M}\longrightarrow\mathcal{N}$ *a smooth mapping. The mapping f is the Lie-Backlund mapping iff it locally maps every integral submanifold of* $\mathcal{M}$ *into an integral submanifold of* $\mathcal{N}$.

Proof. Indeed, let $(U,\varphi,\mathbb{R}^{m+I},\partial_x)$ and $(V,\psi,\mathbb{R}^{n+J},\partial_y)$ be foliating charts with coordinates x,u and y,v and with basis Cartan fields

$$\partial_{x^\mu} = \frac{\partial}{\partial x^\mu}, \ \mu = 1,\dots,m, \ \text{and} \ \partial_{y^\nu} = \frac{\partial}{\partial y^\nu}, \ \nu = 1,\dots,n,$$

on $\mathcal{M}$ and $\mathcal{N}$, respectively, such that $f(U)\subset V$. In these coordinates the mapping f will have the form

$$y^{\nu} = Y^{\nu}(x,u), \ \nu = 1,\dots,n, \ v^{j} = V^{j}(x,u), \ j \in J,$$

where Y^{ν}, V^{j} are smooth functions on U. By Proposition 2.2, the mapping f is Lie-Backlund iff

$$\partial_{x^{\mu}} V^{j} = \frac{\partial V^{j}}{\partial x^{\mu}} = 0, \quad \mu = 1,\dots,m, \ j \in J$$

(note that the coefficients $A^{i}_{\mu}, B^{j}_{\nu}=0$ here), i.e., iff $v^{j}=V^{j}(u)$ for all $j\in J$. Suppose f is a Lie-Backlund mapping and let $S=\{u=u_{0}\}\subset U$ be an integral submanifold of $\mathcal{M}$, then its image

$$f(S) = \left\{y=Y(x,u_{0}), \ v=V(u_{0})\right\} \subset T = \left\{v=v_{0}\right\},$$

where $T=\{v=v_{0}\}\subset V$ is an integral submanifold of $\mathcal{N}$, $v_{0}=V(u_{0})$. On the other side, suppose that f maps integral submanifolds of $\mathcal{M}$ into integral submanifolds of $\mathcal{N}$. Let us fix u_{0} such that $S=\{u=u_{0}\}\subset U$ is an integral submanifold of $\mathcal{M}$ then its image

$$f(S) = \left\{y=Y(x,u_{0}), \ v=V(x,u_{0})\right\} \subset T = \left\{v=v_{0}\right\},$$

where $T=\{v=v_{0}\}\subset V$ is some integral submanifold of $\mathcal{N}$. In particular, $v^{j}=V^{j}(x,u_{0})\equiv v^{j}_{0}$, $j\in J$. Hence, functions $v^{j}=V^{j}(u)$ do not depend on x for all $j\in J$, i.e., the mapping f is Lie-Backlund.

Thus, in the finite-dimensional case there is the nice criterion for a smooth mapping to be Lie-Backlund. In the infinite-dimensional case the situation is more complicated due to the absence of foliating charts. The best we can prove is that in some cases Lie-Backlund mappings locally map integral submanifolds into integral submanifolds.

Let $\mathcal{M}=(M,CTM)$ and $\mathcal{N}=(N,CTN)$ be differential manifolds. An isomorphism $f:M\simeq N$ is called *Lie-Backlund isomorphism*, written $f:\mathcal{M}\simeq\mathcal{N}$, if the mappings f and f^{-1} are both Lie-Backlund. Clear, that in this case the image of every integral submanifold is the integral submanifold.

Example 2.3(c). For the differential manifolds $\mathcal{M}$ and $\mathcal{N}$, defined by the modified Burgers' and the heat equations, we have the Lie-Backlund isomorphism

$$s = t,\ y = x,\ v_i = \partial_x^i e^{u_0},\ i \in \mathbb{N}$$

(cf. Cole-Hopf substitution, see, for example, Whitham [1; Ch.4]). This is a particular case ($p=0$) of the general result proved in Example 2.3(b).

A Lie-Backlund mapping $f:\mathcal{M}\longrightarrow\mathcal{N}$ is said to be *covering* if $f_*:CT_aM\simeq CT_{f(a)}N$ for each point $a\in M$, where f_* is the tangent mapping (cf. Vinogradov, Krasil'shchik, Lychagin [1;Ch.VII, §5]). Clear, that in this case the Cartan dimensions of $\mathcal{M}$ and $\mathcal{N}$ coincide.

Proposition 2.7. *Let* $(U,\varphi,\mathbb{R}^{m+I},\partial_x)$ *and* $(V,\psi,\mathbb{R}^{m+J},\partial_y)$ *be standard charts on* $\mathcal{M}$ *and* $\mathcal{N}$*, respectively. A Lie-Backlund mapping* $f:U\longrightarrow V$ *is covering iff the* $m\times m$*-matrix* $\left\|\partial_{x^\mu}Y^\nu\right\|$ *is nonsingular* (we are keeping the notation of Proposition 2.1).

Proof. By the definition, a Lie-Backlund mapping f be covering iff

$$f_*\left(\partial_{x^\mu}\right) = \sum_{\nu=1}^{m} R_\mu^\nu \, \partial_{y^\nu},$$

where R_μ^ν is a nonsingular matrix, i.e., iff

$$\sum_{\nu=1}^{m} \left(\partial_{x^\mu} Y^\nu\right) \frac{\partial}{\partial y^\nu} + \sum_{j\in J} \left(\partial_{x^\mu} V^j\right) \frac{\partial}{\partial v^j} = \sum_{\nu=1}^{m} R_\mu^\nu \left(\frac{\partial}{\partial y^\nu} + \sum_{j\in J} B_\nu^j(y,v) \frac{\partial}{\partial v^j}\right),$$

i.e., iff $\partial_{x^\mu} Y^\nu = R_\mu^\nu$ is a nonsingular matrix.

In particular, the Lie-Backlund mappings from Example 2.3(b) are covering.

Proposition 2.8. *Let $f:M\to N$ be a covering Lie-Backlund mapping then f locally maps every integral submanifold of M onto an integral submanifold of N.*

Proof. Let $(U,\varphi,\mathbb{R}^{m+I},\partial_x)$ and $(V,\psi,\mathbb{R}^{m+J},\partial_y)$ be standard charts on M and N, respectively, such that $f(U)\subset V$ (we are keeping the notation of Proposition 2.1). Let

$$S = \left\{x,u \; : \; u = s(x)\right\} \subset U$$

be an integral submanifold of M (see Proposition 1.7) and let $f(S)$ be its image. By the definition, $T_b f(S)=f_* T_a S=f_* CT_a M=CTN$ for each point $b=f(a)\in f(S)$. Hence, the only thing we should check is that $f(S)$ is an m-dimensional submanifold of N. Indeed, the image

$$f(S) = \left\{y = Y(x,s(x)), \; v = V(x,s(x))\right\},$$

and, by Propositions 1.8 and 2.7, the $m\times m$-matrix

$$\frac{\partial Y^{\nu}(x,s(x))}{\partial x^{\mu}} = \left(\partial_{x^{\mu}} Y^{\nu}(x,u)\right)\Big|_{u=s(x)}$$

is nonsingular. Thus, by the implicit function theorem, we can locally solve $x^{\mu}=X^{\mu}(y)$, $\mu=1,\dots,m$. Then the image $f(S)$ will take the form

$$f(S) = \left\{v = r(y)\right\},$$

where the smooth functions $r^{j}(y)=V^{j}(X(y),s(X(y)))$, $j\in J$. According to Section 0.4, $f(S)$ is an m-dimensional submanifold of N.

Example 2.9. Let M be the differential manifold with global coordinates x, $\mathbf{u}=(u_i;\, i\in\mathbb{N})$, $u_1>0$, and the basis Cartan field

$$\partial_x = \frac{\partial}{\partial x} + \sum_{i\in\mathbb{N}} u_{i+1}\frac{\partial}{\partial u_i}$$

(cf. Example 0.6.1). In the same way, let N be the differential manifold with coordinates y, $\mathbf{v}=(v_i;\, i\in\mathbb{N})$, $v_0>0$, and the basis Cartan field

$$\partial_y = \frac{\partial}{\partial y} + \sum_{i\in\mathbb{N}} \frac{v_{i+1}}{v_0}\frac{\partial}{\partial v_i}.$$

Consider the Lie-Backlund mapping $f\colon M\longrightarrow N$ defined by the formulas

$$y = Y(x,\mathbf{u}) = u_0, \quad v_i = V_i(x,\mathbf{u}) = u_{i+1}, \quad i\in\mathbb{N}.$$

Then

$$f_*\left(\partial_x\right) = u_1\frac{\partial}{\partial y} + \sum_{i\in\mathbb{N}} u_{i+2}\frac{\partial}{\partial v_i} = v_0\frac{\partial}{\partial y} + \sum_{i\in\mathbb{N}} v_{i+1}\frac{\partial}{\partial v_i} = v_0\partial_y,$$

i.e., the mapping f is covering. Now, let

$$S = \left\{u_i = \varphi^{(i)}(x),\ i \in \mathbb{N}\right\}$$

be an integral submanifold of $\mathcal{M}$, where $\varphi(x)$ is a smooth function, $\varphi^{(i)}(x)=d^i\varphi(x)\big/dx^i$, $\varphi^{(1)}(x)>0$, then its image

$$f(S) = \left\{y = \varphi(x),\ v_i = \varphi^{(i+1)}(x),\ i \in \mathbb{N}\right\}$$

$$= \left\{v_i = \left[\varphi^{(i+1)}(x)\right]\Big|_{x=\varphi^{-1}(y)},\ i \in \mathbb{N}\right\}$$

will be an integral submanifold of $\mathcal{N}$ (note that the inverse function $x=\varphi^{-1}(y)$ exists since $\varphi^{(1)}(x)>0$).

Example 2.3(d). Consider the heat equation $u_t=u_{xx}$ and the Burgers' equation $v_s=v_{yy}+v_xv$. Let $\mathcal{M}$ and $\mathcal{N}$ be the corresponding differential manifolds (here $f=u_2$ and $g=v_2+v_1v_0$, we are keeping the notation of Example 2.3). The Lie-Backlund mapping $F:\mathcal{M}\longrightarrow\mathcal{N}$ given by the formulas

$$s = t,\quad y = x,\quad v_i = \partial_x^i V,\quad i \in \mathbb{N},$$

is covering, where $V(t,x,\mathbf{u})=2u_1\big/u_0$ is the *Cole-Hopf substitution* (see, for example, Whitham [1; Ch.4], Ibragimov [1; §17.5], note that $u_1=u_x$, $u_0=u$ in another notation).

Indeed, here $f_i=u_{i+2}$, the defining equation has the form

$$\partial_t V - g(t,x,\mathbf{V}) = \partial_t V - \left(V_2 + V_1V_0\right)$$

$$= \partial_t V - \left[\partial_x^2 V + \left(\partial_x V\right)V\right] = 0$$

and it is easy to check that $V=2u_1\big/u$ is a solution of it. Moreover, the matrix

$$\left\| \partial_{x^\mu} Y^\nu \right\| = \left\| \begin{matrix} \partial_t S & \partial_t Y \\ \partial_x S & \partial_x Y \end{matrix} \right\| = \left\| \begin{matrix} 1 & 0 \\ 0 & 1 \end{matrix} \right\|$$

is nonsingular and, by Proposition 2.7, the mapping F is covering (here $S=t$, $Y=x$).

A Lie-Backlund mapping $f:\mathcal{M}\longrightarrow\mathcal{N}$ is called a *Lie-Backlund immersion* if $f:M\longrightarrow N$ is an immersion and $f_*\left(CT_aM\right)=CT_{f(a)}N\cap f_*\left(T_aM\right)$ for each point $a\in M$, where f_* is the tangent mapping.

Proposition 2.10. *Let $f:\mathcal{M}\longrightarrow\mathcal{N}$ be a Lie-Backlund immersion. Then for every point $a\in M$ there exist standard charts $(U,\varphi,\mathbb{R}^{m+I},\partial)$ and $(V,\psi,\mathbb{R}^{m+n+I+J},\nabla)$ on $\mathcal{M}$ and $\mathcal{N}$, respectively, $a\in U$, $f(U)\subset V$, with the following properties:*

the chart $(U,\varphi,\mathbb{R}^{m+I},\partial)$ has coordinates x^μ,u^i, $\mu=1,\ldots,m$, $i\in I$, and basis Cartan fields

$$\partial_{x^\mu} = \frac{\partial}{\partial x^\mu} + \sum_{i\in I} a^i_\mu(x,u)\,\frac{\partial}{\partial u^i},\quad \mu = 1,\ldots,m;$$

the chart $(V,\psi,\mathbb{R}^{m+n+I+J},\nabla)$ has coordinates x^μ,u^i,y^ν,v^j, $\mu=1,\ldots,m$, $i\in I$, $\nu=1,\ldots,n$, $j\in J$, and basis Cartan fields

$$\nabla_{x^\mu} = \frac{\partial}{\partial x^\mu} + \sum_{i\in I} A^i_\mu(x,u,y,v)\,\frac{\partial}{\partial u^i} + \sum_{j\in J} B^j_\mu(x,u,y,v)\,\frac{\partial}{\partial v^j},\quad \mu=1,\ldots,m,$$

$$\nabla_{y^\nu} = \frac{\partial}{\partial y^\nu} + \sum_{i\in I} F^i_\nu(x,u,y,v)\,\frac{\partial}{\partial u^i} + \sum_{j\in J} G^j_\nu(x,u,y,v)\,\frac{\partial}{\partial v^j},\quad \nu=1,\ldots,n;$$

the mapping f acts on U by the rule: $(x,u)\mapsto(x,u,0,0)$ and

$$\partial_{x^\mu} = \nabla_{x^\mu}\Big|_{f(U)},\quad \mu = 1,\ldots,m,$$

i.e., $a^i_\mu(x,u)=A^i_\mu(x,u,0,0)$, $B^j_\mu(x,u,0,0)=0$, $\mu=1,\dots,m$, $i\in I$, $j\in J$.

Proof. By the definition of an immersion, there exist a chart $(U,\varphi,\mathbb{R}^K)$ on M with coordinates $z=(u^i;i\in K)$ and a chart $(V,\chi,\mathbb{R}^{K+L})$ on N with coordinates $z=(u^i;i\in K),w=(v^j;j\in L)$, such that $a\in U$ and f acts on U by the rule: $z\mapsto(z,w;w=0)$. Without loss of generality we may assume that the chart $(U,\varphi,\mathbb{R}^K)$ is already standard, i.e., $K=m+I$, $z=(x,u)$, $x=(x^\mu;\mu=1,\dots,m)$, $u=(u^i;i\in I)$ and the Cartan distribution CTM is given on U by the basis fields

$$\partial_{x^\mu} = \frac{\partial}{\partial x^\mu} + \sum_{i\in I} a^i_\mu(x,u)\frac{\partial}{\partial u^i}, \quad \mu = 1,\dots,m.$$

Further, suppose that the Cartan distribution CTN is given on V by the basis fields

$$\zeta_\lambda = \sum_{\mu=1}^{m} X^\mu_\lambda(x,u,w)\frac{\partial}{\partial x^\mu} + \sum_{i\in I} U^i_\lambda(x,u,w)\frac{\partial}{\partial u^i} + \sum_{j\in L} V^j_\lambda(x,u,w)\frac{\partial}{\partial v^j},$$

$\lambda=1,\dots,m+n$. By the definition of a Lie-Backlund mapping, there exist coefficients $R^\lambda_\mu(x,u)$ such that

$$f_*\left(\partial_{x^\mu}\right) = \partial_{x^\mu} = \sum_{\lambda=1}^{m+n} R^\lambda_\mu(x,u)\,\zeta_\lambda\Big|_{f(U)}, \quad \mu=1,\dots,m.$$

In particular,

$$\frac{\partial}{\partial x^\mu}+\dots = \sum_{\sigma=1}^{m}\left(\sum_{\lambda=1}^{m+n} R^\lambda_\mu(x,u)X^\sigma_\lambda(x,u,0)\right)\frac{\partial}{\partial x^\sigma}+\dots,$$

i.e.,

$$\sum_{\lambda=1}^{m+n} R^\lambda_\mu(x,u)X^\sigma_\lambda(x,u,0) = \delta^\sigma_\mu, \quad \sigma,\mu = 1,\dots,m.$$

Hence, shrinking the charts if necessary, we may assume that on V the $(m+n)\times m$-matrix $X^{\mu}_{\lambda}(x,u,w)$ has a left-inverse $m\times(m+n)$-matrix $Y^{\lambda}_{\mu}(x,u,w)$, such that $Y^{\lambda}_{\mu}(x,u,0)=R^{\lambda}_{\mu}(x,u)$. Consider Cartan fields

$$\begin{aligned}\eta_{\mu} &= \sum_{\lambda=1}^{m+n} Y^{\lambda}_{\mu}(x,u,w)\ \zeta_{\lambda} \\ &= \frac{\partial}{\partial x^{\mu}} + \sum_{i\in I} \varphi^{i}_{\mu}(x,u,w)\ \frac{\partial}{\partial u^{i}} + \sum_{j\in L} \psi^{j}_{\mu}(x,u,w)\ \frac{\partial}{\partial v^{j}},\quad \mu = 1,\dots,m,\end{aligned}$$

then $\partial_{x^{\mu}}=\eta_{\mu}\Big|_{f(M)}$, i.e., $\varphi^{i}_{\mu}(x,u,0)=a^{i}_{\mu}(x,u)$, $\psi^{j}_{\mu}(x,u,0)=0$, for all $\mu=1,\dots,m$, $i\in I$, $j\in L$. Among the fields $\zeta_{1},\dots,\zeta_{m+n}$ there are n linearly independent with the fields $\eta_{1},\dots,\eta_{m}$. Suppose that these fields are $\zeta_{1},\dots,\zeta_{n}$, then the fields $\eta_{1},\dots,\eta_{m},\zeta_{1},\dots,\zeta_{n}$ will be a new basis of the Cartan distribution CTN on V. Further,

$$\begin{aligned}\zeta_{\nu} &= \sum_{\mu=1}^{m} X^{\mu}_{\lambda}(x,u,w)\ \frac{\partial}{\partial x^{\mu}} + \sum_{i\in I} U^{i}_{\lambda}(x,u,w)\ \frac{\partial}{\partial u^{i}} + \sum_{j\in L} V^{j}_{\lambda}(x,u,w)\ \frac{\partial}{\partial v^{j}} \\ &= \sum_{\mu=1}^{m} X^{\mu}_{\lambda}(x,u,w)\ \eta_{\mu} + \xi_{\nu},\quad \nu = 1,\dots,n,\end{aligned}$$

where fields

$$\xi_{\nu} = \sum_{i\in I} \tilde{U}^{i}_{\nu}(x,u,w)\ \frac{\partial}{\partial u^{i}} + \sum_{j\in L} \tilde{V}^{j}_{\nu}(x,u,w)\ \frac{\partial}{\partial v^{j}},\quad \nu = 1,\dots,n.$$

Now, the condition $f_{*}\left(CT_{a}M\right)=CT_{f(a)}N\cap f_{*}\left(T_{a}M\right)$ for all $a\in M$ implies that

$$\text{rank}\ \left\|\tilde{V}^{j}_{\nu}\right\| = n\quad \text{on } V$$

(maybe after a shrinking the charts). Hence, there are indices $j_1, \ldots, j_n$ such that $n\times n$-matrix $\tilde{V}_\nu^{j_\sigma}(x,u,w)$ is nonsingular and has the inverse matrix $S_\sigma^\nu(x,u,w)$ on V. Let us put $y^1 = v^{j_1}, \ldots, y^n = v^{j_n}$, $v = (v^j; j\in J)$, $J = L\setminus\{j_1, \ldots, j_n\}$, and pass from the fields ξ_ν to the fields

$$\begin{aligned}\nabla_{y^\nu} &= \sum_{\sigma=1}^{n} S_\nu^\sigma\, \xi_\sigma \\ &= \frac{\partial}{\partial y^\nu} + \sum_{i\in I} F_\nu^i(x,u,y,v)\frac{\partial}{\partial u^i} + \sum_{j\in J} G_\nu^j(x,u,y,v)\frac{\partial}{\partial v^j}, \quad \nu = 1, \ldots, n,\end{aligned}$$

of the required form. In our new notation the fields η_μ have the form

$$\eta_\mu = \frac{\partial}{\partial x^\mu} + \sum_{i\in I}\varphi_\mu^i\frac{\partial}{\partial u^i} + \sum_{\nu=1}^{n}\psi_\mu^{j_\nu}\frac{\partial}{\partial y^\nu} + \sum_{j\in J}\psi_\mu^j\frac{\partial}{\partial v^j}, \qquad \mu = 1, \ldots, m,$$

and to finish the proof it is enough to pass from the fields ξ_μ to the new fields

$$\begin{aligned}\nabla_{x^\mu} &= \eta_\mu - \sum_{\nu=1}^{n}\psi_\mu^{j_\nu}\,\nabla_{y^\nu} \\ &= \frac{\partial}{\partial x^\mu} + \sum_{i\in I} A_\mu^i(x,u,y,v)\frac{\partial}{\partial u^i} + \sum_{j\in J} B_\mu^j(x,u,y,v)\frac{\partial}{\partial v^j}, \qquad \mu = 1, \ldots, m.\end{aligned}$$

Clear that in this case $a_\mu^i(x,u) = A_\mu^i(x,u,0,0)$ and $B_\mu^j(x,u,0,0) = 0$ for all $\mu = 1, \ldots, m$, $i\in I$, $j\in J$.

A Lie-Backlund mapping $f: M\longrightarrow N$ is called a *Lie-Backlund submersion* if $f: M\longrightarrow N$ is a submersion and $f_*\left(CT_aM\right) = CT_{f(a)}N$ for each point $a\in M$, where f_* is the tangent mapping. A Lie-Backlund submersion $f:M\longrightarrow N$ is called a *Lie-Backlund covering submersion* if Cartan dimensions of M and N are equal. In the last case $f\ :CT\ M\simeq CT\quad N$ for each point $a\in M$ (cf. Krasil'shchik, Vinogradov [1]).

Proposition 2.11. *Let $f:M\longrightarrow N$ be a Lie-Backlund submersion. Then for any $a\in M$ there exist standard charts $(V,\psi,\mathbb{R}^{m+n+I+J},\nabla)$ and $(U,\varphi,\mathbb{R}^{m+I},\partial)$ on M and N, respectively, $a\in V$, $f(V)\subset U$, with the following properties:*

the chart $(U,\varphi,\mathbb{R}^{m+I},\partial)$ has coordinates x^{μ},u^{i}, $\mu=1,\dots,m$, $i\in I$, and basis Cartan fields

$$\partial_{x^{\mu}} = \frac{\partial}{\partial x^{\mu}} + \sum_{i\in I} A^{i}_{\mu}(x,u)\,\frac{\partial}{\partial u^{i}},\quad \mu = 1,\dots,m;$$

the chart $(V,\psi,\mathbb{R}^{m+n+I+J},\nabla)$ has coordinates $x^{\mu},u^{i},y^{\nu},v^{j}$, $\mu=1,\dots,m$, $i\in I$, $\nu=1,\dots,n$, $j\in J$, and basis Cartan fields

$$\nabla_{x^{\mu}} = \partial_{x^{\mu}} + \sum_{j\in J} B^{j}_{\mu}(x,u,y,v)\,\frac{\partial}{\partial v^{j}},\quad \mu = 1,\dots,m,$$

$$\nabla_{y^{\nu}} = \frac{\partial}{\partial y^{\nu}} + \sum_{j\in J} C^{j}_{\nu}(x,u,y,v)\,\frac{\partial}{\partial v^{j}},\quad \nu = 1,\dots,n;$$

the mapping f acts on V by the rule: $(x,u,y,v)\mapsto(x,u)$, hence

$$f_{*}\left(\nabla_{x^{\mu}}\right) = \partial_{x^{\mu}},\ \mu = 1,\dots,m,\ \text{and}\ f_{*}\left(\nabla_{y^{\nu}}\right) = 0,\ \nu = 1,\dots,n.$$

In particular, the basis fields ∇_{μ} and ∂_{μ} are f-related on V for $\mu=1,\dots,m$.

Remind that vector fields ξ on a manifold M and ϑ on a manifold N are called *f-related* on M, where a mapping $f:M\longrightarrow N$, if

$f(\xi(x))=\vartheta(y)$ for all $x\in M$, $y=f(x)$ (see, for example, Kobayashi, Nomizu [1] or Sternberg [1]).

Proof. Let $(V,\psi,\mathbb{R}^L,\zeta)$ and $(U,\varphi,\mathbb{R}^K,\chi)$ be charts on M and N, respectively, such that a given point $a\in V$ and $f(V)\subset U$. By Proposition 1.6, we may assume that the chart $(U,\varphi,\mathbb{R}^K,\chi)$ already has the required form, i.e., it has coordinates $x=(x^\mu;\mu=1,\dots,m)$, $u=(u^i;i\in I)$, $K=m+I$, and basis Cartan fields $\chi_\mu=\partial_{x^\mu}$, $\mu=1,\dots,m$. By the definition of a submersion, we may assume that the chart $(V,\psi,\mathbb{R}^L,\zeta)$ has coordinates $x,u,w=(v^j;j\in L)$ and the mapping f takes a point (x,u,w) to the point (x,u). Suppose that basis Cartan fields

$$\zeta_\lambda=\sum_{\mu=1}^{m} X^\mu_\lambda(x,u,w)\,\frac{\partial}{\partial x^\mu}+\sum_{i\in I} U^i_\lambda(x,u,w)\,\frac{\partial}{\partial u^i}+\sum_{j\in L} V^j_\lambda(x,u,w)\,\frac{\partial}{\partial v^j},$$

$\lambda=1,\dots,m+n$, then

$$f_*\left(\zeta_\lambda\right)=\sum_{\mu=1}^{m} X^\mu_\lambda(x,u,w)\,\frac{\partial}{\partial x^\mu}+\sum_{i\in I} U^i_\lambda(x,u,w)\,\frac{\partial}{\partial u^i}$$

$$=\sum_{\mu=1}^{m} X^\mu_\lambda(x,u,w)\,\partial_{x^\mu},\quad \lambda=1,\dots,m+n,$$

by the definition of a Lie-Backlund mapping. It follows from the condition $f_*(CT_aM)=CT_{f(a)}N$ that the rank of the $(m+n)\times n$-matrix X^λ_μ is equal to m. Suppose that, for example, the minor made up from the first m rows is nonzero. Then shrinking the charts if necessary we can pass from the basis fields $\zeta_1,\dots,\zeta_{m+n}$ to the basis fields $\xi_1,\dots,\xi_m,\zeta_{m+1},\dots,\zeta_{m+n}$, where

$$\xi_\mu = \partial_{x^\mu} + \sum_{j\in L} W^j_\mu(x,u,w) \frac{\partial}{\partial v^j}, \quad \mu = 1,\dots,m.$$

Further,

$$\begin{aligned}\zeta_{m+\nu} &= \sum_{\mu=1}^{m} X^\mu_{m+\nu}\, \partial_{x^\mu} + \sum_{j\in L} V^j_{m+\nu} \frac{\partial}{\partial v^j} \\ &= \sum_{\mu=1}^{m} X^\mu_{m+\nu}\, \xi_\mu + \sum_{j\in L}\left(V^j_{m+\nu} - \sum_{\mu=1}^{m} X^\mu_{m+\nu} W^j_\mu\right) \frac{\partial}{\partial v^j}, \quad \nu = 1,\dots,n,\end{aligned}$$

and we can pass to the basis fields $\xi_1,\dots,\xi_m,\eta_1,\dots,\eta_n$, where

$$\eta_\nu = \sum_{j\in L}\left(V^j_{m+\nu} - \sum_{\mu=1}^{m} X^\mu_{m+\nu} W^j_\mu\right) \frac{\partial}{\partial v^j} = \sum_{j\in L} H^j_\nu \frac{\partial}{\partial v^j}, \quad \nu = 1,\dots,n.$$

By the construction, the fields $\eta_1,\dots,\eta_n$ are linearly independent, hence, the rank of the matrix H^j_ν is equal to n. Choosing indices $j_1,\dots,j_n$ so that the corresponding minor is nonzero and setting $y^\nu = v^{j_\nu}$, $\nu=1,\dots,n$, $v=(v^j; j\in J)$, $J=L\setminus\{j_1,\dots,j_n\}$, we pass from the fields $\eta_1,\dots,\eta_n$ to the fields $\nabla_{y^1},\dots,\nabla_{y^n}$ of the required form. At last, subtracting appropriate linear combinations of the fields $\nabla_{y^1},\dots,\nabla_{y^n}$ from the fields $\xi_1,\dots,\xi_m$ we pass to basis fields $\nabla_{x^1},\dots,\nabla_{x^m}$ of the required form, thereby concluding the proof.

Example 2.12. Let $\mathcal{M}$ be the differential manifold defined by an evolution equation $v_s=g$ with global coordinates $s, y, \mathbf{v}=(v_i; i\in\mathbb{N})$ and basis Cartan fields

$$\nabla_y = \frac{\partial}{\partial y} + \sum_{i=0}^{\infty} v_{i+1} \frac{\partial}{\partial v_i}, \quad \nabla_s = \frac{\partial}{\partial s} + \sum_{i=0}^{\infty} g_i \frac{\partial}{\partial v_i}.$$

In the same way, let N be defined by $u_t=f$ with coordinates $t,x,\mathbf{u}=(u_i;i\in\mathbb{N})$ and basis fields

$$\partial_x = \frac{\partial}{\partial x} + \sum_{i=0}^{\infty} u_{i+1}\frac{\partial}{\partial u_i},\quad \partial_t = \frac{\partial}{\partial t} + \sum_{i=0}^{\infty} f_i\frac{\partial}{\partial u_i}$$

(cf. Example 2.3). Consider the submersion $F:M\longrightarrow N$ given by the formulas

$$t = s,\ x = y,\ u_i = v_{i+k},\ i\in\mathbb{N},$$

where a number $k\in\{1,2,\ldots\}$ is called the *index* of the submersion. Then

$$F_*\left(\nabla_y\right) = \partial_x \quad\text{and}\quad F_*\left(\nabla_s\right) = \frac{\partial}{\partial t} + \sum_{i=0}^{\infty} g_{i+k}\frac{\partial}{\partial u_i}.$$

The submersion F will be Lie-Backlund if $F_*\left(\nabla_s\right)=\partial_t$, i.e., if $g_{i+k}=f_i$, $i\in\mathbb{N}$, i.e., if

$$f(t,x,\mathbf{u}) = \left[\nabla_y^k g(t,y,\mathbf{v})\right]\Big|_{y=x,\,v_{i+k}=u_i,\,i\in\mathbb{N}}$$

(in particular, the right side does not depend on $v_0,\ldots,v_{k-1}$). Let us introduce the special coordinates $t,x,\mathbf{u},\mathbf{w}$ on M, where

$$t = s,\ x = y,\ \mathbf{u} = (u_i=v_{i+k};i\in\mathbb{N}),\ \mathbf{w} = (u_i=v_{i+k};i=-k,\ldots,-1).$$

In this coordinates the basis fields take the form

$$\nabla_y \equiv \nabla_x = \frac{\partial}{\partial x} + \sum_{i=-k}^{\infty} u_{i+1}\frac{\partial}{\partial u_i} = \partial_x + \sum_{i=-k}^{-1} u_{i+1}\frac{\partial}{\partial u_i},$$

$$\nabla_s \equiv \nabla_t = \frac{\partial}{\partial t} + \sum_{i=-k}^{\infty} g_{i+k}\frac{\partial}{\partial u_i},\quad g_{i+k} = \nabla_x^{i+k} g(t,x,\mathbf{u},\mathbf{v}).$$

Now,

$$\nabla_t = \partial_t + \sum_{i=-k}^{-1} g_{i+k} \frac{\partial}{\partial u_i} \quad \text{if } f(t,x,\mathbf{u}) = \nabla_x^k g(t,x,\mathbf{u},\mathbf{w}).$$

In particular, for $k=1$ the last condition takes the form

$$f(t,x,\mathbf{u}) = \partial_x g + u_0 g_{u_{-1}}, \quad g = g(t,x,\mathbf{u},u_{-1}).$$

It is fulfilled iff

$$g = h + au_{-1}, \quad f = \partial_x h + au_0,$$

where $h(t,x,\mathbf{u})$ and $a(t)$ are some smooth functions.

Example 2.12(a). The Burgers' equation

$$u_t = u_{xx} + uu_x = \left(u_x + \frac{u^2}{2}\right)_x$$

(see Example 2.3(a)) is covered by the equation

$$v_t = v_{xx} + \frac{v_x^2}{2}$$

(the last equation turns into the modified Burgers' equation, see Example 2.3(b), after dilation $v=2u$) with the index $k=1$. In more detail, let $g=v_2+v_1^2/2$ and $f=u_2+uu_1$, then $f=\partial_x g$ and we have the submersion of the above type with the index $k=1$ (here $h=g$, $a=0$).

If we define the differential manifold $\mathcal{M}$ by modified Burgers' equation with coordinates $s,y,\mathbf{v}=(v_i;i\in\mathbb{N})$ and $g=v_2+u_1^2$ and the differential manifold $\mathcal{N}$ by Burgers' equation with coordinates $t,x,\mathbf{u}=(u_i;i\in\mathbb{N})$ and $f=u_2+uu_1$ then the submersion $F:\mathcal{M}\longrightarrow\mathcal{N}$ will be given by the formulas

$$t = s, \; x = y, \; u_i = \frac{1}{2}v_{i+1}, \; i \in \mathbb{N}.$$

Let $\mathcal{M}=(M,CTM)$ be a differential manifold and Aut$\mathcal{M}$ be the group of all *Lie-Backlund automorphisms* of $\mathcal{M}$, i.e., Lie-Backlund isomorphisms $f:\mathcal{M}\to\mathcal{M}$. Every Lie-Backlund automorphism $f\in$Aut$\mathcal{M}$ takes an integral submanifold of $\mathcal{M}$ into another integral submanifold of $\mathcal{M}$. We denote by Aut$_0\mathcal{M}$ the set of all trivial Lie-Backlund automorphisms, taking every maximal integral submanifold into itself (remind that an integral submanifold is called *maximal* if it is not a proper part of any bigger integral submanifold).

Lemma 2.13. *The set* Aut$_0\mathcal{M}$ *is a normal subgroup of* Aut$\mathcal{M}$.

Proof. Let $f,g\in$Aut$\mathcal{M}$. We should prove that $g^{-1}f\in$Aut$_0\mathcal{M}$ iff $fg^{-1}\in$Aut$_0\mathcal{M}$. But, by the definition, $g^{-1}f\in$Aut$_0\mathcal{M}$ ($fg^{-1}\in$Aut$_0\mathcal{M}$) iff $g^{-1}f(S)=S$ ($fg^{-1}(S)=S$), i.e., iff $f(S)=g(S)$ ($g^{-1}(S)=f^{-1}(S)$), for any maximal integral submanifold S. Now, the last two conditions are clearly equivalent.

Thus, the quotient group Tr$\mathcal{M}$=Aut$\mathcal{M}\big/$Aut$_0\mathcal{M}$ of *Lie-Backlund transformations* of a differential manifold $\mathcal{M}$ is defined.

Example 2.14. Consider an ordinary differential equation $u_x=f$ and the corresponding differential manifold $\mathcal{M}$ with global coordinates x,u and a basis Cartan field

$$\partial_x = \frac{\partial}{\partial x} + f(x,u)\,\frac{\partial}{\partial u}$$

(see Examples 1.1 and 2.2). An automorphism $F:\mathcal{M}\simeq\mathcal{M}$ given by the formulas

$$x' = X(x,u), \quad u' = U(x.u),$$

will be Lie-Backlund iff smooth functions X,U satisfy the defining equation

$$\partial_x U - f(X,U)\, \partial_x X = 0,$$

i.e., iff

$$(U_x + f(x,u)U_u) - f(X,U)(X_x + f(x,u)X_u) = 0.$$

Here we can choose $X(x,u)$ to be an arbitrary smooth function and then find $U(x,u)$ as a solution of the quasi-linear PDE (sure, we should check that the mapping F will be an automorphism).

Example 2.14(a). In particular, let $f(x,u)=u$. For a fixed function $X(x,u)$ a function $U(x,u)$ should satisfy the equation

$$\partial_x U = U\partial_x X.$$

Let us put $U=V(x,u)e^X$ then V should be a solution of the equation

$$\partial_x V = V_x + uV_u = 0.$$

The general solution of the last PDE is $V=\varphi(ue^{-x})$, where $\varphi(t)$ is an arbitrary smooth function. Hence, in this case the group Aut$\mathcal{M}$ consists of all mappings $F:\mathcal{M}\simeq\mathcal{M}$ given by the formulas

$$x' = X(x,u), \quad u' = \varphi(ue^{-x})e^X,$$

where X,φ are arbitrary smooth functions such that F is invertible (note that in this case the Jacobian of the mapping F is $J=e^{X-x}\varphi'\partial_x X$, where φ' is the derivative of the function φ). Such mapping takes an integral curve $\gamma=\{x,u;u=ce^x\}$, where c is an arbitrary constant, into the integral curve $\gamma'=\{x,u;u=\varphi(c)e^{x}\}$ and it is trivial iff $\varphi(t)\equiv t$. Thus, the subgroup $\mathrm{Aut}_0\mathcal{M}$ consists of all trivial mappings

$$x' = X(x,u),\ u' = ue^{X-x}.$$

At last, the quotient group $\mathrm{Tr}\mathcal{M}$ has a parameterization

$$x' = x,\ u' = \varphi(ue^{-x})e^{x}.$$

Example 2.15. Let $\mathcal{M}$ be a differential manifold covered by a global foliating chart with coordinates x^{μ}, u^{i} and basis Cartan fields $\partial_{\mu}=\partial\Big/\partial x^{\mu}$, $\mu=1,\ldots,m$, $i\in I$. The defining equations $\partial_{\mu}U^{i}=0$, $\mu=1,\ldots,m$, $i\in I$, for an automorphism $F:\mathcal{M}\simeq\mathcal{M}$ given by the formulas,

$$x' = X(x,u),\ u' = U(x,u)$$

have the general solution $X=X(x,u)$, $U=U(u)$, where X,U are arbitrary smooth functions, such that the inverse mapping $F^{-1}:\mathcal{M}\simeq\mathcal{M}$ is defined. An integral submanifold $S=\{x,u;u=c\}$, where c is a constant, is mapped by such automorphism into the integral submanifold $S'=\{x,u;u=U(c)\}$. This automorphism is trivial iff $U(u)\equiv u$. Hence, the subgroup $\mathrm{Aut}_0\mathcal{M}$ consists of all trivial mappings

$$x' = X(x,u),\ u' = u,$$

and the quotient group $\mathrm{Tr}\mathcal{M}$ has a parameterization

$$x' = x,\ u' = U(u).$$

Example 2.16. Consider the differential manifold $\mathcal{M}=(\mathbb{R}\times\mathbb{R}_{\mathbb{N}},D_{x})$ with global coordinates x, $\mathbf{u}=(u\ ;i\in\mathbb{N})$ and a basis Cartan field

$$D_x = \frac{\partial}{\partial x} + \sum_{i\in\mathbb{N}} u_{i+1} \frac{\partial}{\partial u_i}$$

(see Example 1.5). The defining equations

$$D_x U_i - U_{i+1} D_x X = 0, \quad i \in \mathbb{N},$$

for an automorphism $F:\mathcal{M}\simeq\mathcal{M}$, given by the formulas

$$x' = X(x,\mathbf{u}), \quad u'_i = U_i(x,\mathbf{u}), \quad i \in \mathbb{N},$$

have the general solution

$$U_i = \left(\frac{1}{D_x X} D_x\right)^i U, \quad i \in \mathbb{N},$$

where $X, U(x,u_0,\ldots,u_p)$ are arbitrary smooth functions (the last representation is of course valid in the points where $D_x X \neq 0$). The inverse mapping $F^{-1}:\mathcal{M}\simeq\mathcal{M}$ is given by the formulas

$$x = Y, \quad u_i = V_i = \left(\frac{1}{D_x Y} D_x\right)^i V, \quad i \in \mathbb{N},$$

where $Y, V(x',u'_0,\ldots,u'_q)$ are smooth functions such that

$$x \equiv Y(X,U_0,\ldots,U_q), \quad u_i \equiv V_i(X,U_0,\ldots,U_q), \quad i \in \mathbb{N}.$$

The simple analysis shows that this is possible only if $X=X(x,u_0)$, $U=U(x,u_0)$ (and then $Y=Y(x',u'_0)$, $V(x',u'_0)$). Now, a mapping $F:\mathcal{M}\rightarrow\mathcal{M}$, defined by smooth functions $X,U(x,u_0)$, takes an integral curve γ, defined by a parameterization $x=f(t)$, $u_0=g(t)$ (see Example 1.5), into the integral curve γ', parameterized by $x'=X(f((t),g(t))$, $u'_0=U(f(t),g(t))$. It is easy to check that $\gamma=\gamma'$ for any maximal integral curve γ iff $X=x$, $U=u_0$, i.e., iff F is the identity mapping.

Example 2.17. Consider the differential manifold $\mathcal{M}=(\mathbb{R}\times\mathbb{R}^2_{\mathbb{N}}, D_x)$ with global coordinates x, $\mathbf{u}=(u_i; i\in\mathbb{N})$, $\mathbf{v}=(v_i; i\in\mathbb{N})$ and basis Cartan field

$$D_x = \frac{\partial}{\partial x} + \sum_{i\in\mathbb{N}} \left(u_{i+1} \frac{\partial}{\partial u_i} + v_{i+1} \frac{\partial}{\partial v_i} \right).$$

The defining equations

$$D_x U_i - U_{i+1} D_x X = 0, \; D_x V_i - V_{i+1} D_x X = 0, \quad i \in \mathbb{N},$$

for an automorphism $F:\mathcal{M}\simeq\mathcal{M}$, given by the formulas

$$x' = X(x,\mathbf{u},\mathbf{v}), \; u'_i = U_i(x,\mathbf{u},\mathbf{v}), \; v'_i = V_i(x,\mathbf{u},\mathbf{v}), \quad i \in \mathbb{N},$$

have the general solution

$$U_i = \left(\frac{1}{D_x X} D_x \right)^i U, \; V_i = \left(\frac{1}{D_x X} D_x \right)^i V, \quad i \in \mathbb{N},$$

where $X,U,V(x,u_0,\ldots,u_p,v_0,\ldots,v_q)$ are arbitrary smooth functions. The inverse mapping $F^{-1}:\mathcal{M}\simeq\mathcal{M}$ is given by the formulas

$$x = Y, \; u_i = R_i = \left(\frac{1}{D_x Y} D_x \right)^i R, \; v_i = S_i = \left(\frac{1}{D_x Y} D_x \right)^i S, \quad i \in \mathbb{N},$$

where $Y,R,S(x',u'_0,\ldots,u'_k,v'_0,\ldots,v'_l)$ are smooth functions such that

$$x \equiv Y(X,\mathbf{U},\mathbf{V}), \; u_i \equiv R_i(X,\mathbf{U},\mathbf{V}), \; v_i \equiv S_i(X,\mathbf{U},\mathbf{V}), \quad i \in \mathbb{N}.$$

In this case there are automorphisms defined by functions X,U,V of higher orders. For example, we may choose

$$X = x, \; U = u_0, \; V = v_0 + f(x,\mathbf{u}),$$

where f is any smooth function of any order, then the inverse mapping will be defined by the functions

$$Y = x', \; R = u'_0, \; S = v'_0 - f(x',\mathbf{u}').$$

3. Lie-Backlund Fields and Infinitesimal Symmetries

Let $\mathcal{M}=(M,CTM)$ be a differential manifold with the tangent bundle TM and its Cartan subbundle CTM. Let $\mathcal{O}$ be an open subset of M. A vector field $\zeta\in\mathcal{T}(\mathcal{O})$ is called a *Lie-Backlund field* if the Lie bracket $[\zeta,\eta]\in C\mathcal{T}(\mathcal{O})$ for any $\eta\in C\mathcal{T}(\mathcal{O})$, where $C\mathcal{T}(\mathcal{O})$ is the Lie subalgebra of Cartan fields of the Lie algebra $\mathcal{T}(\mathcal{O})$ of vector fields on $\mathcal{O}$. We denote by $LB\mathcal{T}(\mathcal{O})$ the set of all Lie-Backlund fields on $\mathcal{O}$. By Jacoby identity $LB\mathcal{T}(\mathcal{O})$ is a Lie subalgebra of $\mathcal{T}(\mathcal{O})$ and by the definition $C\mathcal{T}(\mathcal{O})$ is its ideal. Hence, the quotient algebra Sym$\mathcal{O}$= $LB\mathcal{T}(\mathcal{O})\big/C\mathcal{T}(\mathcal{O})$ of *infinitesimal symmetries* (*symmetries*, for brevity) of $\mathcal{O}$ is defined.

Proposition 3.1. *Let $\mathcal{M}=(M,CTM)$ be a differential manifold and $(U,\varphi,\mathbb{R}^{m+I},\partial)$ a standard chart on $\mathcal{M}$ with coordinates $x=(x^\mu)$, $u=(u^i)$ and basis Cartan fields*

$$\partial_\mu = \frac{\partial}{\partial x^\mu} + \sum_{i\in I} A^i_\mu(x,u)\,\frac{\partial}{\partial u^i},\qquad \mu = 1,\ldots,m.$$

A vector field (see Proposition 1.10)

$$\zeta = \sum_{\mu=1}^{m} \xi^\mu(x,u)\,\partial_\mu + \sum_{i\in I} \eta^i(x,u)\,\frac{\partial}{\partial u^i} = \xi + \eta \in \mathcal{T}(U)$$

will be Lie-Backlund iff smooth functions η^i satisfy the defining equations

$$\partial_\mu \eta^i - \sum_{j\in I} \frac{\partial A^i_\mu}{\partial u^j} \eta^j = 0, \quad \mu = 1,\dots,m, \; i \in I.$$

Note that there are no restrictions on smooth functions ξ^μ.

Proof. Indeed, the Lie bracket $[\xi,\partial_\mu]\in C\mathcal{T}(U)$ for any smooth functions ξ^μ and $\mu=1,\dots,m$. Hence, the vector field ζ will be Lie-Backlund iff the field η is Lie-Backlund. Now, the Lie bracket

$$\left[\partial_\mu,\eta\right] = \sum_{i\in I} \left(\partial_\mu \eta^i\right) \frac{\partial}{\partial u^i} + \sum_{i\in I} \eta^i \, \partial_\mu \frac{\partial}{\partial u^i} - \sum_{j\in I} \eta^j \frac{\partial}{\partial u^j} \partial_\mu$$

$$= \sum_{i\in I} \left(\partial_\mu \eta^i - \sum_{j\in I} \eta^j \frac{\partial A^i_\mu}{\partial u^j}\right) \frac{\partial}{\partial u^i} \in C\mathcal{T}(U), \quad \mu = 1,\dots,m,$$

iff the defining equations are valid.

Note that the field η will be Lie-Backlund iff $[\partial_\mu,\eta]=0$ for $\mu=1,\dots,m$.

Proposition 3.2. *Let Γ be a local one-parameter group of Lie-Backlund transformations of a differential manifold $\mathcal{M}=(M,CTM)$. Then its vector field ζ* (see Section 0.3) *is Lie-Backlund.*

Proof. Let $(U,\varphi,\mathbb{R}^{m+I},\partial)$ be a standard chart on $\mathcal{M}$ with coordinates $x=(x^\mu)$, $u=(u^i)$ and basis Cartan fields ∂_μ (we are keeping the notation of Proposition 3.1). Suppose that the group Γ is given on U by the formulas

$$x^\mu(t) = X^\mu(t;x,u), \; X^\mu(0;x,u) = x^m, \; \mu = 1,\dots,m,$$

$$u^i(t) = U^i(t;x,u), \; U^i(0;x,u) = u^i, \; i \in I,$$

where X^μ, U^i are smooth functions. Then the vector field ζ will has the form

$$\zeta = \sum_{\mu=1}^{m} \xi^{\mu} \frac{\partial}{\partial x^{\mu}} + \sum_{i\in I} \vartheta^{i} \frac{\partial}{\partial u^{i}} = \sum_{\mu=1}^{m} \xi^{\mu} \partial_{\mu} + \sum_{i\in I} \eta^{i} \frac{\partial}{\partial u^{i}},$$

where

$$\xi^{\mu}(x,u) = \left.\frac{\partial X^{\mu}(t,x,u)}{\partial t}\right|_{t=0}, \quad \vartheta^{i}(x,u) = \left.\frac{\partial U^{i}(t,x,u)}{\partial t}\right|_{t=0},$$

$$\eta^{i} = \vartheta^{i} - \sum_{\mu=1}^{m} \xi^{\mu} A^{i}_{\mu}.$$

Now, the mappings $(x,u)\mapsto(x(t),u(t))$ are Lie-Backlund. Hence, by Proposition 2.1, the functions X^{μ},U^{i} satisfy the defining equations

$$\partial_{\mu}U^{i}(t;x,u) - \sum_{\nu=1}^{m} A^{i}_{\nu}(X(t;x,u),U(t;x,u))\, \partial_{\mu}X^{\nu}(t;x,u) = 0,$$

for all $\mu=1,\dots,m$, $i\in I$. Taking derivatives of these equations with respect to t at the point $t=0$ we get

$$\partial_{\mu}\vartheta^{i} - \sum_{\nu=1}^{m}\left(\sum_{\lambda=1}^{m} \frac{\partial A^{i}_{\nu}(x,u)}{\partial x^{\lambda}} \xi^{\lambda} + \sum_{j\in I} \frac{\partial A^{i}_{\nu}(x,u)}{\partial u^{j}} \vartheta^{j}\right) \delta^{\nu}_{\mu} - \sum_{\nu=1}^{m} A^{i}_{\nu}(x,u)\, \partial_{\mu}\xi^{\nu} = 0,$$

or

$$\partial_{\mu}\left(\eta^{i} + \sum_{\nu=1}^{m} A^{i}_{\nu}\, \xi^{\nu}\right) - \sum_{\nu=1}^{m} \frac{\partial A^{i}_{\mu}}{\partial x^{\nu}} \xi^{\nu} - \sum_{j\in I} \frac{\partial A^{i}_{\mu}}{\partial u^{j}} \left(\eta^{j} + \sum_{\nu=1}^{m} A^{j}_{\nu}\, \xi^{\nu}\right)$$
$$- \sum_{\nu=1}^{m} A^{i}_{\nu}\, \partial_{\mu}\xi^{\nu} = 0,$$

or

$$\partial_{\mu}\eta^{i} + \sum_{\nu=1}^{m} \left(\partial_{\mu}A^{i}_{\nu}\right) \xi^{\nu} - \sum_{\nu=1}^{m} \left(\frac{\partial A^{i}_{\mu}}{\partial x^{\nu}} + \sum_{j\in I} A^{j}_{\nu} \frac{\partial A^{i}_{\mu}}{\partial u^{j}}\right) \xi^{\nu} - \sum_{j\in I} \frac{\partial A^{i}_{\mu}}{\partial u^{j}} \eta^{j} = 0,$$

i.e.,

$$\partial_\mu \eta^i \mu - \sum_{j\in I} \frac{\partial A^i_\mu}{\partial u^j} \eta^j + \sum_{\nu=1}^{m} \left(\partial_\mu A^i_\nu - \partial_\nu A^i_\mu \right) \xi^\nu = 0.$$

Thus, taking into account the equalities $\partial_\mu A^i_\nu = \partial_\nu A^i_\mu$ characterizing involutive distributions (see the definition of standard charts in Section 1), we come to the defining equations of Proposition 3.1. This ends the proof.

If the manifold $\mathcal{M}$ is finite-dimensional the converse result is also true.

Proposition 3.3. *Let $\mathcal{M}=(M,CTM)$ be a finite-dimensional differential manifold and ζ a Lie-Backlund field on $\mathcal{M}$. Then the local one-parameter transformation group Γ on M generated by ζ will be Lie-Backlund.*

Proof. By the classical result of Lie the vector field ζ indeed generates a local one-parameter transformation group Γ and the only fact we should check is that this group will be Lie-Backlund. So, let $(U,\varphi,\mathbb{R}^{m+I},\partial)$ be a foliating chart on $\mathcal{M}$ with coordinates $x=(x^\mu),u=(u^i)$ and basis Cartan fields $\partial_\mu=\partial/\partial x^\mu$. The defining equations for the vector field

$$\zeta = \sum_{\mu=1}^{m} \xi^\mu(x,u)\, \partial_\mu + \sum_{i\in I} \eta^i(x,u) \frac{\partial}{\partial u^i}$$

(we use the notation of Proposition 3.2) take the form

$$\frac{\partial \eta^i}{\partial x^\mu} = 0, \quad \mu = 1,\dots,m, \ i \in I,$$

(note that $A^i_\mu=0$, hence, $\vartheta^i=\eta^i$ here) i.e., components $\eta^i=\eta^i(u)$ do not depend on x. Further, the Lie equations for the group Γ have the form

$$\frac{dX^\mu(t;x,u)}{dt} = \xi^\mu(X,U), \quad X^\mu(0,x,u) = x^\mu, \quad \mu = 1,\dots,m,$$
$$\frac{dU^i(t;x,u)}{dt} = \eta^i(U), \qquad U^i(0;x,u) = u^i, \quad i \in I.$$

In the considered here finite-dimensional case this Cauchy problem has an unique solution. Clear that functions $U^i=U^i(t;u)$ do not depend on x. Hence, the mappings $(x,u)\mapsto(x(t),u(t))$ are Lie-Backlund (see the proof of Proposition 2.6), q.e.d.

There are no reasons to expect this result to be true in the infinite-dimensional case considering the fact that far not every field can be integrated to a local one-parameter group in this case (see Section 0.6).

Let us study the correspondence between trivial transformations and Cartan fields.

Proposition 3.4. *Let $\mathcal{M}=(M,CTM)$ be a differential manifold with a foliating atlas (i.e., an atlas consisting of foliating charts). Let Γ be a local one-parameter group of trivial transformations of $\mathcal{M}$. Then its vector field ζ is Cartan.*

Proof. Let $(U,\varphi,\mathbb{R}^{m+I},\partial)$ be a foliating chart on $\mathcal{M}$ with coordinates $x=(x^\mu),u=(u^i)$ and basis Cartan fields $\partial_\mu=\partial/\partial x^\mu$. The group Γ given on U by the formulas

$$x^{\mu}(t) = X^{\mu}(t;x,u),\ X^{\mu}(0;x,u) = x^{m},\ \mu = 1,\dots,m,$$

$$u^{i}(t) = U^{i}(t;x,u),\ U^{i}(0;x,u) = u^{i},\ i \in I,$$

will be trivial iff $U^{i}(t;x,u)=U^{i}(0;x,u)=u^{i}$, $i \in I$. Now, the vector field ζ generated by Γ will has the form

$$\zeta = \sum_{\mu=1}^{m} \xi^{\mu}\, \partial_{\mu} + \sum_{i\in I} \eta^{i} \frac{\partial}{\partial u^{i}},$$

where

$$\eta^{i} = \vartheta^{i} = \frac{\partial U^{i}}{\partial t} = 0,\quad i \in I.$$

Hence,

$$\zeta = \sum_{\mu=1}^{m} \xi^{\mu}\, \partial_{\mu} \in C\mathcal{T}(U).$$

Proposition 3.5. *Let ζ be a Cartan field on a finite-dimensional differential manifold $\mathcal{M}=(M,CTM)$. Then the local one-parameter group Γ generated by ζ will be trivial.*

Proof. In the notation of Proposition 3.3 the field ζ has the form

$$\zeta = \sum_{\mu=1}^{m} \xi^{\mu}(x,u)\, \partial_{\mu}.$$

Hence, the defining equations for the group Γ will take the form

$$\frac{dX^{\mu}(t;x,u)}{dt} = \xi^{\mu}(X,U),\ X^{\mu}(0,x,u) = x^{\mu},\ \mu = 1,\dots,m,$$

$$\frac{dU^{i}(t;x,u)}{dt} = 0,\qquad U^{i}(0;x,u) = u^{i},\ i \in I,$$

because $\eta^{i}=0$ here. Thus, $U^{i}(t;x,u)=u^{i}$, $i\in I$, and the group Γ is trivial.

We see that, though there is no direct correspondence between local one-parameter symmetry groups of a differential manifold $\mathcal{M}$ and elements of the quotient algebra Sym$\mathcal{M}$, there are enough reasons to call the latter infinitesimal symmetries of $\mathcal{M}$. Note that in a standard chart on $\mathcal{M}$ every equivalence class $[\zeta]\in$Sym$\mathcal{M}$ has the unique representative $\zeta\in LB\mathcal{T}(\mathcal{M})$ of the form

$$\zeta = \sum_{i\in I} \eta^{i}(x,u) \frac{\partial}{\partial u^{i}}.$$

Note, also, that the defining equations for components η^{i} are always linear while the defining equations for Lie-Backlund automorphisms are in the general case nonlinear.

Example 3.6. Consider the differential manifold $\mathcal{M}$ defined by an ordinary differential equation $u_{t}=f(t,u)$ (see Example 1.1), with the global coordinates t,u and the basis Cartan field

$$\partial_{t} = \frac{\partial}{\partial t} + f(t,u) \frac{\partial}{\partial u}.$$

A vector field

$$\zeta = \eta(t,u) \frac{\partial}{\partial u}$$

will be Lie-Backlund iff the smooth function $\eta(t,u)$ satisfies the defining equation

$$\partial_{t}\eta - \frac{\partial f}{\partial u} \eta = 0, \text{ or } \eta_{t} + f\eta_{u} - f_{u}\eta = 0.$$

Note that the last equation can be also rewritten in the form

$$\left(\frac{1}{\eta}\right)_{t} + \left(\frac{f}{\eta}\right)_{u} = 0.$$

Suppose, for example, that $f(t,u)=\varphi(t)\psi(u)$, where φ,ψ are some smooth functions, and let us look for solutions $\eta(t,u)=\alpha(t)\beta(u)$ then for smooth functions α,β we get the equation

$$\alpha_t\beta + \varphi\psi\alpha\beta_u - \varphi\psi_u\alpha\beta = 0$$

implying

$$\frac{\alpha_t}{\alpha\varphi} = \psi_u - \frac{\psi\beta_u}{\beta} = c = const.$$

The last two equations have the general solutions

$$\alpha(t) = \exp\left(c\int\varphi(t)dt\right),\ \beta(u) = \psi(u)\exp\left(-c\int\frac{du}{\psi(u)}\right).$$

Hence, in this case

$$\eta(t,u) = \psi(u)\exp\left[c\left(\int\varphi(t)dt - \int\frac{du}{\psi(u)}\right)\right].$$

Example 3.7. Let $\mathcal{M}=(\mathbb{R}\times\mathbb{R}_{\mathbb{N}},\partial_x)$ be the differential manifold with the global coordinates $x,\mathbf{u}=(u_i;i\in\mathbb{N})$ and the basis Cartan field

$$\partial_x = \frac{\partial}{\partial x} + \sum_{i\in\mathbb{N}} u_{i+1}\frac{\partial}{\partial u_i}$$

(see Example 0.6.1). A vector field

$$\zeta = \sum_{i\in\mathbb{N}} \eta_i(x,\mathbf{u})\frac{\partial}{\partial u_i}$$

will be Lie-Backlund iff the smooth functions η_i satisfy the defining equations

$$\partial_x\eta_i - \eta_{i+1} = 0,\ i \in \mathbb{N},$$

i.e., iff $\eta_i=\partial_x^i\varphi$, $i\in\mathbb{N}$, where $\varphi(x,\mathbf{u})$ is a smooth function. Hence, in this case infinitesimal symmetries are in fact *evolution differentiations* (see, for example, Vinogradov, Krasil'shchik, Lychagin [1; Ch.3, §2])

$$\zeta = \mathrm{E}_\varphi = \sum_{i\in\mathbb{N}} \partial_x^i\varphi \frac{\partial}{\partial u_i}.$$

By the definition, $[\partial_x,\mathrm{E}_\varphi]=0$, hence,

$$\partial_x^i\mathrm{E}_\varphi(f) = \mathrm{E}_\varphi(\partial_x^i f)$$

for all smooth functions $\varphi, f(x,\mathbf{u})$ and $i\in\mathbb{N}$. Moreover, let E_φ and E_ψ be evolution differentiations then their Lie bracket is also evolution differentiation,

$$\left[\mathrm{E}_\varphi,\mathrm{E}_\psi\right] = \mathrm{E}_\chi,$$

where

$$\chi = \{\varphi,\psi\} = \mathrm{E}_\varphi(\psi) - \mathrm{E}_\psi(\varphi)$$

is the *Jacoby bracket* of the smooth functions $\varphi,\psi(x,\mathbf{u})$. To prove this it is sufficient to use the previous equality. Now, the set $\mathcal{C}^\infty(\mathbb{R}\times\mathbb{R}_\mathbb{N})$ of all smooth functions $\varphi(x,\mathbf{u})$ forms the Lie algebra with the Jacoby bracket as the multiplication. Moreover, the Lie algebra Sym$\mathcal{M}$ of infinitesimal symmetries of the differential manifold $\mathcal{M}=(\mathbb{R}\times\mathbb{R}_\mathbb{N},\partial_x)$ is isomorphic to $\mathcal{C}^\infty(\mathbb{R}\times\mathbb{R}_\mathbb{N})$.

Example 3.8. (cf. Ibragimov [1; §19]) Let $\mathcal{M}$ be the differential manifold defined by an evolution equation $u_t=f$ (see Example

1.4) with the global coordinates $t, x, \mathbf{u}=(u\ ; i\in\mathbb{N})$ and the basis Cartan fields

$$\partial_x = \frac{\partial}{\partial x} + \sum_{i\in\mathbb{N}} u_{i+1} \frac{\partial}{\partial u_i},$$

$$\partial_t = \frac{\partial}{\partial t} + \sum_{i\in\mathbb{N}} \partial_x^i f \frac{\partial}{\partial u_i}$$

$$= \frac{\partial}{\partial t} + \mathrm{E}_f.$$

A vector field

$$\zeta = \sum_{i\in\mathbb{N}} \eta_i(t,x,\mathbf{u}) \frac{\partial}{\partial u_i}$$

will be Lie-Backlund iff the smooth functions η_i satisfy the defining equations

$$\partial_x \eta_i - \eta_{i+1} = 0, \quad i \in \mathbb{N},$$

$$\partial_t \eta_i - \sum_{k\in\mathbb{N}} \frac{\partial}{\partial u_k}\left(\partial_x^i f\right) \eta_k = 0, \quad i \in \mathbb{N}.$$

The first subsystem easily yields $\eta_i=\partial_x^i\varphi$, $i\in\mathbb{N}$, where $\varphi(t,x,\mathbf{u})$ is a smooth function (i.e., ζ is an evolution differentiation). Then the second one takes the form

$$\partial_t \partial_x^i \varphi - \mathrm{E}_\varphi(\partial_x^i f) = 0, \quad i \in \mathbb{N},$$

i.e.,

$$\partial_x^i\left(\partial_t \varphi - \mathrm{E}_\varphi(f)\right) = 0, \quad i \in \mathbb{N},$$

where we used the equalities $\partial_t\partial_x^i=\partial_x^i\partial_t$ and $\mathrm{E}_\varphi(\partial_x^i f)=\partial_x^i \mathrm{E}_\varphi(f)$. Thus, infinitesimal symmetries of an evolution equation $u_t=f$ are in fact evolution differentiations

$$\mathrm{E}_\varphi = \sum_{i\in\mathbb{N}} \partial_x^i\varphi \frac{\partial}{\partial u_i},$$

where smooth functions $\varphi(t,x,\mathbf{u})$ satisfy the single defining equation

$$\partial_t\varphi - f_*\varphi = 0.$$

The linear differential operator

$$f_* = \sum_{k\in\mathbb{N}} f_{u_k}\partial_x^k$$

acting on smooth functions $\chi(x,\mathbf{u})$ is called the *universal linearization operator* of a smooth function f (see, for example, Vinogradov, Krasil'shchik, Lychagin [1; Ch.3, §2], t is a parameter). By the definition, $f_*(\varphi)=\mathrm{E}_\varphi(f)$. Note that the Jacoby bracket $\{\varphi,\psi\}$ of any two solutions φ,ψ of the defining equation is also the solution of this equation. Now, the set $\mathcal{C}^\infty(\mathbb{R}^2\times\mathbb{R}_\mathbb{N})$ of all smooth functions of $\varphi(t,x,\mathbf{u})$ forms the Lie algebra with the Jacoby bracket $\{\cdot,\cdot\}$ as the multiplication and the set of all solutions of the defining equation $\partial_t\varphi-f_*\varphi=0$ is its subalgebra. Moreover, the Lie algebra $\mathrm{Sym}\mathcal{M}$ of infinitesimal symmetries of the differential manifold $\mathcal{M}$ (i.e., of the evolution equation $u_t=f$) is isomorphic to this subalgebra. We shall identify them.

Example 3.8(a). Consider the heat equation $u_t=u_{xx}$. Here, $f=u_2$ and

$$\mathrm{E}_f = \sum_{i\in\mathbb{N}} u_{i+2}\frac{\partial}{\partial u_i}, \quad f_* = \partial_x^2.$$

Hence, the defining equation for a function $\varphi(t,x,u_0,\dots,u_p)$ of an order p has the form

$$\left(\partial_t - \partial_x^2\right)\varphi = \varphi_t - \varphi_{xx} - 2\sum_{i=0}^{p} u_{i+1}\varphi_{xu_i} - \sum_{i,k=0}^{p} u_{i+1}u_{k+1}\varphi_{u_{i+1}u_{k+1}} = 0.$$

To solve this equation we shall use the technics of Vinogradov, Krasil'shchik [1]. First note that a function $\varphi(t,x)$ (so to say of the order $p=-1$) will be a solution of the defining equation iff it is a solution of the original heat equation. Moreover, the Jacoby bracket $\{\varphi,\psi\}=0$ for any functions $\varphi,\psi(t,x)$. Thus, the set $\mathcal{Sol}$ of all solution of the heat equation is the commutative Lie subalgebra of the Lie algebra Sym$\mathcal{M}$ of all solutions of the defining equation. Further, by direct computations one can check that there are the following solutions of lower orders (see, for example, Ibragimov [1; §18.1])

$$\varphi_0^0 = u_0,$$

$$\varphi_1^0 = u_1,\ \varphi_1^1 = tu_1+\tfrac{x}{2}u_0,$$

$$\varphi_2^0 = u_2,\ \varphi_2^1 = tu_2+\tfrac{x}{2}u_1,\ \varphi_2^2 = t^2u_2+txu_1+\tfrac{1}{4}(2t+x^2)u_0,$$

and obvious solutions

$$\varphi_3^0 = u_3,\ \varphi_4^0 = u_4,\dots .$$

It is easy to check that any solution $\varphi_p=\varphi_p(\dots,u_p)$ of the defining equation has the representation

$$\varphi_p = a(t)u_p+(\tfrac{1}{2}a'x+b(t))u_{p-1}+\psi_{p-2}$$

(cf. Examples 2.3(a) and 2.3(b)), where $a(t),b(t),\psi_{p-2}(\ldots,u_{p-2})$ are smooth functions. Now, suppose that φ_p is a solution of an order p. As we know, the Jacoby bracket of any two solutions is also the solution, hence,

$$\left\{\varphi_1^0,\varphi_p\right\} = -\left(\varphi_p\right)_x = -\tfrac{1}{2}a'u_{p-1}+\left(\psi_{p-2}\right)_x$$

is the solution of the order $p-1$, $a'=da\big/dt$. Repeating this procedure $p-1$ more times we get the solution of the order 0

$$\varphi_0 = 2^{-p}a^{(p)}u_0+\left(\psi_{p-2}\right)_{x\ldots x} = cu_0+\psi(t,x)$$

(see the solutions of lower orders listed above). Thus, $a(t)$ is a polynomial of the pth degree. On the other side, suppose that

$$\varphi_p^k = t^k u_p+\psi^k(\ldots,u_{p-1})$$

is a solution of the defining equation. Then the Jacoby bracket

$$\left\{\varphi_2^2,\varphi_p^k\right\} = (p-k)t^{k+1}u_p+\ldots = (p-k)\varphi_p^{k+1},$$

where $\varphi_p^{k+1}=t^{k+1}u_p+\psi^{k+1}(\ldots,u_{p-1})$ is the solution of the same type, $p=1,2,\ldots,$ $k=0,\ldots,p-1$. Hence, starting from the obvious solutions $\varphi_p^0=u_p$ we can get a solution $\varphi_p=a(t)u_p+\psi_{p-1}$, where $a(t)$ is any given polynomial of a degree $\le p$ and ψ_{p-1} is some smooth function of an order $\le p-1$. This provides us with the method of construction of a general solution of a given order by induction. Further, if φ_p^k is a solution of the above type then the Jacoby brackets

$$\left\{\varphi_p^k, \varphi_1^0\right\} = \frac{k}{2}\varphi_{p-1}^{k-1} \text{ and } \left\{\varphi_p^k, \varphi_3^0\right\} = \frac{3k}{2}\varphi_{p+1}^{k-1} \text{ for } p=1,2,\dots,\ k=1,\dots,p,$$

where φ_{p-1}^{k-1} and φ_{p+1}^{k-1} are solutions of the same type. This leads to the following description.

The Lie algebra Sym$\mathcal{M}$ *of infinitesimal symmetries of the heat equation has the representation*

$$\text{Sym}\mathcal{M} = \text{Sym}_{ess}\mathcal{M} \oplus \mathcal{S}ol,$$

where $\mathcal{S}ol$ *is the commutative ideal of solutions of the heat equation and* Sym$_{ess}\mathcal{M}$ *is the Lie subalgebra, generated by the elements* φ_1^0, φ_2^2 *and* φ_3^0. *In particular,*

$$\varphi_p^k = \sum_{i=0}^{p} a_i^k u_i,$$

where a_i^k *are polynomials in* t,x, $a_p^k=t^k$.

Let $\mathcal{M}=(M,CTM)$ and $\mathcal{N}=(N,CTN)$ be two differential manifold and $f:\mathcal{M}\simeq\mathcal{N}$ be a Lie-Backlund isomorphism. Then the tangent mapping f_* will isomorphically map the Lie algebra $\mathcal{T}(M)$ onto the Lie algebra $\mathcal{T}(N)$, $C\mathcal{T}(M)$ onto $C\mathcal{T}(N)$ and $LB\mathcal{T}(M)$ onto $LB\mathcal{T}(N)$. Hence, the quotient isomorphism $\mathbf{f}_*:\text{Sym}\mathcal{M}\simeq\text{Sym}\mathcal{N}$ is defined.

Example 3.9. Let $\mathcal{M}=(\mathbb{R}\times\mathbb{R}_{\mathbb{N}},\partial_x)$ and $\mathcal{N}=(\mathbb{R}\times\mathbb{R}_{\mathbb{N}},\partial_x)$ be the differential manifolds with coordinates $x,\mathbf{u}=(u_i)$ and $y,\mathbf{v}=(v_i)$ and basis Cartan fields

$$\partial_x = \frac{\partial}{\partial x} + \sum_{i\in\mathbb{N}} u_{i+1}\frac{\partial}{\partial u_i} \text{ and } \partial_y = \frac{\partial}{\partial y} + \sum_{i\in\mathbb{N}} v_{i+1}\frac{\partial}{\partial v_i},$$

respectively (see Example 2.16). Let $f:\mathcal{M}\simeq\mathcal{N}$ be a Lie-Backlund isomorphism given by the formulas

$$y = Y(x,\mathbf{u}),\ \ \mathbf{v} = \mathbf{V}(x,\mathbf{u}) \text{ and } x = X(y,\mathbf{v}),\ \ \mathbf{u} = \mathbf{U}(y,\mathbf{v}),$$

where

$$V_i = \left(\frac{1}{\partial_x Y}\,\partial_x\right)^i V \text{ and } U_i = \left(\frac{1}{\partial_y X}\,\partial_y\right)^i U,\quad i \in \mathbb{N},$$

$Y, V(x,u_0)$ and $X, U(y,v_0)$ are smooth functions (cf. Example 2.16). Let

$$\zeta = \mathrm{E}_\varphi = \sum_{i\in\mathbb{N}} \partial_x^i\varphi\,\frac{\partial}{\partial u_i}$$

be an evolution differentiation on $\mathcal{M}$, then its image

$$\eta = f_*(\zeta) = \chi(y,\mathbf{v})\,\frac{\partial}{\partial y} + \sum_{i\in\mathbb{N}} \eta_i(y,\mathbf{v})\,\frac{\partial}{\partial v_i},$$

where

$$\chi = \mathrm{E}_\varphi(Y)\Big|_{x=X(y,\mathbf{v}),\,\mathbf{u}=\mathbf{U}(y,\mathbf{v})} \text{ and } \eta_i = \mathrm{E}_\varphi(V_i)\Big|_{x=X(y,\mathbf{v}),\,\mathbf{u}=\mathbf{U}(y,\mathbf{v})}.$$

In particular, let $y=x$ (i.e., $Y(x,\mathbf{u})=x$, $X(y,\mathbf{v})=y$), so

$$V_i = \partial_x^i V \text{ and } U_i = \partial_y^i U,\quad i \in \mathbb{N}.$$

Then $\chi=0$ and

$$\eta_i = \mathrm{E}_\varphi\left(\partial_x^i V\right)\Big|_{x=y,\,\mathbf{u}=\mathbf{U}(y,\mathbf{v})} = \left(\partial_x^i \mathrm{E}_\varphi(V)\right)\Big|_{x=y,\,\mathbf{u}=\mathbf{U}(y,\mathbf{v})} = \partial_y^i\psi(y,\mathbf{v}),$$

where we used the equality $\mathrm{E}_\varphi\partial_x^i=\partial_x^i\mathrm{E}_\varphi$ (see Example 3.8) and Lemma 2.4. Hence, in this case

$$f_*\left(\mathrm{E}_\varphi\right) = \mathrm{E}_\psi,\quad \psi(y,\mathbf{v}) = \left[\mathrm{E}_\varphi(V)(x,\mathbf{u})\right]\Big|_{x=y,\,\mathbf{u}=\mathbf{U}(y,\mathbf{v})},$$

that is, an evolution differentiation is mapped into an evolution differentiation.

Example 3.10. Let $\mathcal{M}$ and $\mathcal{N}$ be the differential manifolds defined by evolution equations $u_t=f$ and $v_s=g$, respectively (see Example 2.3, we shall the notation). Let $F:\mathcal{M}\simeq\mathcal{N}$ be a Lie-Backlund isomorphism given by the formulas

$$s = S,\ y = Y,\ \mathbf{v} = \mathbf{V} \text{ and } t = T,\ x = X,\ \mathbf{u} = \mathbf{U},$$

where smooth functions $S,Y,\mathbf{V}(t,x,\mathbf{u})$ and $T,X,\mathbf{U}(s,y,\mathbf{v})$ satisfy the corresponding defining equations. Let an evolution differentiation

$$\zeta = \mathrm{E}_\varphi,\ \partial_t\varphi = f_*\varphi,$$

be an infinitesimal symmetry of $\mathcal{M}$ (see Example 3.8), then its image

$$\eta = F_*(\zeta) = \sigma(s,y,\mathbf{v})\,\frac{\partial}{\partial s} + \vartheta(s,y,\mathbf{v})\,\frac{\partial}{\partial y} + \sum_{i\in\mathbb{N}} \eta_i(s,y,\mathbf{v})\,\frac{\partial}{\partial v_i},$$

where

$$\sigma = \mathrm{E}_\varphi(S),\ \vartheta = \mathrm{E}_\varphi(Y),\ \eta_i = \mathrm{E}_\varphi(V_i)\Big|_{t=T,x=X,\mathbf{u}=\mathbf{U}},\quad i\in\mathbb{N}.$$

In particular, suppose that $s=t$, $y=x$, then

$$F_*\left(\mathrm{E}_\varphi\right) = \mathrm{E}_\psi,\quad \psi(s,y,\mathbf{v}) = \left[\mathrm{E}_\varphi(V)(t,x,\mathbf{u})\right]\Big|_{t=s,x=y,\mathbf{u}=\mathbf{U}(s,y,\mathbf{v})},$$

like in the previous example.

Example 3.10(a). Let $\mathcal{M}$ and $\mathcal{N}$ be the differential manifolds defined by the heat equation and by the modified Burgers' equation, respectively (see Example 2.3(b), note that we switched the places of $\mathcal{M}$ and $\mathcal{N}$). We have the Lie-Backlund isomorphism $F:\mathcal{M}\simeq\mathcal{N}$ given by the formulas

$$s = t,\ y = x,\ v_i = \partial_x^i \log u_0,\ u_i = \partial_y^i e^{v_0},\ i \in \mathcal{N},$$

(see Example 2.3(c)). By this isomorphism an infinitesimal symmetry $\varphi \in \mathrm{Sym}\mathcal{M}$ is mapped into the infinitesimal symmetry $\psi \in \mathrm{Sym}\mathcal{N}$,

$$\psi = \Big[\mathrm{E}_\varphi(\log u_0)\Big]\Big|_{\mathbf{u}=\mathbf{U}(\mathbf{v})} = \varphi(t,x,\mathbf{U}(\mathbf{v}))e^{-v_0}$$

(we identify t with s and x with y, for simplicity). Here,

$$U_0 = e^{v_0},\ U_1 = v_1 e^{v_0},\ U_2 = (v_2+v_1^2)e^{v_0},\ U_3 = (v_3+3v_2v_1+v_1^3)e^{v_0}$$

et cetera. In general, $U_i = P_i e^{v_0}$, where a polynomial $P_i = v_i + \ldots + v_1^i$, $i \in \mathbb{N}$. Hence,

$$\begin{aligned}
\mathcal{S}ol \ni \varphi &\mapsto \psi = \varphi(t,x)e^{-v_0},\\
\mathrm{Sym}_{ess}\mathcal{M} \ni \varphi_0^0 &\mapsto \psi_0^0 = 1,\\
\varphi_1^0 &\mapsto \psi_1^0 = v_1,\\
\varphi_1^1 &\mapsto \psi_1^1 = tv_1 + \tfrac{x}{2},\\
\varphi_2^0 &\mapsto \psi_2^0 = v_2 + v_1^2,\\
\varphi_2^1 &\mapsto \psi_2^1 = t(v_2+v_1^2) + \tfrac{x}{2}v_1,\\
\varphi_2^2 &\mapsto \psi_2^2 = t^2(v_2+v_1^2) + txv_1 + \tfrac{1}{4}(x^2+2t),\\
\varphi_3^0 &\mapsto \psi_3^0 = v_3 + 3v_2v_1 + v_1^3, \ldots
\end{aligned}$$

(see Example 3.8(a)). Thus, the Lie algebra

$$\mathrm{Sym}\mathcal{N} = \mathrm{Sym}_{ess}\mathcal{N} \oplus \mathcal{J},$$

where the ideal $\mathcal{J}$ is the image of the ideal $\mathcal{S}ol$ and the subalgebra $\mathrm{Sym}_{ess}\mathcal{N}$ is the image of the subalgebra $\mathrm{Sym}_{ess}\mathcal{M}$. In particular, every element of $\mathrm{Sym}_{ess}\mathcal{N}$ is a polynomial in $t,x,v_1,\ldots,v_p$, where p

is the order of this element and subalgebra $\mathrm{Sym}_{ess}\mathcal{N}$ is generated by ψ_1^0, ψ_2^2 and ψ_3^0.

Proposition 3.11. *Let $f:\mathcal{M}\longrightarrow\mathcal{N}$ be a Lie-Backlund submersion. If vector fields $\xi\in\mathcal{T}(M)$ and $\vartheta\in\mathcal{T}(N)$ are f-related and the field ξ is Lie-Backlund then the field ϑ is also Lie-Backlund.*

Proof. Remind first that if vector fields $\zeta_i\in\mathcal{T}(M)$ and $\eta_i\in\mathcal{T}(N)$ are f-related, $i=1,2$, then their Lie brackets $[\zeta_1,\zeta_2]$ and $[\eta_1,\eta_2]$ are also f-related (see, for example, Kobayashi, Nomizu [1; Vol.I, Ch.I] or Sternberg [1; Th.8.3]; in the infinite-dimensional case the proof is the same). Now, to prove the proposition it is enough to note that the statement has a local character and to use Proposition 2.11.

This proposition establishes a relation between infinitesimal symmetries of differential manifolds $\mathcal{M}$ and $\mathcal{N}$. Moreover, in the case of Lie-Backlund covering submersion symmetries of $\mathcal{M}$ are sometimes called *nonlocal* symmetries of $\mathcal{N}$ (see Krasil'shchik, Vinogradov [1] and the discussion there). One can also consider Lie-Backlund covering submersions as a geometrical realization of prolongation structures introduced in Wahlquist, Eastabrook [1].

Example 3.12. Let $\mathcal{M}$ and $\mathcal{N}$ be the differential manifolds defined by evolution equations $v_s=g$ and $u_t=f$, respectively (in the notation of Example 2.12). Let the submersion $F:\mathcal{M}\longrightarrow\mathcal{N}$, given by the formulas

$$t = s,\ x = y,\ u_i = v_{i+k},\ i \in \mathbb{N},$$

be Lie-Backlund (i.e., $f=\nabla_y^k g$, k is the index). Let an evolution differentiation

$$\zeta = \mathrm{E}_\varphi,\ \varphi = \varphi(s,y,\mathbf{v}),\ \nabla_s\varphi = g_*\varphi,$$

be an infinitesimal symmetry of $\mathcal{M}$. In the new coordinates on $\mathcal{M}$ (see Example 2.12) we shall have

$$\zeta = \sum_{i=-k}^{\infty} \left(\nabla_x^{i+k}\varphi\right) \frac{\partial}{\partial u_i},\ \varphi = \varphi(t,x,\mathbf{u},\mathbf{w}).$$

The mapping $F\colon M \longrightarrow N$ is acting as the projection $(t,x,\mathbf{u},\mathbf{w}) \mapsto (t,x,\mathbf{u})$, hence, the image

$$F_*(\zeta) = \sum_{i=0}^{\infty} \left(\nabla_x^{i+k}\varphi\right) \frac{\partial}{\partial u_i} = \sum_{i\in 0}^{\infty} \left(\partial_x^i\psi\right) \frac{\partial}{\partial u_i} = \mathrm{E}_\psi$$

will be an infinitesimal symmetry of $\mathcal{N}$ if

$$\nabla_x^k\varphi(t,x,\mathbf{u},\mathbf{w}) = \psi(t,x,\mathbf{u})$$

does not depend on $\mathbf{w}$.

Example 3.12(a). Let the differential manifolds $\mathcal{M}$ and $\mathcal{N}$ be defined by the modified Burgers' and by the Burgers' equations, i.e., $g=v_2+v_1^2$ and $f=u_2+u_0u_1$, respectively (see Examples 2.3(b) and 2.12(a)). Consider the mapping

$$t = s,\ x = y,\ u_i = 2v_{i+1},\ i \in \mathbb{N},$$

including the dilation and having the index $k=1$ (see Example 2.12(a)). Let us introduce the new coordinates

t, x, u_i on M, where $t = s$, $x = y$, $u_i = 2v_{i+1}$, $i = -1, 0, 1, \ldots,$

then the basis Cartan fields will take the form

$$\nabla_y = \frac{\partial}{\partial x} + \sum_{i=-1}^{\infty} u_{i+1} \frac{\partial}{\partial u_i} = \partial_x + u_0 \frac{\partial}{\partial u_{-1}},$$

$$\nabla_s = \frac{\partial}{\partial t} + \sum_{i=-1}^{\infty} \tfrac{1}{2} g_{i+1} \frac{\partial}{\partial u_i} = \frac{\partial}{\partial t} + \sum_{i=0}^{\infty} f_i \frac{\partial}{\partial u_i} + \left(u_1 + \tfrac{1}{2} u_0^2\right) \frac{\partial}{\partial u_{-1}}$$

$$= \partial_t + \left(u_1 + \tfrac{1}{2} u_0^2\right) \frac{\partial}{\partial u_{-1}}.$$

Let $\zeta = E_\varphi \in \mathrm{Sym}\mathcal{M}$ be an infinitesimal symmetry of the modified Burgers' equation then its image

$$F_*(\zeta) = \sum_{i \in \mathbb{N}} \left(\nabla_y^i \psi\right) \frac{\partial}{\partial u_i} + 2\varphi, \text{ where } \psi = 2\nabla_y \varphi.$$

Thus (see Example 3.10(a)),

$$\mathcal{J} \ni \varphi = \chi(t,x) e^{-v_0} \mapsto \psi = \left[2\chi - \chi u_0\right] \exp\left(-u_{-1}\big/2\right), \ \chi_t = \chi_{xx},$$

hence, elements of $\mathcal{J}$ are nonlocal infinitesimal symmetries of the Burgers' equation (cf. Ibragimov [1; §17.5]). Further,

$$\mathrm{Sym}_{ess}\mathcal{M} \ni \varphi_0^0 \mapsto \psi_0^0 = 0,$$

$$\varphi_1^0 \mapsto \psi_1^0 = u_1,$$

$$\varphi_1^1 \mapsto \psi_1^1 = t u_1 + 1,$$

$$\varphi_2^0 \mapsto \psi_2^0 = u_2 + u_0 u_1,$$

$$\varphi_2^1 \mapsto \psi_2^1 = t(u_2 + u_0 u_1) + \tfrac{x}{2} u_1 + \tfrac{1}{2} u_0,$$

$$\varphi_2^2 \mapsto \psi_2^2 = t^2(u_2 + u_0 u_1) + t x u_1 + t u_0 + x,$$

$$\varphi_3^0 \mapsto \psi_3^0 = u_3 + \tfrac{3}{2} u_0 u_2 + \tfrac{3}{2} u_1^2 + \tfrac{3}{4} u_0^2 u_1, \ldots$$

(see Example 3.10(a)), hence, elements of $\mathrm{Sym}_{ess}\mathcal{M}$ generate infinitesimal symmetries of the Burgers' equation. Moreover, one can show that the image $F_*\left(\mathrm{Sym}_{ess}\mathcal{M}\right)=\mathrm{Sym}\mathcal{N}$, i.e., that every infinitesimal symmetry of the Burgers' equation is generated by some infinitesimal symmetry of the modified Burgers' equation. In particular, $\psi_p^k=2\partial_x\varphi_p^k$ and the Lie algebra $\mathrm{Sym}\mathcal{N}$ is generated by the elements ψ_1^0, ψ_2^2 and ψ_3^0 (cf. Vinogradov A.M., Krasil'shchik I.S [1]).

4. Cartan Forms, Currents and Conservation Laws

Let $\mathcal{M}=(M,CTM)$ be a differential manifold and $\mathcal{O}$ be an open subset of M.

A q-form $\omega\in\Lambda^q(\mathcal{O})$ is called *Cartan* if $\omega(\zeta_1,\ldots,\zeta_q)(a)=0$ for all points $a\in\mathcal{O}$ and vectors $\zeta_1,\ldots,\zeta_q\in CT_aM$. We denote by $C\Lambda^q(\mathcal{O})$ the linear space of all Cartan forms on $\mathcal{O}$, $q=1,2,\ldots,$ and put $C\Lambda^0(\mathcal{O})=0$ by the definition. Note that $C\Lambda^q(\mathcal{O})=\Lambda^q(\mathcal{O})$ if $q>m$, where m is the Cartan dimension of $\mathcal{M}$.

It is clear that if a q-form $\omega\in\Lambda^q(\mathcal{O})$ is Cartan then the restriction $\omega\big|_S=0$ for every integral submanifold $S\subset\mathcal{O}$ ($\omega\big|_S\equiv i^*\omega$, where $i:S\longrightarrow\mathcal{O}$ is the natural imbedding). In the finite dimensional case this property characterizes Cartan forms. Namely, the following proposition is valid.

Proposition 4.1. *Let a differential manifold $\mathcal{M}$ be finite-dimensional. Then a q-form $\omega\in\Lambda^q(\mathcal{O})$ is Cartan iff $\omega\big|_S=0$ for every integral submanifold $S\subset\mathcal{O}$.*

Proof. It is enough to remind that by Frobenius' theorem (see Section 0.5) in this case through each point $a\in M$ there passes an integral submanifold S with $T_aS=CT_aM$.

In the infinite-dimensional case this property is no longer valid due to the possible absence of an integral submanifold passing through a given point. Nevertheless, it is convenient to look at Cartan forms as forms vanishing on integral submanifolds.

The involutivity of the Cartan distribution implies that $d\omega \in C\Lambda^{q+1}(\mathcal{O})$ if $\omega \in C\Lambda^{q}(\mathcal{O})$ (see the definition of the exterior derivative in Section 0.2, in this case $[\zeta_\lambda, \zeta_\mu]$ is Cartan if ζ_λ and ζ_μ are Cartan). Hence, the linear mappings $d: C\Lambda^{q}(\mathcal{O}) \longrightarrow C\Lambda^{q+1}(\mathcal{O})$ are defined for $q=0,1,\ldots$.

A form $\omega \in \Lambda^{q}(M)$ is called a *current* if $d\omega \in C\Lambda^{q+1}(M)$ and the number $q=0,1,2,\ldots$ is called a *dimension* of a current ω in this case. The linear space of all currents on M of a dimension q we denote by $\mathrm{Curr}^{q}\mathcal{M}$. If a q-form ω is exact, i.e., $\omega = d\theta$, where θ is a $(q-1)$-form, then $d\omega = 0$ is Cartan, so exact forms are currents; such currents carry no information about $\mathcal{M}$, they are called *trivial*. As we saw above, $d\omega \in C\Lambda^{q+1}(M)$ if $\omega \in C\Lambda^{q}(M)$, so Cartan forms are also currents, and also trivial. Elements of the quotient spaces

$$\mathrm{Cons}^{q}\mathcal{M} = \frac{d^{-1}C\Lambda^{q+1}(M)}{C\Lambda^{q}(M) + d\Lambda^{q-1}(M)}, \quad q = 0,1,\ldots,$$

are called *conservation laws of the dimension* q, where

$$d^{-1}C\Lambda^{q+1}(M) = \left\{\omega \in \Lambda^{q}(M) \;:\; d\omega \in C\Lambda^{q+1}(M)\right\}$$

is the inverse image of the space $C\Lambda^{q+1}(M)$ and

$$d\Lambda^{q-1}(M) = \left\{\omega = d\theta \;;\; \theta \in \Lambda^{q-1}(M)\right\}$$

is the image of the space $\Lambda^{q-1}(M)$. Let $S\subset M$ be a submanifold of M then

$$d\left(\omega\Big|_S\right) = \left(d\omega\right)\Big|_S \text{ for any form } \omega\in\Lambda^q(M).$$

Hence, we can look on conservation laws as on forms nontrivially closed on integral submanifolds. Consider the limit cases. First, $C\Lambda^0(M)=d\Lambda^{-1}(M)=0$, so

$$\mathrm{Cons}^0\mathcal{M} = \left\{f\in\mathcal{C}^\infty(M)\ :\ df\in C\Lambda(M)\right\}.$$

On the other hand, $C\Lambda^q(M)=\Lambda^q(M)$ for $q>m$, thus

$$\mathrm{Cons}^m\mathcal{M} = \frac{\Lambda^m(M)}{C\Lambda^m(M)+d\Lambda^{m-1}(M)},$$

$$\mathrm{Cons}^q\mathcal{M} = 0 \text{ for } q = m+1, m+2, \ldots.$$

Note that $\Omega\Big|_S=0$ for any form $\Omega\in\Lambda^{m+1}(M)$ and any m-dimensional submanifold $S\subset M$. From this point of view all m-dimensional currents can be considered trivial.

Suppose that a differential manifold $\mathcal{M}$ is covered by a single standard chart $(U,\varphi,\mathbb{R}^{m+I},\partial)$ with coordinates x^μ, u^i, $\mu=1,\ldots,m$, $i\in I$, has basis Cartan fields

$$\partial_\mu = \frac{\partial}{\partial x^\mu} + \sum_{i\in I} A^i_\mu(x,u)\,\frac{\partial}{\partial u^i},\quad \mu = 1,\ldots,m,$$

and basis Cartan 1-forms

$$\delta u^i = du^i - \sum_{\mu=1}^{m} A^i_\mu(x,u)\,dx^\mu,\quad i\in I$$

(see Section 1). As the direct corollary of Proposition 1.11 we shall have the following

Proposition 4.2. *Every q-form $\omega\in\Lambda^q(U)$ has the unique representation*

$$\omega = \sum_{k+r=q} \sum_{(i)_k} \sum_{(\mu)_r} \omega_{(i)_k;(\mu)_r} \left(\delta u^i\right)^k \wedge \left(dx^\mu\right)^r,$$

where indices $(i)_k = i_1 \dots i_k \in I^k$, $(\mu)_r = \mu_1 \dots \mu_r \in \{1,\dots,m\}^r$, coefficients $\omega_{(i)_k;(\mu)_r}$ are smooth functions on U skew-symmetric in indices $i_1,\dots,i_k$ and $\mu_1,\dots,\mu_r$, separately,

$$\left(\delta u^i\right)^k = \delta u^{i_1} \wedge \dots \wedge \delta u^{i_k} \text{ and } \left(dx^\mu\right)^r = dx^{\mu_1} \wedge \dots \wedge dx^{\mu_r}.$$

Note that in this representation there are two limit terms

$$\sum_{(\mu)_q} \omega_{(\mu)_q} \left(dx^\mu\right)^q \quad\text{and}\quad \sum_{(i)_q} \omega_{(i)_q} \left(\delta u^i\right)^q.$$

It is easy to check that a form ω is Cartan iff the first limit term is absent. Hence, every q-form $\omega\in\Lambda^q(U)$ has the unique representation

$$\omega = \sum_{\mu_1 \dots \mu_q} \omega_{\mu_1 \dots \mu_q}(x,u)\, dx^{\mu_1} \wedge \dots \wedge dx^{\mu_q} \mod C\Lambda^q(U).$$

Moreover, the exterior derivative $d\omega = \partial\omega \mod C\Lambda^{q+1}(U)$, where

$$\partial\omega = \sum_{\mu_0 \dots \mu_q} \left(\partial_{[\mu_0} \omega_{\mu_1 \dots \mu_q]}\right) dx^{\mu_0} \wedge \dots \wedge dx^{\mu_q},$$

brackets [...] denote the alternation in indices $\mu_0 \dots \mu_q$; in particular,

$$\partial f = \sum_{\mu=1}^{m} \partial_\mu f \, dx^\mu \text{ for any } f \in \mathscr{C}^\infty(U)$$

and $\partial\omega=0$ for any $\omega\in\Lambda^m(U)$.

Proposition 4.3. *The following representations are valid:*

$$\mathrm{Cons}^0\mathcal{M} = \left\{ f\in\mathscr{C}^\infty(U) \ : \ \partial f=0 \right\},$$

$$\mathrm{Cons}^q\mathcal{M} = \frac{\left\{ \omega\in\Lambda^q(U) \ : \ \partial\omega=0 \right\}}{\left\{ \omega=\partial\theta \ ; \ \theta\in\Lambda^{q-1}(U) \right\}}, \quad q = 1,\ldots,m-1,$$

$$\mathrm{Cons}^m\mathcal{M} = \Lambda^m(U) \Big/ \left\{ \omega=\partial\theta \ ; \ \theta\in\Lambda^{m-1}(U) \right\},$$

where

$$\omega = \sum_{(\mu)_q} \omega_{(\mu)_q} \left(dx^\mu\right)^q \quad \text{and} \quad \theta = \sum_{(\mu)_{q-1}} \theta_{(\mu)_{q-1}} \left(dx^\mu\right)^{q-1}.$$

Proof. To prove this proposition it is enough to use the above formulas.

Example 4.4. Suppose that a chart $(U,\varphi,\mathbb{R}^{m+I},\partial)$ is foliating, i.e., the basis fields $\partial_\mu=\partial \big/ \partial x^\mu$, $\mu=1,\ldots,m$, and that the domain U is convex. Then $\partial f=0$ iff a smooth function $f=f(u)$ does not depend on x. Hence, in this case

$$\mathrm{Cons}^0\mathcal{M} = \left\{ f\in\mathscr{C}^\infty(U) \ : \ f=f(u) \right\}.$$

Further, by Poincare's lemma (in the variables x's, the variables u's are parameters), $\partial\omega=0$ iff $\omega=\partial\theta$, where q-form ω and $(q-1)$-form θ are of the above type, $q=1,\ldots,m$. Hence, here

$$\mathrm{Cons}^q\mathcal{M} = 0 \text{ for } q = 1,\dots,m.$$

Remind that any finite-dimensional differential manifold can be covered by foliating charts. Thus, in this case conservation laws are purely global characteristics like de Rham cohomologies for smooth manifolds.

Example 4.5. Let $\mathcal{M}=(M,\partial_t)$ be a differential manifold with global coordinates $(t,x)\in M$ and basis Cartan field

$$\partial_t = \frac{\partial}{\partial t} + f(t,u)\,\frac{\partial}{\partial u},$$

where $f\in\mathcal{C}^\infty(M)$, defined by an ordinary differential equation $u_t=f$ (see Example 1.1). Here,

$$\mathrm{Cons}^0\mathcal{M} = \left\{F(t,u) \in \mathcal{C}^\infty(M) \ :\ \partial_t F = 0\right\}.$$

The equality $\partial_t F=0$ on M is valid iff $F(t,s(t))=const$ for any local solution $u=s(t)$ of the equation $u_t=f$ (see Proposition 1.9). Thus, in this case 0-dimensional conservation laws are nothing but integrals of the original equation $u_t=f$. Further,

$$\mathrm{Cons}^1\mathcal{M} = \frac{\left\{\omega = \phi dt \in \Lambda^1(M)\right\}}{\left\{\omega = \partial\chi \ ;\ \chi \in \mathcal{C}^\infty(M)\right\}} = \frac{\left\{\phi \in \mathcal{C}^\infty(M)\right\}}{\left\{\phi = \partial_t\chi \ ;\ \chi \in \mathcal{C}^\infty(M)\right\}}.$$

In other words, $\mathrm{Cons}^0\mathcal{M}$ and $\mathrm{Cons}^1\mathcal{M}$ are the kernel and the cokernel, respectively, of the linear mapping $\partial_t:\mathcal{C}^\infty(M)\longrightarrow\mathcal{C}^\infty(M)$.

Example 4.6. Let $\mathcal{M}=(M=\mathbb{R}\times\mathbb{R}_{\mathbb{N}},D_x)$ be a differential manifold with global coordinates $x,\mathbf{u}=(u_i;i\in\mathbb{N})$ and a basis Cartan field

$$D_x = \frac{\partial}{\partial x} + \sum_{i\in\mathbb{N}} u_{i+1} \frac{\partial}{\partial u_i}$$

(see Example 1.5). Then

$$\mathrm{Cons}^0 \mathcal{M} = \left\{F(x,\mathbf{u}) \in \mathcal{C}^\infty(M) \ : \ D_x F = 0\right\} = \mathbb{R}$$

(see Lemma 2.5) and

$$\mathrm{Cons}^1 \mathcal{M} = \frac{\left\{\omega = L dx \in \Lambda^1(M)\right\}}{\left\{\omega = \partial F \ ; \ F \in \mathcal{C}^\infty(M)\right\}} = \frac{\left\{L \in \mathcal{C}^\infty(M)\right\}}{\left\{L = D_x F \ ; \ F \in \mathcal{C}^\infty(M)\right\}}.$$

Let us denote by $\mathcal{D}(\mathbb{R})=\mathcal{C}_0^\infty(\mathbb{R})$ the linear space of all smooth functions on $\mathbb{R}$ with compact supports with the natural topology of a locally convex space (this is the space of test-functions for Schwartz distributions, see any book on distributions). Now, let a 1-form $\omega=L(x,\mathbf{u})dx\in\Lambda^1(M)$ and let $s(x)\in\mathcal{D}(\mathbb{R})$, then the integral

$$\mathcal{L}(s) = \int \omega\Big|_S = \int L(x,\mathbf{u})\Big|_{\mathbf{u}=\mathbf{s}(x)} dx$$

is determined, where

$$S = \left\{x,\mathbf{u} \ : \ \mathbf{u} = \mathbf{s}(x), \ \mathbf{s}(x) = \left(s^{(i)}(x)\right)\right\}$$

is the integral submanifold of $\mathcal{M}$ defined by $s(x)$ (see Example 1.5). Moreover, if the form $\omega=\partial F=(D_x F)dx$, then

$$\int \omega\Big|_S = \int \left[D_x F(x,\mathbf{u})\right]\Big|_{\mathbf{u}=\mathbf{s}(x)} dx = \int \frac{d}{dx}\Big[F(x,\mathbf{s}(x))\Big] dx$$

$$= F(x,\mathbf{s}(x))\Big|_{x=-\infty}^{x=+\infty} = 0,$$

where we used the compactness of the support of the function $s(x)$. Hence, elements of the space $\mathrm{Cons}^1 \mathcal{M}$ are in fact functionals on the

Schwartz space $\mathcal{D}(\mathbb{R})$. Note that the compactness of the support of a function $s(x)$ can be replaced by a suitable growth conditions on F and s, implying the existence of the limit

$$\lim_{x\to\pm\infty} F(x,\mathbf{s}(x)) = 0.$$

Example 4.7. Let $\mathcal{M}=(M=\mathbb{R}^2\times\mathbb{R}_{\mathbb{N}},\partial)$ be the differential manifold with global coordinates $t,x,\mathbf{u}$ and basis Cartan fields

$$\partial_x = \frac{\partial}{\partial x} + \sum_{i=0}^{\infty} u_{i+1}\frac{\partial}{\partial u_i},\quad \partial_t = \frac{\partial}{\partial t} + \sum_{i=0}^{\infty} f_i\frac{\partial}{\partial u_i},$$

defined by an evolution equation $u_t=f$, $f(t,x,\mathbf{u})\in\mathcal{C}^\infty(M)$ (see Example 1.4).

In this case

$$\mathrm{Cons}^0\mathcal{M} = \left\{F(t,x,\mathbf{u}) \in \mathcal{C}^\infty(M) : \partial_x F = 0,\ \partial_t F = 0\right\} = \mathbb{R},$$

because the equation $\partial_x F=0$ yields $F=F(t)$ (see Lemma 2.5) and then the equation $\partial_t F=0$ gives $F=const$.

Further,

$$\mathrm{Cons}^1\mathcal{M} = \frac{\left\{\omega = Tdx-Xdt : \partial_t T+\partial_x X=0\right\}}{\left\{\omega = \partial_x\phi dx+\partial_t\phi dt\right\}},$$

where $T,X,\phi\in\mathcal{C}^\infty(M)$, the vector $J=(T,X)$ is called the *current*, the component T is called the *conserved density*, the component X is called the *associated flux* and the equation $\partial_t T+\partial_x X=0$ is called the *divergence equation*. Let $J=(T,X)$ be a current, $s=s(t,x)$ be a solution of the evolution equation, satisfying appropriate growth conditions, and $S=\{\mathbf{u}=\mathbf{s}(t,x)\}$ be the integral submanifold defined by s. Consider the *charge*

$$Q(t) = \int T(t,x,\mathbf{u})\Big|_{\mathbf{u}=\mathbf{s}(t,x)} dx$$

then, by Proposition 1.9,

$$\frac{d}{dt}Q(t) = \int \frac{\partial}{\partial t}T(t,x,\mathbf{s}(t,x))dx = \int \Big[\partial_t F(t,x,\mathbf{u})\Big]\Big|_{\mathbf{u}=\mathbf{s}(t,x)} dx$$

$$= \int \Big[\partial_x X(t,x,\mathbf{u})\Big]\Big|_{\mathbf{u}=\mathbf{s}(t,x)} dx = \int \frac{\partial}{\partial x}X(t,x,\mathbf{s}(t,x))dx$$

$$= X(t,x,\mathbf{s}(t,x))\Big|_{x=-\infty}^{x=+\infty} = 0$$

if the flux X and the solution s satisfy appropriate growth conditions implying the existence of the limit

$$\lim_{x\to\pm\infty} X(t,x,\mathbf{s}(t,x)) = 0$$

(cf. Example 4.6). Thus, the charge $Q=const$, i.e. it is conserved. If the current $J=(\partial_x\phi,-\partial_t\phi)$ is trivial then

$$Q(t) = \int \Big[\partial_x \phi(t,x,\mathbf{u})\Big]\Big|_{\mathbf{u}=\mathbf{s}(t,x)} dx = \int \frac{\partial}{\partial x}\phi(t,x,\mathbf{s}(t,x))dx$$

$$= \phi(t,x,\mathbf{s}(t,x))\Big|_{x=-\infty}^{x=+\infty} = 0$$

under the same growth conditions, i.e., trivial currents generate zero charges. Hence, every equivalence class $[\omega]\in \mathrm{Cons}^1 M$ defines a conserved quantity and this property allows us to call it the conservation law.

At last, the space

$$\mathrm{Cons}^2\mathcal{M} = \frac{\left\{\omega = L\,dt\wedge dx\right\}}{\left\{\omega = (\partial_t T+\partial_x X)\,dt\wedge dx\right\}},$$

where $L,T,X\in\mathcal{C}^\infty(M)$. Let 2-form $\omega=L\,dt\wedge dx\in\Lambda^2(M)$ and $s=s(t,x)$ be a solution of the evolution equation, satisfying appropriate growth conditions, and S the integral submanifold defined by s. Consider the integral

$$\mathcal{L}(s) = \int\omega\Big|_S = \int L(t,x,\mathbf{s}(t,x))\,dt\wedge dx$$

(we assume that it does exist due to the implied growth conditions). If the form ω is trivial, i.e., if $\omega=(\partial_t T+\partial_x X)dt\wedge dx$ with some $T,X\in\mathcal{C}^\infty(M)$, then $\mathcal{L}(s)=0$ by Proposition 1.9, Green's theorem and the growth conditions. Hence, elements of $\mathrm{Cons}^2\mathcal{M}$ are functionals on the space of solutions of the studied evolution equation, satisfying appropriate growth conditions.

Let us study a bit more closely the space

$$\mathrm{Cons}^1\mathcal{M} = \mathrm{Curr}^1\mathcal{M}\Big/\mathrm{TrCurr}^1\mathcal{M},$$

where

$$\mathrm{Curr}^1\mathcal{M} = \left\{J=(T,X)\ :\ \partial_t T+\partial_x X=0\right\},\quad \mathrm{TrCurr}^1\mathcal{M} = \left\{J=(\partial_x\phi,-\partial_t\phi)\right\},$$

functions $T,X,\phi(t,x,\mathbf{u})\in\mathcal{C}^\infty(M)$. We shall say that a current $J=(T,X)$ has the *order* p if the density T has the order p, i.e., if $T=T(\ldots.,u_p)$. Two currents J_1 and J_2 are called *equivalent* if they define the same conservation law, i.e., if $[J_1]=[J_2]\in\mathrm{Cons}^1\mathcal{M}$.

Lemma 4.8. *A current $J=(T,X)\in\mathrm{Curr}^1 M$ of an order $p\geq 1$ is equivalent to a current of an order $\leq(p-1)$ iff $T_{u_p u_p}=0$.*

Proof. Suppose, first, that a current $J=(T,X)$ of the order p is equivalent to a current $J'=(T',X')$ of an order $\leq(p-1)$. Then $T=\partial_x\phi+T'$, where ϕ is a smooth function of the order $p-1$, i.e., $T=u_p\phi_{u_{p-1}}+\ldots$, where low-order terms are omitted, and hence, $T_{u_p u_p}=\phi_{u_{p-1}u_p}=0$. On the other hand, let $J=(T,X)$ be a current of the order p and suppose that $T_{u_p u_p}=0$. Then $T=Au_p+B$, where A,B are smooth functions of the order $p-1$. Let ϕ be a primitive of A in the variable u_{p-1}, i.e., $A=\phi_{u_{p-1}}$, ϕ is a smooth function of the order $p-1$. Then $T=\partial_x\phi+T'$, where the smooth function $T'=B-\partial'_x\phi$ has an order $\leq(p-1)$, $\partial'_x\phi=u_{p-1}\phi_{u_{p-2}}+\ldots+\phi_x$. Thus, the current J is equivalent to the current $J'=(T',X')$ of an order $\leq(p-1)$, $X'=X+\partial_t\phi$.

Example 4.7(a). Let $\mathcal{M}=(M=\mathbb{R}^2\times\mathbb{R}_{\mathbb{N}},\partial)$ be the differential manifold with the global coordinates $t,x,\mathbf{u}$ and the basis Cartan fields

$$\partial_x=\frac{\partial}{\partial x}+\sum_{i=0}^{\infty}u_{i+1}\frac{\partial}{\partial u_i},\quad \partial_t=\frac{\partial}{\partial t}+\sum_{i=0}^{\infty}u_{i+2}\frac{\partial}{\partial u_i},$$

defined by the heat equation $u_t=u_2$ (see Example 3.8(a)). Let $J=(T,X)$ be a current of an order p then the divergence equation has the form

$$\partial_t T+\partial_x X=\left(u_{p+2}T_{u_p}+u_{p+1}T_{u_{p-1}}+\ldots\right)+\left(u_{q+1}X_{u_q}+u_q X_{u_{q-1}}+\ldots\right)=0,$$

where q is an order of the flux X. Without loss of generality, we can assume $q=p+1$, so

$$u_{p+2}\left(T_{u_p}+X_{u_{p+1}}\right) + \ldots = 0,$$

implying $X_{u_{p+1}}=-T_{u_p}$, i.e., $X=-u_{p+1}T_{u_p}+A$, where A is a smooth function of an order $\le p$. Now, the divergency equation yields

$$-u_{p+1}^2T_{u_pu_p}+u_{p+1}\left(T_{u_{p-1}}+A_{u_p}\right) + \ldots . = 0.$$

Hence, $T_{u_pu_p}=0$ and, by Lemma 4.8, the current J is equivalent to a current of an order $\le(p-1)$. By the induction we derive that every current is equivalent to a current of the order 0. So, let $J=(T,X)$ be a current of the order 0, i.e., $T=T(t,x,u_0)$, $X=X(t,x,u_0,u_1)$ and

$$u_2T_{u_0}+T_t+u_2X_{u_1}+u_1X_{u_0}+X_x = 0.$$

The general solution of the last equation has the form

$$T = au_0+\phi_x,\ X = -au_1+a_xu_0-\phi_t,$$

where $a(t,x),\phi(t,x)$ are smooth functions on $\mathbb{R}^2$, $a_t+a_{xx}=0$. Thus, for the heat equation the space

$$\mathrm{Cons}^1\mathcal{M} = \left\{J=\left(au_0,-au_1+a_xu_0\right)\ :\ a(t,x)\in\mathcal{C}^\infty(\mathbb{R}^2),\ a_t+a_{xx}=0\right\},$$

i.e., in this case the space $\mathrm{Cons}^1\mathcal{M}$ is isomorphic to the linear space of all global smooth solutions of the equation $a_t+a_{xx}=0$, dual to the heat equation.

Example 4.7(b). For the differential manifold $\mathcal{M}$ defined by the Burgers' equation (see Example 2.3(a), here $f=u_2+u_1u_0$) by the same procedure we get

$$\mathrm{Cons}^1\mathcal{M} = \left\{J=\left(cu_0, -c\left(u_1+\frac{u_0^2}{2}\right)\right) \ : \ c=const\right\},$$

i.e., in this case the space $\mathrm{Cons}^1\mathcal{M}$ is one-dimensional.

Example 4.7(c). Consider the differential manifold $\mathcal{M}$ defined by the evolution equation

$$u_t = u_{xx}^2$$

(see Abellanas, Galindo [1]). Here $f=u_2^2$ and for $T=T(\ldots,u_p)$, $X=X(\ldots,u_q)$ the divergency equation reads

$$\partial_t T+\partial_x X = 2u_{p+2}u_2T_{u_p}+u_{q+1}X_{u_q}+\ldots = 0.$$

This yields $q=p+1$, and then $X_{u_{p+1}}=-2u_2T_{u_p}$, i.e., $X=-2u_{p+1}u_2T_{u_p}+Y$, where $Y=Y(\ldots,u_p)$ is a smooth function of an order $\leq p$. Further,

$$\begin{aligned}\partial_t T+\partial_x X &= (2u_{p+2}u_2+(p+1)u_{p+1}u_3)T_{u_p}+2u_{p+1}u_2T_{u_{p-1}}\\ &\quad -2u_{p+2}u_2T_{u_p}-2u_{p+1}^2u_2T_{u_pu_p}+u_{p+1}Y_{u_p}+\ldots\\ &= -2u_{p+1}^2u_2T_{u_pu_p}+u_{p+1}\left((p+1)u_3T_{u_p}+2u_2T_{u_{p-1}}+Y_{u_p}\right)+\ldots\\ &= 0.\end{aligned}$$

This implies $T_{u_pu_p}=0$, and, by Lemma 4.8 and by the induction, we come to the case $T=T(t,x,u_0)$, $X=X(t,x,u_0,u_1)$ with the divergency equation

$$u_2^2T_{u_0}+T_t+u_2X_{u_1}+u_1X_{u_0}+X_x = 0,$$

yielding $T_{u_0}=0$, $X_{u_1}=0$, $X_{u_0}=0$. Thus, $T,X(t,x)$ and $T_t+X_x=0$. The last equation has only trivial solutions $T=\phi_x$, $X=-\phi_t$, $\phi\in\mathcal{C}^\infty(\mathbb{R}^2)$. Hence,

$$\mathrm{Cons}^1\mathcal{M} = 0,$$

i.e., *the equation* $u_t=u_{xx}^2$ *has no conservation laws.*

Example 4.7(d). For the differential manifold $\mathcal{M}$ defined by the evolution equation $u_t=0$ the function $f=0$, hence,

$$\partial_t = \frac{\partial}{\partial t}, \ \partial_x = \frac{\partial}{\partial x} + \sum_{i\in\mathbb{N}} u_{i+1}\frac{\partial}{\partial u_i}.$$

The divergency equation for functions $T,X(t,x,\mathbf{u})\in\mathscr{C}^\infty(\mathbb{R}^2\times\mathbb{R}_\mathbb{N})$ takes the form

$$T_t + \partial_x X = 0.$$

This equation has evident solutions $T=T(x,\mathbf{u})\in\mathscr{C}^\infty(\mathbb{R}\times\mathbb{R}_\mathbb{N})$, $X=0$. On the other hand, let T,X be a solution and let ϕ be a primitive of X in the variable t, i.e., $X=\phi_t$. Then, $T=-\partial_x\phi+T'$, where a smooth function T' does not depend on t, i.e., $T'=T'(x,\mathbf{u})$. Hence, the current $J=(T,X)$ is equivalent to the current $J'=(T',0)$, because $\phi_t=\partial_t\phi$. Thus, in this case

$$\mathrm{Cons}^1\mathcal{M} = \frac{\left\{T(x,\mathbf{u}) \in \mathscr{C}^\infty(\mathbb{R}\times\mathbb{R}_\mathbb{N})\right\}}{\left\{T=\partial_x\phi \ : \ \phi(x,\mathbf{u}) \in \mathscr{C}^\infty(\mathbb{R}\times\mathbb{R}_\mathbb{N})\right\}}.$$

Let us return to the general case. Let $\mathcal{M}=(M,CTM)$ be a differential manifold of the Cartan dimension m. Consider a current $\omega\in\mathrm{Curr}^q\mathcal{M}$ and an integral submanifold $S\subset M$. The restriction $\omega\big|_S$ will be an exact q-form on the m-dimensional smooth manifold S, so,

$$\int_{\partial\mathcal{D}} \omega\Big|_S = \int_{\mathcal{D}} d\Big(\omega\Big|_S\Big) = \int_{\mathcal{D}} \Big(d\omega\Big)\Big|_S = 0,$$

by Stokes' theorem, for every compact domain $\mathcal{D}\subset S$ with a piece-wise smooth boundary $\partial\mathcal{D}$. This property is widely used in applications, especially in the case $q=m-1$. For example, it allows to derive *a priori* estimates for PDE and conserved charges for evolution systems (see, for example, Olver [1] and Vinogradov [1]).

Let $f:\mathcal{M}\longrightarrow\mathcal{N}$ be a Lie-Backlund mapping of a differential manifold $\mathcal{M}=(M,CTM)$ into a differential manifold $\mathcal{N}=(N,CTN)$. Consider the cotangent mapping $f^*:\Lambda^q(N)\longrightarrow\Lambda^q(M)$, $q=0,1,2,\ldots$. Let a q-form $\omega\in\Lambda^q(N)$ be Cartan, then its image $f^*\omega\in\Lambda^q(M)$ will be also Cartan. Indeed, let vectors $\zeta_1,\ldots,\zeta_q\in CTM$ then images $f_*\zeta_1,\ldots,f_*\zeta_q\in CTN$ and $f^*\omega(\zeta_1,\ldots,\zeta_q)=\omega(f_*\zeta_1,\ldots,f_*\zeta_q)=0$. Thus, $f^*:\Lambda^q(N)\longrightarrow\Lambda^q(M)$. Remind that $f^*\circ d=d\circ f^*$ for any smooth mapping $f:M\longrightarrow N$ (see Section 0.3). Hence, the quotient mapping

$$\mathbf{f}^* \ : \ \mathrm{Cons}^q\mathcal{N} \longrightarrow \mathrm{Cons}^q\mathcal{M}, \ q=0,1,2,\ldots,$$

is induced. Clear that $\mathbf{f}^*$ is isomorphism if f is Lie-Backlund isomorphism.

Example 4.9. Let $\mathcal{M}$ and $\mathcal{N}$ be the differential manifolds defined by evolution equations $u_t=f$ and $v_s=g$, respectively. Let $f:\mathcal{M}\longrightarrow\mathcal{N}$ be a Lie-Backlund mapping given by the formulas

$$s = t, \ y = x, \ v_i = \partial_{\mathbf{x}}^i V(t,x,\mathbf{u}), \ i \in \mathbb{N}, \ \partial_t V = g(t,x,\mathbf{V})$$

(see Example 2.3, we are keeping the notation). Let $K=(S,Y)$ be a current on the manifold $\mathcal{N}$, then its image $J=f^{*}K=(T,X)\in\mathrm{Curr}^{1}\mathcal{M}$ can be calculated by the formulas

$$T = S\Big|_{s=t,\,y=x,\,\mathbf{v}=\mathbf{V}},\quad X = Y\Big|_{s=t,\,y=x,\,\mathbf{v}=\mathbf{V}}.$$

Indeed, in this case

$$\partial_{t}T+\partial_{x}X = \left[\partial_{s}S+\partial_{y}Y\right]\Big|_{s=t,\,y=x,\,\mathbf{v}=\mathbf{V}} = 0,$$

by Proposition 2.1. Note that the current J can be trivial even if the current K is nontrivial.

Example 4.9(a). Let $\mathcal{M}$ and $\mathcal{N}$ be the differential manifolds defined by the modified Burgers' and the heat equations, respectively (see Example 2.3(c)). The Lie-Backlund isomorphism $f:\mathcal{M}\longrightarrow\mathcal{N}$, given by the formulas

$$s = t,\ y = x,\ v_{i} = \partial_{x}^{i}e^{u_0},\ i\in\mathbb{N},$$

takes a current $K=\left(av_{0},-av_{1}+a_{y}v_{0}\right)$, $a_{s}+a_{yy}=0$, of the heat equation $\mathcal{N}$ (see Example 4.7(a)) into the current $J=\left(ae^{u_0},(-au_{1}+a_{x})e^{u_0}\right)$ of the modified Burgers' equation $\mathcal{M}$. Hence,

$$\mathrm{Cons}^{1}\mathcal{M} = \left\{J=\left(ae^{u_0},(-au_{1}+a_{x})e^{u_0}\right)\ :\ a(t,x)\in\mathcal{C}^{\infty}(\mathbb{R}^{2}),\ a_{t}+a_{xx}=0\right\}.$$

Example 4.9(b). Let $\mathcal{M}$ and $\mathcal{N}$ be the differential manifolds defined by the modified Burgers' and the Burgers' equations, respectively (see Example 3.12(a)). The Lie-Backlund covering submersion $f:\mathcal{M}\longrightarrow\mathcal{N}$, given by the formulas

$$s = t,\ y = x,\ v_{i} = 2u_{i+1},\ i\in\mathbb{N},$$

takes the current $K=\left(v_0, -v_1-\frac{v_0^2}{2}\right)$ of the Burgers' equation (see Example 4.7(b)) into the trivial current $J=\left(\partial_x\phi, -\partial_t\phi\right)$, where $\phi=2u_0$. Hence, in this case the quotient mapping $\mathbf{f}^*:\mathrm{Cons}^1\mathcal{N}\longrightarrow\mathrm{Cons}^1\mathcal{M}$ has the trivial image, $\mathrm{Im}\mathbf{f}^*=0$.

Let $\mathcal{M}=(M,CTM)$ be a differential manifold. Let $\mathcal{O}$ be an open subset of M and ζ a vector field on $\mathcal{O}$, $\zeta\in\mathcal{T}(\mathcal{O})$, then the Lie derivative L_ζ will define the linear mapping $L_\zeta:\Lambda^q(\mathcal{O})\longrightarrow\Lambda^q(\mathcal{O})$ for $q=0,1,2,\ldots$ (see Section 0.2). Remind that $L_\zeta\circ d=d\circ L_\zeta$, i.e.,

$$L_\zeta \ : \ d\Lambda^q(\mathcal{O}) \longrightarrow d\Lambda^q(\mathcal{O}).$$

Now, let the vector field ζ be Lie-Backlund, then the Lie bracket $[\zeta,\eta]$ is Cartan if a vector field η is Cartan, hence,

$$L_\zeta \ : \ C\Lambda^q(\mathcal{O}) \longrightarrow C\Lambda^q(\mathcal{O}).$$

Indeed,

$$L_\zeta\omega(\zeta_1,\ldots,\zeta_q) = \zeta\left(\omega(\zeta_1,\ldots,\zeta_q)\right) - \sum_{\lambda=1}^{q} \omega(\zeta_1,\ldots,[\zeta,\zeta_\lambda],\ldots,\zeta_q) = 0$$

for any Cartan q-form ω and Cartan fields $\zeta_1,\ldots,\zeta_q$. These properties imply that $L_\zeta\omega$ is the current if ω is a current and the current $L_\zeta\omega$ is trivial if the current ω is trivial. Further, let the field ζ be Cartan and ω be a current, then the Lie derivative $L_\zeta\omega$ will be the trivial current, because

$$L_\zeta\omega = \zeta\rfloor(d\omega) + d(\zeta\rfloor\omega)$$

(see Section 0.2), where the form $\zeta \rfloor (d\omega)$ is Cartan (it is easy to check that the interior product of a Cartan field and a Cartan form is a Cartan form) and the form $d(\zeta \rfloor \omega)$ is exact. Thus, the following lemma is valid.

Lemma 4.10. *Let* $[\zeta]$ *be a symmetry of a differential manifold* $\mathcal{M}$, *i.e.,* $[\zeta] \in \mathrm{Sym}\mathcal{M}$, *then the quotient mapping*

$$L_{[\zeta]} : \mathrm{Cons}^q\mathcal{M} \longrightarrow \mathrm{Cons}^q\mathcal{M}, \; q = 0, 1, \ldots, m,$$

is defined.

5. C-Spectral Sequence. Further Properties of Conservation Laws

The spaces of conservation laws $\mathrm{Cons}^q\mathcal{M}$ of a differential manifold $\mathcal{M}$ can be considered as elements of a spectral sequence associated with $\mathcal{M}$. Namely, let $\mathcal{M}=(M,CTM)$ be a differential manifold and let $\mathcal{O}$ be an open subset of M. A form $\omega\in\Lambda^q(\mathcal{O})$, $q=0,1,\ldots,$ is called *p-Cartan*, $p=0,1,\ldots,$ if $\omega(\zeta_1,\ldots,\zeta_q) = 0$ when at least $q-p+1$ of the fields $\zeta_1,\ldots,\zeta_q\in\mathcal{T}(\mathcal{O})$ are Cartan, i.e., belong to $C\mathcal{T}(\mathcal{O})$. The linear space of all p-Cartan q-forms on $\mathcal{O}$ is denoted by $C^p\Lambda^q(\mathcal{O})$. Clear,

$$C^0\Lambda^q(\mathcal{O}) = \Lambda^q(\mathcal{O}), \text{ and } C^1\Lambda^q(\mathcal{O}) = C\Lambda^q(\mathcal{O}).$$

Moreover,

$$C^p\Lambda^q(\mathcal{O}) = 0 \text{ if } p>q, \text{ and } \Lambda^q(\mathcal{O}) = C^{q-m}\Lambda^q(\mathcal{O}) \text{ if } q>m,$$

where m is the Cartan dimension of $\mathcal{M}$. For further convenience we also denote $\Lambda^q(\mathcal{O})=0$ for $q<0$ and $C^p\Lambda^q(\mathcal{O})=\Lambda^q(\mathcal{O})$ for $p<0$. Now, $C^p\Lambda^q(\mathcal{O})\supset C^{p+1}\Lambda^q(\mathcal{O})$, by the definition, and $d\omega\in C^p\Lambda^{q+1}(\mathcal{O})$ if $\omega\in C^p\Lambda^q(\mathcal{O})$, by the involutivity of the Cartan distribution (see the analogous statement in Section 4). Hence, the de Rham complex

$$0 \longrightarrow \mathbb{R} \longrightarrow \Lambda^0(M) \xrightarrow{d} \Lambda^1(M) \xrightarrow{d} \Lambda^2(M) \xrightarrow{d} \Lambda^3(M) \xrightarrow{d} \ldots$$

of the differential manifold $\mathcal{M}$ has the filtration

$$C^0\Lambda^q(M) \supset C^1\Lambda^q(M) \supset C^2\Lambda^q(M) \supset \ldots \supset C^q\Lambda^q(M) \supset 0$$

compatible with the exterior differentiations, i.e.,

$$d: C^p\Lambda^q(M) \longrightarrow C^p\Lambda^{q+1}(M),$$

where $q=0,1,\ldots,$ $p=0,\ldots,q$. The standard technique of the algebraic topology (see Godeman [1]) allows now to associate with $\mathcal{M}$ a spectral sequence $\left\{E_r^{p,q}(\mathcal{M}), d_r^{p,q}\right\}$ which is called the *C-spectral sequence associated with* $\mathcal{M}$ (Vinogradov [2], see also [4]). We give the inductive description of this sequence.

The zero-order terms are equal to

$$E_0^{p,q}(\mathcal{M}) = \frac{C^p\Lambda^{p+q}(M)}{C^{p+1}\Lambda^{p+q}(M)}, \quad p,q = 0,1,\ldots.$$

It is easy to see that $C^p\Lambda^{p+q}(M) = C^{p+q-m}\Lambda^{p+q}(M)$ if $q>m$, where m is the Cartan dimension of $\mathcal{M}$. Hence, $E_0^{p,q}(\mathcal{M})=0$ when $q>m$. The quotient linear mappings

$$d_0^{p,q} : E_0^{p,q}(\mathcal{M}) \longrightarrow E_0^{p,q+1}(\mathcal{M}), \quad p,q = 0,1,\ldots,$$

are induced by exterior differentiations d's, i.e., $d_0^{p,q}[\omega]=[d\omega]$ for any equivalence class $[\omega]\in E_0^{p,q}(\mathcal{M})$.

The first-order terms are equal to

$$E_1^{p,q}(\mathcal{M}) = \frac{\operatorname{Ker} d_0^{p,q}}{\operatorname{Im} d_0^{p,q-1}} = \frac{C^p\Lambda^{p+q}(M)\cap d^{-1}C^{p+1}\Lambda^{p+q+1}(M)}{C^{p+1}\Lambda^{p+q}(M)+dC^p\Lambda^{p+q-1}(M)}, \quad p,q = 0,1,\ldots,$$

where

$$d^{-1}C^{p+1}\Lambda^{p+q+1}(M) = \left\{\omega\in C^{p}\Lambda^{p+q}(M) \;:\; d\omega\in C^{p+1}\Lambda^{p+q+1}(M)\right\}$$

and

$$dC^{p}\Lambda^{p+q-1}(M) = \left\{\omega=d\theta \;;\; \theta\in C^{p}\Lambda^{p+q-1}(M)\right\}.$$

Again $E_1^{p,q}(\mathcal{M})=0$ for q>m, and the quotient linear mappings

$$d_1^{p,q} \;:\; E_1^{p,q}(\mathcal{M}) \longrightarrow E_1^{p+1,q}(\mathcal{M}),\quad p,q = 0,1,\dots,$$

are induced by the exterior differentiations, i.e., $d_1^{p,q}[\omega]=[d\omega]$ for an equivalence class $[\omega]\in E_1^{p,q}(\mathcal{M})$.

The second-order terms are equal to

$$E_2^{p,q}(\mathcal{M}) = \frac{\operatorname{Ker} d_1^{p,q}}{\operatorname{Im} d_1^{p,q-1}}$$

$$= \frac{C^{p}\Lambda^{p+q}(M)\cap d^{-1}C^{p+2}\Lambda^{p+q+1}(M)}{C^{p+1}\Lambda^{p+q}(M)\cap d^{-1}C^{p+2}\Lambda^{p+q+1}(M)+C^{p}\Lambda^{p+q}(M)\cap dC^{p-1}\Lambda^{p+q-1}(M)},$$

$p,q=0,1,\dots$.

By the induction, the r-order terms are equal to

$$E_r^{p,q}(\mathcal{M}) = \frac{Z_r^{p,q}}{Z_{r-1}^{p+1,q-1}+B_{r-1}^{p,q}},\quad p,q,r = 0,1,\dots,$$

where

$$Z_r^{p,q} = C^{p}\Lambda^{p+q}(M)\cap d^{-1}C^{p+r}\Lambda^{p+q+1}(M)$$

and

$$B_r^{p,q} = C^{p}\Lambda^{p+q}(M)\cap dC^{p-r}\Lambda^{p+q-1}(M).$$

The exterior differentiations induce the linear mappings

$$d_r^{p,q} \;:\; E_r^{p,q}(\mathcal{M}) \longrightarrow E_r^{p+r,q-r+1}(\mathcal{M}).$$

The infinite-order terms are

$$E_{\infty}^{p,q}(\mathcal{M}) = \frac{Z_{\infty}^{p,q}}{Z_{\infty}^{p+1,q-1}+B_{\infty}^{p,q}}, \quad p,q = 0,1,\ldots,$$

where

$$Z_{\infty}^{p,q} = C^p\Lambda^{p+q}(M)\cap d^{-1}\{0\} = \left\{\omega\in C^p\Lambda^{p+q}(M) \; : \; d\omega=0\right\}$$

and

$$B_{\infty}^{p,q} = C^p\Lambda^{p+q}(M)\cap d\Lambda^{p+q-1}(M) = \left\{\omega\in C^p\Lambda^{p+q}(M) \; : \; \omega=d\theta, \; \theta\in\Lambda^{p+q-1}(M)\right\}.$$

Proposition 5.1. *The equalities*

$$E_{r}^{p,q}(\mathcal{M}) = E_{\infty}^{p,q}(\mathcal{M})$$

are valid for all $p,q=0,1,\ldots,$ *and* $r>\max\left\{p,q+1\right\}$.

Proof. Indeed, $C^{p+r}\Lambda^{p+q+1}(M)=0$ if $p+r>p+q+1$, hence, $Z_{r}^{p,q}=Z_{\infty}^{p,q}$ for $r>q+1$. Also, $C^{p-r}\Lambda^{p+q-1}(M)=\Lambda^{p+q-1}(M)$ if $p-r\le 0$, hence $B_{r}^{p,q}=B_{\infty}^{p,q}$ for $r\ge p$. Thus, $Z_{r}^{p,q}=Z_{\infty}^{p,q}$, $Z_{r-1}^{p+1,q-1}=Z_{\infty}^{p+1,q-1}$, $B_{r-1}^{p,q}=B_{\infty}^{p,q}$ and the equality is valid for all $r>\max\left\{p,q+1\right\}$.

Note that $E_{r}^{p,q}(\mathcal{M})=0$ for $q>m$, because $C^p\Lambda^{p+q}(M)=C^{p+q-m}\Lambda^{p+q}(M)$ in this case. Also, it is easy to check that

$$E_{0}^{0,0}(\mathcal{M}) = \mathcal{C}^{\infty}(M),$$

$$E_{1}^{0,0}(\mathcal{M}) = \mathcal{C}^{\infty}(M)\cap d^{-1}C\Lambda(M) = \left\{f\in\mathcal{C}^{\infty}(M) \; : \; \zeta(f)=0, \; \forall\zeta\in C\mathcal{T}(M)\right\},$$

$$E_{r}^{0,0}(\mathcal{M}) = \mathcal{C}^{\infty}(M)\cap d^{-1}\{0\} = \left\{f\in\mathcal{C}^{\infty}(M) \; : \; df=0\right\}, \quad r = 2,3,\ldots.$$

Remind that $C\mathcal{T}(M)$ is the Lie algebra of all Cartan fields on M. Thus, elements of $E_{1}^{0,0}(\mathcal{M})$ can be considered as constants for the Lie algebra of Cartan differentiations of the algebra $\mathcal{C}^{\infty}(M)$. Elements of $E_r^{0,0}(\mathcal{M})$, $r\geq 2$, are locally constant functions on M, in particular, $E_r^{0,0}(\mathcal{M})=\mathbb{R}$ if the manifold M is connected.

Proposition 5.2. *The operations* $d\omega$, $\zeta\rfloor\omega$ *and* $L_\zeta\omega$ (see the definitions in Section 0.2) *have the following properties on the spaces* $C^p\Lambda^q(M)$:

(a) $d\omega \in C^p\Lambda^{q+1}(M)$ *for all* $\omega \in C^p\Lambda^q(M)$;

(b) $\zeta\rfloor\omega \in C^{p-1}\Lambda^{q-1}(M)$ *for all* $\zeta \in \mathcal{T}(M)$, $\omega \in C^p\Lambda^q(M)$;

(c) $\zeta\rfloor\omega \in C^p\Lambda^{q-1}(M)$ *for all* $\zeta \in C\mathcal{T}(M)$, $\omega \in C^p\Lambda^q(M)$;

(d) $L_\zeta\omega \in C^{p-1}\Lambda^q(M)$ *for all* $\zeta \in \mathcal{T}(M)$, $\omega \in C^p\Lambda^q(M)$;

(e) $L_\zeta\omega \in C^p\Lambda^q(M)$ *for all* $\zeta \in LB\mathcal{T}(M)$, $\omega \in C^p\Lambda^q(M)$;

(f) $L_\zeta\omega = d(\zeta\rfloor\omega) + \zeta\rfloor(d\omega)$ *for all* $\zeta \in \mathcal{T}(M)$, $\omega \in C^p\Lambda^q(M)$;

$p,q=0,1,\ldots$.

Proof. Some of these properties we already proved and used above (for example, in Section 4), other are simple exercises. Basically, we listed all these formulas only for convenience of further references.

Proposition 5.3. *The interior product* $\zeta\rfloor\omega$ *induces the* interior product $[\zeta]\rfloor_0[\omega]\in E_0^{p-1,q}(\mathcal{M})$ *for all* $[\zeta]\in\mathrm{Sym}\mathcal{M}$ *and* $[\omega]\in E_0^{p,q}(\mathcal{M})$, $p,q=0,1,\ldots$.

Proof. Let $[\zeta]\in\mathrm{Sym}\mathcal{M}$ and $[\omega]\in E_0^{p,q}(\mathcal{M})$ then (we are omitting *(M)* for brevity)

$$
\begin{aligned}
[\zeta]\rfloor[\omega] &= \left(\zeta + C\mathcal{T}\right)\rfloor\left(\omega + C^{p+1}\Lambda^{p+q}\right) \\
&= \zeta\rfloor\omega + C\mathcal{T}\rfloor\omega + \zeta\rfloor C^{p+1}\Lambda^{p+q} + C\mathcal{T}\rfloor C^{p+1}\Lambda^{p+q} \\
&\subset \zeta\rfloor\omega + C^{p}\Lambda^{p+q-1} + C^{p}\Lambda^{p+q-1} + C^{p+1}\Lambda^{p+q-1} \\
&\subset \zeta\rfloor\omega + C^{p}\Lambda^{p+q-1} = [\zeta\rfloor\omega] \in E_0^{p-1,q}(\mathcal{M}),
\end{aligned}
$$

where $\zeta\in LB\mathcal{T}(M)$, $\omega\in C^{p}\Lambda^{p+q}(M)$, and we used Proposition 5.2.

Proposition 5.4. *The interior product $\zeta\rfloor\omega$ induces the interior product $[\zeta]\rfloor_1[\omega]\in E_1^{p-1,q}(\mathcal{M})$ for all $[\zeta]\in\mathrm{Sym}\mathcal{M}$ and $[\omega]\in E_1^{p,q}(\mathcal{M})$, $p,q=0,1,\ldots$.*

Proof. Let $[\zeta]\in\mathrm{Sym}\mathcal{M}$ and $[\omega]\in E_1^{p,q}(\mathcal{M})$, then we shall have (we denote $X=C^{p+1}\Lambda^{p+q}(M)$ and $Y=C^{p}\Lambda^{p+q-1}(M)$ for brevity)

$$
\begin{aligned}
[\zeta]\rfloor[\omega] &= \left(\zeta + C\mathcal{T}\right)\rfloor\left(\omega + X + dY\right) \\
&= \zeta\rfloor\omega + C\mathcal{T}\rfloor\omega + \zeta\rfloor X + \zeta\rfloor dY + C\mathcal{T}\rfloor\left(X + dY\right) \\
&\subset \zeta\rfloor\omega + C^{p}\Lambda^{p+q-1} + C^{p}\Lambda^{p+q-1} + \left(C^{p}\Lambda^{p+q-1} - dC^{p-1}\Lambda^{p+q-2}\right) \\
&\quad + C^{p+1}\Lambda^{p+q-1} + C^{p}\Lambda^{p+q-1} \subset \zeta\rfloor\omega + C^{p}\Lambda^{p+q-1} + dC^{p-1}\Lambda^{p+q-2} \\
&= [\zeta\rfloor\omega] \in E_1^{p-1,q}(\mathcal{M}),
\end{aligned}
$$

because

$$
d\left(\zeta\rfloor\omega\right) = L_{\zeta}\omega - \zeta\rfloor d\omega \in C^{p}\Lambda^{p+q}(M),
$$

where $\zeta\in LB\mathcal{T}(M)$, $\omega\in C^{p}\Lambda^{p+q}(M)$, $d\omega\in C^{p+1}\Lambda^{p+q+1}(M)$, and we used Proposition 5.2, in particular, the equality *(f)*.

Proposition 5.5. *The Lie derivative $L_{\zeta}\omega$ induces the* Lie derivative $L_{[\zeta]}[\omega]\in E_1^{p,q}(\mathcal{M})$ *for all* $[\zeta]\in\mathrm{Sym}\mathcal{M}$ *and* $[\omega]\in E_1^{p,q}(\mathcal{M})$, *where* $p,q=0,1,\ldots$ (cf. Lemma 4.10).

Proof. Let $[\zeta]\in\mathrm{Sym}\mathcal{M}$ and $[\omega]\in E_1^{p,q}(\mathcal{M})$, then (we are keeping the notation of the previous proof)

$$L_{[\zeta]}[\omega] = \left(L_\zeta + L_{C\mathcal{T}}\right)\left(\omega + X + dY\right)$$

$$= L_\zeta\omega + L_{C\mathcal{T}}\omega + L_\zeta X + L_{C\mathcal{T}}X + L_\zeta dY + L_{C\mathcal{T}}dY$$

$$\subset L_\zeta\omega + \left(dC^p\Lambda^{p+q-1} + C^{p+1}\Lambda^{p+q}\right) + C^{p+1}\Lambda^{p+q} + C^{p+1}\Lambda^{p+q}$$

$$+ dC^p\Lambda^{p+q-1} + dC^p\Lambda^{p+q-1} \subset L_\zeta\omega + X + dY$$

$$= \left[L_\zeta\omega\right] \in E_0^{p,q}(\mathcal{M}),$$

because

$$d\left(L_\zeta\omega\right) = L_\zeta d\omega \in C^{p+1}\Lambda^{p+q+1}(M),$$

where $\zeta\in LB\mathcal{T}(M)$, $\omega\in C^p\Lambda^{p+q}(M)$, $d\omega\in C^{p+1}\Lambda^{p+q+1}(M)$, we used Proposition 5.2 and the commutation relation $L_\zeta d\theta = dL_\zeta\theta$ (see Section 0.2).

Proposition 5.6. *The formulas*

(a) $L_{[\zeta]}[\omega] = d_1\left([\zeta]\rfloor_1[\omega]\right) + [\zeta]\rfloor_1\left(d_1[\omega]\right)$,

(b) $L_{[\zeta]}\left(d_1[\omega]\right) = d_1\left(L_{[\zeta]}[\omega]\right)$,

are valid for all $[\zeta]\in\mathrm{Sym}\mathcal{M}$ *and* $[\omega]\in E_1^{p,q}(\mathcal{M})$ (we omit the superscribes p,q in $d_1^{p,q}$ for brevity).

Proof. These formulas directly follow from the formula *(f)* of Proposition 5.2 and the equality $L_\zeta(d\omega)=d(L_\zeta\omega)$ (see Section 0.2).

Now, by the definition,

$$E_1^{0,q}(\mathcal{M}) = \frac{\Lambda^q(M)\cap d^{-1}C\Lambda^{q+1}(M)}{C\Lambda^q(M)+dC\Lambda^{q-1}(M)} = \mathrm{Cons}^q\mathcal{M},\ q = 0,1,\ldots,m,$$

hence, the spaces of conservation laws $\mathrm{Cons}^q\mathcal{M}$ of a differential manifold $\mathcal{M}$ are elements of the spectral sequence $\left\{E_r^{p,q}(\mathcal{M}), d_r^{p,q}\right\}$. By Proposition 5.5 (see also Section 4), the Lie derivative $L_{[\zeta]}[\omega]$ is the conservation law if $[\zeta]$ is a symmetry and $[\omega]$ is a conservation law. On the other hand, by Proposition 5.4, the interior product $[\zeta]\rfloor_1[\omega]$ is the conservation law if $[\zeta]$ is a symmetry and $[\omega]\in E_1^{1,q}(\mathcal{M})$. Note, that by Proposition 5.6, in this case $(p=0)$ we have the equality

$$L_{[\zeta]}[\omega] = [\zeta]\rfloor_1\left(d_1[\omega]\right),$$

because $[\zeta]\rfloor_1[\omega]\in E_1^{-1,q}(\mathcal{M})=0$ here. The linear mapping

$$d_1 : E_1^{0,q}(\mathcal{M}) \longrightarrow E_1^{1,q}(\mathcal{M})$$

is called *characteristic* when $q=m-1$ (cf. Lax [1], Zharinov [2], Vinogradov [5] and Olver [1]), and *variational* or *Euler-Lagrange* when $q=m$ (see, for example, Vinogradov [1] and Olver [1]). In many cases this mapping is injective and its image has simple description.

We shall call a differential manifold $\mathcal{M}$ *star-shaped* if it can be covered by one chart $(M,\varphi,\mathbb{R}^J)$ (not necessary standard) with a star-shaped domain $\varphi(M)\subset\mathbb{R}^J$ and global coordinates u^i, $i\in J$. In this case the Poincaré lemma (see, for example, Cartan [1]) can be refined in the following way.

Proposition 5.7. *Let a differential manifold $\mathcal{M}$ be star-shaped and $\omega\in C^p\Lambda^q(\mathcal{M})$, $p=1,2,\ldots$, $q=p,p+1,\ldots$. The form ω is closed, $d\omega=0$,* *iff* $\omega=d\theta$, *where a form* $\theta\in C^{p-1}\Lambda^{q-1}(M)$.

Proof. Consider the proof of the Poincaré lemma based on the homotopy formula

$$d(h(\omega)) + h(d(\omega)) = \omega,$$

where the homotopy operator

$$h(\omega)(u;\zeta_1,\dots,\zeta_{q-1}) = \int_0^1 \omega(tu;\upsilon,\zeta_1,\dots,\zeta_{q-1})t^{q-1}dt,$$

a point $u=(u^i)\in M$, vector fields $\zeta_1,\dots,\zeta_{q-1}\in\Lambda^{q-1}(M)$ and the scaling vector field

$$\upsilon = \sum_{i\in J} u^i\frac{\partial}{\partial u^i}.$$

Now, in order to get the desired refinement it is enough to note that $h(\omega)\in C^{p-1}\Lambda^{q-1}(M)$ if $\omega\in C^p\Lambda^q(M)$.

Proposition 5.8. *If a differential manifold* $\mathcal{M}$ *is star-shaped, then* $E_r^{p,q}(\mathcal{M})=0$ *for all* $p,q=0,1,\dots$, $p+q>0$, *and* $r=2,3,\dots$.

Proof. Clear that it is sufficient to prove the proposition for $r=2$. So, let an equivalence class $[\omega]\in E_2^{p,q}(\mathcal{M})$, where a form $\omega\in C^p\Lambda^{p+q}(M)$ and $d\omega\in C^{p+2}\Lambda^{p+q+1}(M)$. Then the form $\chi=d\omega\in C^{p+2}\Lambda^{p+q+1}(M)$ and $d\chi=dd\omega=0$. By Proposition 5.2, $\chi=d\theta$ with $\theta\in C^{p+1}\Lambda^{p+q}(M)$. Hence, the form $\omega-\theta\in C^p\Lambda^{p+q}(M)$ and $d(\omega-\theta)=0$ thus $\omega-\theta=d\phi$, $\phi\in C^{p-1}\Lambda^{p+q}(M)$, by Proposition 5.2 again. Thus, the form $\omega=\theta+d\phi$ belongs to the trivial subspace, and the equivalence class $[\omega]=0$.

Corollary 5.9. *If a differential manifold* $\mathcal{M}$ *is star-shaped, then the mappings* $d_1:E_1^{0,q}(\mathcal{M})\longrightarrow E_1^{1,q}(\mathcal{M})$ *induce the exact sequence*

$$0 \longrightarrow \mathbb{R} \longrightarrow E_1^{0,0}(\mathcal{M}) \longrightarrow \mathrm{Ker}\ d_1^{1,0} \longrightarrow 0$$

and the isomorphisms

$$E_1^{0,q}(\mathcal{M}) \simeq \mathrm{Ker}\ d_1^{1,q}, \quad q = 1,\dots,m,$$

where

$$\mathrm{Ker}\ d_1^{1,0} = C\Lambda \cap d^{-1}\{0\}$$

and

$$\mathrm{Ker}\ d_1^{1,q} = \frac{C\Lambda^{q+1} \cap d^{-1}C^3\Lambda^{q+2}}{C^2\Lambda^{q+1} \cap d^{-1}C^3\Lambda^{q+2} + dC\Lambda^q}, \quad q = 1,\dots,m.$$

Proof. Indeed, in this case we have the exact sequences (we omit (M) for brevity)

$$0 \longrightarrow \mathbb{R} \longrightarrow E_1^{0,0} \xrightarrow{d_1^{0,0}} E_1^{1,0} \xrightarrow{d_1^{1,0}} E_1^{2,0} \dots$$

and

$$0 \longrightarrow E_1^{0,q} \xrightarrow{d_1^{0,q}} E_1^{1,q} \xrightarrow{d_1^{1,q}} E_1^{2,q} \dots, \quad q = 1,\dots,m,$$

so one should only to check the expressions for Ker $d_1^{1,q}$.

Let us assume now that a differential manifold $\mathcal{M}$ is covered by one standard chart $(M,\varphi,\mathbb{R}^{m+I},\partial)$ with global coordinates x^μ, u^i, $\mu=1,\dots,m$, $i \in I$, basis Cartan fields

$$\partial_\mu = \frac{\partial}{\partial x^\mu} + \sum_{i\in I} A_\mu^i(x,u)\ \frac{\partial}{\partial u^i}, \quad \mu = 1,\dots,m,$$

and basis Cartan 1-forms

$$\delta u^i = du^i - \sum_{\mu=1}^m A_\mu^i(x,u)\ dx^\mu, \quad i \in I.$$

Consider linear spaces

$$E^{p,q} = \left\{\omega = \sum_{(i)_p} \sum_{(\mu)_q} \omega_{(i)_p;(\mu)_q} \left(\delta u^i\right)^p \wedge \left(dx^\mu\right)^q\right\},$$

(we use the notation of Proposition 4.2, clear that $E^{p,q}=0$ for $q>m$). One can check that the exterior differentiation d acts from $E^{p,q}$ into $E^{p+1,q}+E^{p,q+1}$. Moreover, $d=\delta+\partial$, where the linear mappings $\delta: E^{p,q} \longrightarrow E^{p+1,q}$ and $\partial: E^{p,q} \longrightarrow E^{p,q+1}$ are defined by the formula

$$df = \delta f + \partial f = \sum_{i\in I} \frac{\partial f}{\partial u^i} \delta u^i + \sum_{\mu=1}^{m} \partial_\mu f \, dx^\mu, \quad f \in \mathcal{C}^\infty(M) = E^{0,0}$$

(cf. Proposition 1.13). In particular,

$$du^i = \sum_{k\in I} \frac{\partial u^i}{\partial u^k} \delta u^k + \sum_{\mu=1}^{m} \partial_\mu u^i \, dx^\mu = \delta u^i + \sum_{\mu=1}^{m} A^i_\mu \, dx^\mu,$$

hence,

$$\delta u^i = \delta u^i \quad \text{and} \quad \partial u^i = \sum A^i_\mu \, dx^\mu, \quad i \in I.$$

Also,

$$d\delta u^i = \delta\delta u^i + \partial\delta u^i = d\left(du^i - \sum_{\mu=1}^{m} A^i_\mu \, dx^\mu\right) = -\sum_{\mu=1}^{m} \sum_{k\in I} \frac{\partial A^i_\mu}{\partial u^k} \delta u^k \wedge dx^\mu,$$

so,

$$\delta\delta u^i = 0 \quad \text{and} \quad \partial\delta u^i = -\sum_{\mu=1}^{m} \sum_{k\in I} \frac{\partial A^i_\mu}{\partial u^k} \delta u^k \wedge dx^\mu$$

(cf. Proposition 1.14). The composition equality $dd=0$ implies the formulas

$$\delta\delta = 0, \; \delta\partial + \partial\delta = 0, \; \partial\partial = 0.$$

Note that $\partial\omega=0$ for any $\omega\in E^{p,m}$, $p=0,1,\ldots$.

Every equivalence class $[\omega]\in E_0^{p,q}(\mathcal{M})$, $p,q=0,1,\dots$, has one and only one representative $\omega\in E^{p,q}$, hence, we can identify the spaces $E_0^{p,q}(\mathcal{M})$ and $E^{p,q}$. Clear that $d_0^{p,q}[\omega] = [\partial\omega]$ for any $\omega\in E^{p,q}$ in this case. Further,

$$E_1^{p,q}(\mathcal{M}) = \frac{\operatorname{Ker} d_0^{p,q}}{\operatorname{Im} d^{p,q-1}} = \frac{E^{p,q}\cap\partial^{-1}\{0\}}{\partial E^{p,q-1}}, \quad p,q = 0,1,\dots,$$

where,

$$E^{p,q}\cap\partial^{-1}\{0\} = \left\{\omega\in E^{p,q} \;:\; \partial\omega=0\right\}, \quad \partial E^{p,q-1} = \left\{\partial\chi \;;\; \chi\in E^{p,q-1}\right\}$$

(cf. Proposition 4.3). The linear operators $d_1^{p,q}$ are defined now by the formula

$$d_1^{p,q}[\omega] = [\delta\omega], \; [\omega] \in E^{p,q}\cap d^{-1}\{0\}, \; p,q = 0,1,\dots.$$

In particular, the spaces from Corollary 5.9 have the following descriptions:

$$E_1^{0,0}(\mathcal{M}) = \left\{f\in\mathcal{C}^\infty(M) \;:\; \partial f=0\right\};$$

$$\operatorname{Ker} d_1^{1,0} = \left\{\omega = \sum_{i\in I} \omega_i \;\delta u^i \;:\; \partial\omega=0\right\},$$

where

$$\partial\omega = \sum_{\mu=1}^{m} \sum_{i\in I} \left\{\partial_\mu\omega_i + \sum_{k\in I} \frac{\partial A_\mu^k}{\partial u^i}\,\omega_k\right\} dx^\mu\wedge\delta u^i;$$

$$E_1^{0,q}(\mathcal{M}) = \frac{\left\{\omega\in E^{0,q} \;:\; \partial\omega=0\right\}}{\left\{\partial\theta \;;\; \theta\in E^{0,q-1}\right\}},$$

and, for example,

$$\omega = \sum_{(\mu)_q} \omega_{(\mu)_q} \left(dx^{\mu}\right)^q, \quad \partial\omega = \sum_{(\mu)_{q+1}} \partial_{[\mu_0}\omega_{\mu_1\ldots\mu_q]} \left(dx^{\mu}\right)^{q+1};$$

$$\operatorname{Ker} d_1^{1,q} = \frac{\left\{\omega\in E^{1,q} \;:\; \partial\omega=0,\ \delta\omega=\partial\chi,\ \chi\in E^{2,q-1}\right\}}{\left\{\partial\theta \;;\; \theta\in E^{1,q-1}\right\}},$$

where, for example,

$$\omega = \sum_{i\in I} \sum_{(\mu)_q} \omega_{i;(\mu)_q} \delta u^i \wedge \left(dx^{\mu}\right)^q$$

and

$$\delta\omega = \sum_{i,k\in I} \sum_{(\mu)_q} \left(\frac{\partial}{\partial u^{[i}} \omega_{k];(\mu)_q}\right) \delta u^i \wedge \delta u^k \wedge \left(dx^{\mu}\right)^q.$$

Let $f:\mathcal{N}\longrightarrow\mathcal{M}$ a Lie Backlund mapping of a differential manifold $\mathcal{M}=(M,CTM)$ into a differential manifold $\mathcal{N}=(N,CTN)$. The cotangent mapping $f^*:\Lambda^q(M)\longrightarrow\Lambda^q(N)$ has the following properties: $f^*\circ d=d\circ f^*$, and $f^*:C^p\Lambda^q(M)\longrightarrow C^p\Lambda^q(N)$. Hence, the quotient mapping

$$f^* \;:\; E_r^{pq}(\mathcal{M}) \longrightarrow E_r^{pq}(\mathcal{N})$$

is induced (we keep the same notation for it), and $f^*\circ d_r^{pq}=d_r^{pq}\circ f^*$, for all $r,p,q=0,1,\ldots$.

6. TRIVIAL EQUATIONS. THE FORMAL VARIATIONAL CALCULUS

In this section we consider differential manifolds $\mathcal{M}=(M,CTM)$, having the following property: every m-dimensional submanifold of M is integral, where m is a Cartan dimension of $\mathcal{M}$ (see Examples 1.5 and 1.6 above). Such manifolds correspond to trivial differential equations, i.e., differential equations that don't impose any restrictions on its solutions.

We shall use the standard multi-index notation:

$\mathbb{N}=\{0,1,2,\ldots\}$ is the set of all natural numbers, including zero;

$|i|=i_1+\ldots+i_m$ and $i!=i_1!\ldots i_m!$ for $i=(i_1,\ldots,i_m)\in\mathbb{N}^m$;

the multi-index $(\mu)=(0,\ldots,0,1,0,\ldots,0)$, where the unity stands in the μ-th position, $\mu=1,\ldots,m$;

$i+k=(i_1+k_1,\ldots,i_m+k_m)$ for $i,k\in\mathbb{N}^m$;

$i\le k$ means $i_1\le k_1,\ldots,i_m\le k_m$ for $i,k\in\mathbb{N}^m$;

multinomial coefficient $\binom{k}{i} = \dfrac{k!}{i!(k-i)!}$ for $i,k\in\mathbb{N}^m$.

$\left(\dfrac{\partial}{\partial x}\right)^i = \left(\dfrac{\partial}{\partial x^1}\right)^{i_1}\ldots\left(\dfrac{\partial}{\partial x^m}\right)^{i_m}$ for $i\in\mathbb{N}^m$ and $x=(x^1,\ldots,x^m)\in\mathbb{R}^m$,

in particular,

$$\left(\frac{\partial}{\partial x}\right)^i\varphi(x) = \frac{\partial^{|i|}\varphi(x)}{(\partial x)^i}, \text{ where } i\in\mathbb{N}^m, \quad \left(\partial x\right)^i = \left(\partial x^1\right)^{i_1}\ldots\left(\partial x^m\right)^{i_m}$$

and a smooth function $\varphi(x)\in\mathscr{C}^{\infty}(O)$, O is an open subset of $\mathbb{R}^m$.

We shall also denote by $\mathscr{C}^{\infty}(U)^n$ the direct product of n copies of the space $\mathscr{C}^{\infty}(U)$, where U is some open set; a function $\varphi\in\mathscr{C}^{\infty}(U)^n$ iff $\varphi=(\varphi^{\nu})$, where $\varphi^{\nu}\in\mathscr{C}^{\infty}(U)$, $\nu=1,\dots,n$.

A differential manifold $\mathcal{M}=(M,D)$ is called *trivial* if the manifold $M=\mathbb{R}^m\times\mathbb{R}^a_I$ has global coordinates $x=(x^{\mu})\in\mathbb{R}^m$, $\mathbf{u}=(u^{\alpha}_i)\in\mathbb{R}^a_I$ (here $\mu=1,\dots,m$, $I=\mathbb{N}^m$, $\alpha=1,\dots,a$, $i\in I$), and the Cartan distribution D on M is determined by basis vector fields

$$D_{\mu} = \frac{\partial}{\partial x^{\mu}} + \sum_{\alpha=1}^{a}\sum_{i\in I} u^{\alpha}_{i+(\mu)}\frac{\partial}{\partial u^{\alpha}_i}, \quad \mu = 1,\dots,m.$$

One can easily check that $\left[D_{\mu},D_{\nu}\right]=0$ for $\mu,\nu=1,\dots,m$. Moreover, the differentiations D_{μ} are characterized by the following property:

$$\left[D_{\mu}f(x,\mathbf{u})\right]\Big|_{u=\varphi(x)} = \frac{\partial}{\partial x^{\mu}}\left[f(x,\mathbf{u})\Big|_{u=\varphi(x)}\right], \quad \mu = 1,\dots,m,$$

for all functions $f(x,\mathbf{u})\in\mathscr{C}^{\infty}(M)$ and functions $\varphi(x)\in\mathscr{C}^{\infty}(\mathbb{R}^m)^a$, where

$$u=\varphi(x) \text{ stands for } u^{\alpha}_i = \left(\frac{\partial}{\partial x}\right)^i\varphi^{\alpha}(x),\ \alpha = 1,\dots,a,\ i\in I$$

(cf. Proposition 1.9). The basis Cartan 1-forms are

$$\delta u^{\alpha}_i = du^{\alpha}_i - \sum_{\mu=1}^{m} u^{\alpha}_{i+(\mu)}\,dx^{\mu}, \quad \alpha = 1,\dots,a,\ i\in I,$$

and in this case

$$\partial\delta u^{\alpha}_i = -\sum_{\mu=1}^{m}\delta u^{\alpha}_{i+(\mu)}\wedge dx^{\mu}$$

(see the analogous formulas in Section 5, we use the global standard chart notation, introduced there).

We begin our study of the differential manifold $\mathcal{M}=(M,D)$ with the description of its integral submanifolds.

Proposition 6.1. *A m-dimensional smooth submanifold S of M is integral iff it has a local representation*

$$S = \left\{x,\mathbf{u} \; : \; u_i^\alpha=\left(\frac{\partial}{\partial x}\right)^i\varphi^\alpha(x), \;\; \alpha=1,\ldots,a, \;\; i\in I\right\},$$

where $\varphi(x)\in\mathcal{C}^\infty(\mathcal{O})^a$, $\mathcal{O}$ is an open subset of $\mathbb{R}^m$.

Note that exactly this property enables us to call the manifold $\mathcal{M}$ trivial.

Proof. Indeed, in this case the contact equations for functions $s_i^\alpha(x)$ (see Proposition 1.8) take the form

$$\frac{\partial s_i^\alpha}{\partial x^\mu} = s_{i+(\mu)}^\alpha, \quad \alpha = 1,\ldots,m, \;\; i \in I,$$

and have the evident general solution

$$s_i^\alpha = \left(\frac{\partial}{\partial x}\right)^i\varphi^\alpha(x), \quad \alpha=1,\ldots,a, \;\; i\in I,$$

where $\varphi^\alpha(x)$ are arbitrary smooth functions.

The next step of our study is the description of infinitesimal symmetries of $\mathcal{M}$.

Proposition 6.2. *A vector field*

$$\zeta = \sum_{\alpha=1}^{a} \sum_{i\in I} \zeta_i^\alpha(x,\mathbf{u}) \frac{\partial}{\partial u_i^\alpha}$$

is Lie-Backlund iff $\zeta_i^\alpha=D_i\varphi^\alpha(x,\mathbf{u})$, where $\varphi\in\mathcal{C}^\infty(M)^a$, $\alpha=1,\ldots,a$, $i\in I$.

Here and below we denote $D_i=D_1^{i_1}\ldots D_m^{i_m}$.

Proof. By Proposition 3.1, a vector field ζ will be Lie-Backlund iff its components ζ_i^α satisfy the defining equations

$$D_\mu \zeta_i^\alpha - \zeta_{i+(\mu)}^\alpha = 0, \ \mu = 1,\dots,m, \ \alpha = 1,\dots,a, \ i \in I.$$

Clear that these equations have the general solution we claim.

In other words, infinitesimal symmetries of $\mathcal{M}$ are in fact *evolution differentiations*

$$\mathrm{E}_\varphi = \sum_{\alpha=1}^{a} \sum_{i\in I} \left(D_i \varphi^\alpha(x,\mathbf{u})\right) \frac{\partial}{\partial u_i^\alpha}, \quad \varphi \in \mathcal{C}^\infty(M)^a$$

(cf. Example 3.7, see also Vinogradov, Krasil'shchik, Lychagin [1; Ch.3, §2]).

Remark 6.3. Note that

$$D_i \mathrm{E}_\varphi(f) = \mathrm{E}_\varphi(D_i f)$$

for all $\varphi\in\mathcal{C}^\infty(M)^a$, $f\in\mathcal{C}^\infty(M)$, $i\in I$, and if E_φ and E_ψ are evolution differentiations then their Lie bracket is also evolution differentiation,

$$\left[\mathrm{E}_\varphi, \mathrm{E}_\psi\right] = \mathrm{E}_\chi,$$

where

$$\chi^\alpha = \{\varphi,\psi\}^\alpha = \mathrm{E}_\varphi(\psi^\alpha) - \mathrm{E}_\psi(\varphi^\alpha), \quad \alpha = 1,\dots,a,$$

is the *Jacoby bracket* of functions $\varphi,\psi\in\mathcal{C}^\infty(M)^a$ (cf. Example 3.7). Thus, the set $\mathcal{C}^\infty(M)^a$ of all smooth functions $\varphi(x,\mathbf{u})$ forms the Lie algebra with the Jacoby bracket as the multiplication. Moreover, the Lie algebra Sym$\mathcal{M}$ of infinitesimal symmetries of the differential manifold $\mathcal{M}$ is isomorphic to $\mathcal{C}^\infty(M)^a$. It is worth to note that in applications it is convenient sometimes to use Lie-Backlund fields that are not evolution differentiations. For example, space translations $(x,\mathbf{u})\mapsto(x-l\tau,\mathbf{u})$ of the manifold $M=\mathbb{R}^m\times\mathbb{R}^a_I$, where $l\in\mathbb{R}^m$ is a fixed vector, are usually described by the Lie-Backlund field

$$\zeta = -\sum_{\mu=1}^{m} l^\mu \frac{\partial}{\partial x^\mu},$$

which is equivalent to the evolution differentiation E_φ, where

$$\varphi = \left(\varphi^\alpha = \sum_\mu l^\mu u^\alpha_{(\mu)}\right) \in \mathcal{C}^\infty(M)^a,$$

because

$$-\sum_{\mu=1}^{m} l^\mu \frac{\partial}{\partial x^\mu} = -\sum_{\mu=1}^{m} l^\mu D_\mu + \mathrm{E}_\varphi .$$

Consider now the C-spectral sequence $\left\{E_r^{p,q}(\mathcal{M}), d_r^{p,q}\right\}$, associated with $\mathcal{M}$. By Proposition 5.8, terms $E_r^{p,q}(\mathcal{M})=0$ for all $p+q>0$ and $r>1$. Moreover, let

$$E^{p,q} = \left\{\omega = \sum_{(i)_p} \sum_{(\mu)_q} \omega_{(i)_p;(\mu)_q} \left(\delta u^i\right)^p \wedge \left(dx^\mu\right)^q\right\}$$

be the linear spaces introduced in Section 5, $p,q=0,1,\ldots,$ then the following theorem is true.

Theorem 6.4 (Vinogradov [2], Tsujishita [1]). *The only nontrivial terms of the C-spectral sequence, associated with the differential manifold $\mathcal{M}=(M,D)$ are:*

$$E_0^{p,q}(\mathcal{M}) = E^{p,q}, \quad p = 0,1,\ldots, \quad q = 0,\ldots,m$$

(see the notation of Section 5);

$$E_r^{0,0}(\mathcal{M}) = \mathbb{R}, \quad r = 1,2,\ldots;$$

$$E_1^{p,m}(\mathcal{M}) = \frac{E^{p,m}}{\partial E^{p,m-1}}, \quad p = 0,1,\ldots.$$

I don't know an elementary proof of this theorem, appropriate to reproduce it here, and refer an interested reader to the original papers of Vinogradov and Tsujishita.

The spaces $E_1^{p,m}(\mathcal{M})$, $p=0,1$, and the mapping

$$d_1^{0,m} : E_1^{0,m}(\mathcal{M}) \longrightarrow E_1^{1,m}(\mathcal{M})$$

are the main objects of the formal variational calculus (*formal* means that no restrictions like smoothness or boundary conditions are imposed). We give a brief exposition of it here (for more details and the bibliography see Tsujishita [1], Olver [1], Vinogradov [1,2,4], Galindo, Martinez [1], Gelfand, Manin, Shubin [1] Martinez [1] and others).

By the previous theorem

$$E_1^{0,m}(\mathcal{M}) = \frac{E^{0,m}}{\partial E^{0,m-1}},$$

where

$$E^{0,m} = \left\{\omega = Ld^m x \; ; \; L(x,\mathbf{u}) \in \mathcal{C}^\infty(\mathcal{M})\right\},$$

$$E^{0,m-1} = \left\{\theta = \sum_{\mu=1}^{m} J^\mu d^{m-1} x_\mu \; ; \; J^\mu(x,\mathbf{u}) \in \mathcal{C}^\infty(\mathcal{M})\right\}$$

and

$$\partial E^{0,m-1} = \left\{\omega = \partial\theta = \left(\sum_{\mu=1}^{m} D_\mu J^\mu\right) d^m x\right\}.$$

Here m-forms ω are called *Lagrangians*, smooth functions $L(x,\mathbf{u})$ are called *Lagrangian densities*,

$$d^m x = dx^1\wedge\ldots\wedge dx^m$$

is the volume element in $\mathbb{R}^m$,

$$d^{m-1}x_\mu = (-1)^{\mu-1}dx^1\wedge\ldots\wedge\widehat{dx^\mu}\wedge\ldots\wedge dx^m, \quad \mu = 1,\ldots,m$$

(the differential dx^μ is omitted), m-forms $\partial\theta$ are called *divergencies*. Let $\omega=Ld^m x$ be a Lagrangian. Suppose that $\varphi(x)\in\mathcal{C}^\infty(\mathbb{R}^m)^a$ satisfies suitable decrease conditions as $|x|\longrightarrow\infty$ (has a compact support, for example) then the integral

$$L[\varphi] = \int L(x,\mathbf{u})\Big|_{u=\varphi(x)} d^m x$$

will be defined. Further, let a Lagrangian $\omega=\partial\theta$ be a divergency, then

$$\int\left(\sum_{\mu=1}^{m} D_\mu J^\mu\right)\Big|_{u=\varphi(x)} d^m x = \int\sum_{\mu=1}^{m}\frac{\partial}{\partial x^\mu}\left(J^\mu(x,\mathbf{u})\Big|_{u=\varphi(x)}\right)d^m x = 0,$$

by Stokes' theorem (cf. Example 4.6). Hence, elements of the quotient space $E_1^{0,m}(\mathcal{M})$ are in fact *functionals* on a space of smooth decreasing functions.

Let us turn to the quotient space

$$E_1^{1,m}(\mathcal{M}) = \frac{E^{1,m}}{\partial E^{1,m-1}}.$$

Here

$$E^{1,m} = \left\{\Omega = \sum_{\alpha=1}^{a} \sum_{i\in I} \Omega^i_\alpha \ \delta u^\alpha_i \wedge d^m x \ ; \ \Omega^i_\alpha(x,\mathbf{u}) \in \mathscr{C}^\infty(M)\right\},$$

$$E^{1,m-1} = \left\{\chi = \sum_{\alpha=1}^{a} \sum_{i\in I} \sum_{\mu=1}^{m} \Omega^{i;\mu}_\alpha \ \delta u^\alpha_i \wedge d^{m-1} x_\mu \ ; \ \Omega^{i;\mu}_\alpha(x,\mathbf{u}) \in \mathscr{C}^\infty(M)\right\}.$$

Lemma 6.5. *Let* $\Omega=\Sigma\Omega^i_\alpha \delta u^\alpha_i \wedge d^m x \in E^{1,m}$, *then*

$$\Omega = \sum_{\alpha=1}^{a} \left(\sum_{i\in I} (-1)^{|i|} D_i \Omega^i_\alpha \right) \ \delta u^a \wedge d^m x \quad \mathrm{mod}\ \partial E^{1,m-1},$$

where $u^\alpha = u^\alpha_0$, $0=(0,\ldots,0)\in I$ (remind that $I=\mathbb{N}^m$).

Proof. To prove the lemma it is enough to use the evident formula

$$\partial\left(f \ \delta u^\alpha_i \wedge d^{m-1} x_\mu\right) = -\left\{\left(D_\mu f\right) \ \delta u^\alpha_i + f \ \delta u^\alpha_{i+(\mu)}\right\} \wedge d^m x,$$

implying the representation

$$f \ \delta u^\alpha_{i+(\mu)} \wedge d^m x = -\left(D_\mu f\right) \ \delta u^\alpha_i \wedge d^m x - \partial\left(f \ \delta u^\alpha_i \wedge d^{m-1} x_\mu\right),$$

where $f\in\mathscr{C}^\infty(M)$, $\alpha=1,\ldots,a$, $i\in I$.

Lemma 6.6. *Let*

$$\Omega = \sum_{\alpha=1}^{a} \Omega_\alpha \delta u^\alpha \wedge d^m x \in E^{1,m},$$

then the representation $\Omega=\partial\chi$, $\chi\in E^{1,m-1}$, *is valid iff* $\Omega=0$.

Proof. It is convenient to use another multi-index notation, than above. Namely, every multi-index $i=(i_1,\ldots,i_m)\in\mathbb{N}^m$, $|i|=k$, has the unique representation $i=(\mu_1)+\ldots+(\mu_k)$, where the unordered family $\mu_1\ldots\mu_k\in\{1,\ldots,m\}^k$, and we write $i=(i_1,\ldots,i_m)=\mu_1\ldots\mu_k$ in this case. Now, let

$$\chi=\sum_{\alpha=1}^{a}\sum_{|i|=0}^{r}\sum_{\mu=1}^{m}\chi_\alpha^{i;\mu}\,\delta u_i^\alpha\wedge d^{m-1}x_\mu\in E^{1,m-1},$$

then

$$\partial\chi=-\sum_{\alpha=1}^{a}\sum_{|i|=0}^{r}\left\{\left(\sum_{\mu=1}^{m}D_\mu\chi_\alpha^{i;\mu}\right)\delta u_i^\alpha+\sum_{\mu=1}^{m}\chi_\alpha^{\{i;\mu\}}\,\delta u_{i+(\mu)}^\alpha\right\}\wedge d^m x,$$

where $\{i;\mu\}$ denotes the symmetrization in indices $\mu_1,\ldots,\mu_k,\mu$, for $i=\mu_1\ldots\mu_k\in\mathbb{N}^m$, $|i|=k$. Hence, the equality

$$\begin{aligned}\partial\chi=&-\sum_{\alpha=1}^{a}\left\{\sum_{\mu=1}^{m}\left(D_\mu\chi_\alpha^{0;\mu}\right)\delta u^\alpha\right.\\&+\sum_{k=1}^{r}\sum_{|i|=k}\left(\sum_{\mu=1}^{m}D_\mu\chi_\alpha^{\mu_1\ldots\mu_k;\mu}+\chi_\alpha^{\{\mu_1\ldots\mu_{k-1};\mu_k\}}\right)\delta u_{\mu_1\ldots\mu_k}^\alpha\\&\left.+\sum_{|i|=r+1}\chi_\alpha^{\{\mu_1\ldots\mu_r;\mu_{r+1}\}}\,\delta u_{\mu_1\ldots\mu_{r+1}}^\alpha\right\}\wedge d^m x\\=&\ \Omega=\sum_{\alpha=1}^{a}\Omega_\alpha\delta u^\alpha\wedge d^m x\end{aligned}$$

implies

(a) $\displaystyle\sum_{\mu=1}^{m}\left(D_\mu\chi_\alpha^{0;\mu}\right)=-\Omega_\alpha,$

(b) $\displaystyle\sum_{\mu=1}^{m}D_\mu\chi_\alpha^{\mu_1\ldots\mu_k;\mu}+\chi_\alpha^{\{\mu_1\ldots\mu_{k-1};\mu_k\}}=0,\quad k=1,\ldots,r,$

(c) $\chi_\alpha^{\{\mu_1\ldots\mu_r;\mu_{r+1}\}}=0,$

where $\alpha=1,\ldots,a$. So,

$$\begin{aligned}
\Omega_\alpha &= (-1)^1 \sum_{\mu=1}^{m} \left(D_\mu \chi_\alpha^{0;\mu} \right) \\
&= (-1)^2 \sum_{|i|=2} D_{\mu_1} D_{\mu_2} \chi_\alpha^{\{\mu_1;\mu_2\}} \\
&= (-1)^3 \sum_{|i|=3} D_{\mu_1} D_{\mu_2} D_{\mu_3} \chi_\alpha^{\{\mu_1\mu_2;\mu_3\}} \\
&\cdots\cdots\cdots\cdots\cdots\cdots \\
&= (-1)^{r+1} \sum_{|i|=r+1} D_{\mu_1} \ldots D_{\mu_{r+1}} \chi_\alpha^{\{\mu_1\ldots\mu_r;\mu_{r+1}\}} = 0,
\end{aligned}$$

and the proof is completed.

Proposition 6.7. *The following representation*

$$E_1^{1,m}(\mathcal{M}) = \left\{ \Omega = \sum_{\alpha=1}^{a} \Omega_\alpha \delta u^\alpha \wedge d^m x \ ; \ \Omega_\alpha(x,\mathbf{u}) \in \mathcal{C}^\infty(M) \right\}$$

is valid.

Proof. It is enough to combine Lemmas 6.5 and 6.6.

The next step is the linear mapping $d_1^{0,m}: E_1^{0,m}(\mathcal{M}) \rightarrow E_1^{1,m}(\mathcal{M})$. Let a Lagrangian $\omega=L(x,\mathbf{u})d^m x \in E^{0,m}$, then $d_1^{0,m}[\omega]=[\delta\omega]$ and (see Section 5)

$$\begin{aligned}
\delta\left(L d^m x \right) &= \sum_{\alpha=1}^{a} \sum_{i\in I} \frac{\partial L}{\partial u_i^\alpha} \delta u_i^\alpha \wedge d^m x \\
&= \sum_{\alpha=1}^{a} \frac{\delta L}{\delta u^\alpha} \delta u^\alpha \wedge d^m x \mod \partial E^{1,m-1},
\end{aligned}$$

by lemma 6.5, where

$$\frac{\delta L}{\delta u^\alpha} = \sum_{i\in I} (-1)^{|i|} D_i \left(\frac{\partial L}{\partial u_i^\alpha}\right), \quad \alpha = 1,\dots,a.$$

Thus, the mapping $d_1^{0,m}$ is acting by the rule

$$E_1^{0,m}(\mathcal{M}) \ni \left[L d^m x\right] \mapsto \sum_{\alpha=1}^{a} \frac{\delta L}{\delta u^\alpha} \delta u^\alpha \wedge d^m x \in E_1^{1,m}(\mathcal{M}).$$

From the general considerations, it is clear that this action is correct, i.e., if $L=\sum D_\mu J^\mu$ is a divergency, then $\delta L \big/ \delta u^\alpha=0$, $\alpha=1,\dots,a$. Moreover, from the equality $E_2^{0,m}(\mathcal{M})=0$ it follows that converse is also true, i.e., if $\delta L \big/ \delta u^\alpha=0$, $\alpha=1,\dots,a$, then there exists $J=(J^\mu)\in\mathcal{C}^\infty(M)^m$ such that $L=\sum D_\mu J^\mu$. Both statements can be proved directly also (see, for example, Galindo, Martinez [1]).

Consider a Lagrangian $\omega=L(x,\mathbf{u})d^m x$, and let $\varphi(x)\in\mathcal{C}^\infty(\mathbb{R}^m)^a$ be a smooth function, and $\delta\varphi(x)\in\mathcal{C}^\infty(\mathbb{R}^m)^a$ be its variation with a compact support. Using Stokes' theorem one can easily check that the variation of the functional $L[\varphi]$ is equal to

$$\delta L[\varphi] = \int \sum_{\alpha=1}^{a} \left[\frac{\delta L}{\delta u^\alpha}\bigg|_{u=\varphi(x)}\right] \delta\varphi^\alpha(x)\; d^m x.$$

Thus, $\delta L \big/ \delta u^\alpha$ are *variational derivatives* of a function $L\in\mathcal{C}^\infty(M)$ and the linear mapping

$$\mathcal{C}^\infty(M) \ni L \mapsto \frac{\delta L}{\delta u} = \left(\frac{\delta L}{\delta u^\alpha}\right) \in \mathcal{C}^\infty(M)_a$$

is the *Euler-Lagrange operator* of the variational calculus. The above discussion leads to the following

Proposition 6.8. *The kernel of the Euler-Lagrange operator is equal to*

$$\operatorname{Ker} \frac{\delta}{\delta u} = \left\{ L = \sum_{\mu=1} D_\mu J^\mu \ ; \ J^\mu \in \mathcal{C}^\infty(M) \right\}.$$

Let us now describe the image of this operator. First, consider the quotient space

$$E_1^{2,m}(M) = \frac{E^{2,m}}{\partial E^{2,m-1}},$$

where

$$E^{2,m} = \left\{ \omega = \sum_{\alpha,\beta} \sum_{i,j} \omega_{\alpha,\beta}^{i,j} \ \delta u_i^\alpha \wedge \delta u_j^\beta \wedge d^m x \right\},$$

$$E^{2,m-1} = \left\{ \chi = \sum_{\alpha,\beta} \sum_{i,j} \sum_{\mu} \chi_{\alpha,\beta}^{i,j;\mu} \ \delta u_i^\alpha \wedge \delta u_j^\beta \wedge d^{m-1} x_\mu \right\}$$

(we omit the summation limits for brevity). Let us put $D_\mu \omega = L_{D_\mu} \omega$, $\mu=1,\dots,m$, $\omega \in \Lambda^q(M)$, for convenience. In particular, $D_\mu dx^\nu = 0$ and $D_\mu \delta u_i^\alpha = \delta u_{i+(\mu)}^\alpha$. Also the action $D_i \omega$ is defined for all $i \in I$ and $\omega \in \Lambda^q(M)$, in particular, $\delta u_i^\alpha = D_i \delta u^\alpha$.

Lemma 6.9. *Let $\omega \in E^{2,m}$ and $\mu=1,\dots,m$, then*

$$D_\mu \omega \in \partial E^{2,m-1}.$$

Proof. It is enough to prove the lemma for

$$\omega = f(x,\mathbf{u}) \ \delta u_i^\alpha \wedge \delta u_j^\beta \wedge d^m x,$$

but in this case

$$D_\mu \left(f(x,\mathbf{u}) \ \delta u_i^\alpha \wedge \delta u_j^\beta \wedge d^m x \right)$$

$$= \left\{ \left(D_\mu f \right) \ \delta u_i^\alpha \wedge \delta u_j^\beta + f \ \delta u_{i+(\mu)}^\alpha \wedge \delta u_j^\beta + f \ \delta u_i^\alpha \wedge \delta u_{j+(\mu)}^\beta \right\} \wedge d^m x$$

$$= \partial \left(f \ \delta u_i^\alpha \wedge \delta u_j^\beta \wedge d^{m-1} x_\mu \right) \in \partial E^{2,m-1}.$$

Corollary 6.10. *Let*

$$\omega = \sum_{\alpha,\beta} \sum_{i,j} \omega^{i,j}_{\alpha,\beta}\, \delta u^{\alpha}_{i} \wedge \delta u^{\beta}_{j} \wedge d^m x \in E^{2,m},$$

then

$$\omega = \sum_{\alpha,\beta} \sum_{i,j} (-1)^{|i|} \left[D_j \left(\omega^{i,j}_{\alpha,\beta}\, D_i \delta u^{\alpha} \right) \right] \wedge \delta u^{\beta} \wedge d^m x \mod \partial E^{2,m-1}$$

$$= \sum_{\alpha,\beta} \sum_{i} f^{i}_{\alpha,\beta}\, D_i \delta u^{\alpha} \wedge \delta u^{\beta} \wedge d^m x \mod \partial E^{2,m-1}.$$

Proof. To prove this representation one should repeatedly apply the above lemma. Note that the coefficients

$$f^{i}_{\alpha,\beta} = \sum_{j} \sum_{k\le i} (-1)^{|j+k|} \binom{j+k}{k} D_j \omega^{i-k,\,j+k}_{\alpha,\beta}$$

are calculated using the Leibniz' rule.

Consider class of linear operators $F:\mathcal{C}^{\infty}(M)\longrightarrow\mathcal{C}^{\infty}(M)$ which are differential polynomials (i.e., polynomials in D's),

$$F = \sum_{i\in I} f^{i} D_{i},$$

where only a finite number of coefficients $f^{i}\in\mathcal{C}^{\infty}(M)$ differ from zero. For such differential operator F we define the *adjoint* operator $F^{*}:\mathcal{C}^{\infty}(M)\longrightarrow\mathcal{C}^{\infty}(M)$ by the formula

$$F^{*} = \sum_{i\in I} (-1)^{|i|} D_{i}\circ f^{i},$$

where $\circ$ denotes the composition of operators.

Lemma 6.11. *Let F be a differential polynomial. Then*

(a) $F^{**} = F$;

(b) $F^* = \sum_{i\in I} f^{*i} D_i$, where $f^{*i} = \sum_{k\in I} (-1)^{|i+k|} \binom{i+k}{k} D_k f^{i+k}$.

Proof. The property (a) follows from the definition, (b) can be easily checked by direct calculations using the Leibniz' rule.

By Corollary 6.10, every element $[\omega]\in E_1^{2,m}(M)$ has a representative

$$\omega = \sum_{\alpha,\beta} F_{\alpha\beta}\left(\delta u^\alpha\right)\wedge\delta u^\beta\wedge d^m x, \quad \text{where } F_{\alpha\beta} = \sum_{i\in I} f^i_{\alpha\beta} D_i .$$

We shall call a form of this type *standard*. For a standard form ω let us define the *adjoint* form

$$\omega^* = \sum_{\alpha,\beta} F^*_{\alpha\beta}\left(\delta u^\alpha\right)\wedge\delta u^\beta\wedge d^m x \text{ where } F^*_{\alpha\beta} = \sum_{i\in I} (-1)^{|i|} D_i \circ f^i_{\beta\alpha} .$$

Lemma 6.11 have the

Corollary 6.12. *Let ω be a standard form. Then the adjoint form ω^* is also standard and*

(a) $\omega^{**} = \omega$;

(b) $F^*_{\alpha\beta} = \sum_{i\in I} f^{*i}_{\alpha\beta} D_i$, where $f^{*i}_{\alpha\beta} = \sum_{k\in I} (-1)^{|i+k|} \binom{i+k}{k} D_k f^{i+k}_{\beta\alpha}$.

Lemma 6.13. *Let ω be a standard form. Then*

$$\omega + \omega^* \in \partial E^{2,m-1}.$$

Proof. We have

$$\omega = \sum_{\alpha,\beta} F_{\alpha\beta}\left(\delta u^\alpha\right)\wedge\delta u^\beta\wedge d^m x = \sum_{\alpha,\beta}\sum_{i\in I} f^i_{\alpha\beta}\left(D_i\delta u^\alpha\right)\wedge\delta u^\beta\wedge d^m x$$

$$= \sum_{\alpha,\beta} \sum_{i\in I} \left(D_i \delta u^\alpha\right) \wedge \left(f^i_{\alpha\beta} \delta u^\beta\right) \wedge d^m x$$

$$= \sum_{\alpha,\beta} \sum_{i\in I} (-1)^{|i|} \delta u^\alpha \wedge D_i \left(f^i_{\alpha\beta} \delta u^\beta\right) \wedge d^m x \mod \partial E^{2,m-1}$$

$$= \sum_{\alpha,\beta} \delta u^\beta \wedge F^*_{\alpha\beta} \left(\delta u^\alpha\right) \wedge d^m x = - \sum_{\alpha,\beta} F^*_{\alpha\beta} \left(\delta u^\alpha\right) \wedge \delta u^\beta \wedge d^m x = -\omega^* .$$

Corollary 6.14. *Let a standard form* ω *be self-adjoint* ($\omega^*=\omega$), *then* $\omega\in\partial E^{2,m-1}$.

Every function $f=(f_\alpha)\in\mathcal{C}^\infty(M)_a$ defines the differential operator $f_*:\mathcal{C}^\infty(M)^a\longrightarrow\mathcal{C}^\infty(M)_a$, acting by the rule

$$f_*(\varphi)_\alpha = \sum_j \sum_\beta \frac{\partial f_\alpha}{\partial u^\beta_j} D_j \varphi^\beta, \quad \varphi \in \mathcal{C}^\infty(M)^a, \ \alpha = 1,\dots,a,$$

and the adjoint operator $f^*:\mathcal{C}^\infty(M)^a\longrightarrow\mathcal{C}^\infty(M)_a$, acting by the rule

$$f^*(\varphi)_\alpha = \sum_j \sum_\beta (-1)^{|j|} D_j\left(\frac{\partial f_\beta}{\partial u^\alpha_j} \varphi^\beta\right), \quad \varphi \in \mathcal{C}^\infty(M)^a, \ \alpha = 1,\dots,a.$$

Lemma 6.15. *Let a function* $f=(f_\alpha)$ *belong to the image of Euler-Lagrange operator, i.e.,*

$$f_\alpha = \frac{\delta L}{\delta u^\alpha}, \ \alpha = 1,\dots,a, \quad \text{for some } L \in \mathcal{C}^\infty(M).$$

Then the differential operator f_* *defined by* f *is self-adjoint, i.e.,* $f_*=f^*$.

Proof. Indeed, in this case

$$f_*(\varphi)_\alpha = \sum_j \sum_\beta \frac{\partial}{\partial u_j^\beta}\left(\sum_i (-1)^{|i|} D_i \frac{\partial L}{\partial u_i^\alpha}\right) D_j \varphi^\beta$$

$$= \sum_{i,j} \sum_\beta (-1)^{|i|} \frac{\partial}{\partial u_j^\beta}\left(D_i \frac{\partial L}{\partial u_i^\alpha}\right) D_j \varphi^\beta$$

$$= \sum_i (-1)^{|i|} E_\varphi\left(D_i \frac{\partial L}{\partial u_i^\alpha}\right) = \sum_i (-1)^{|i|} D_i E_\varphi\left(\frac{\partial L}{\partial u_i^\alpha}\right)$$

$$= \sum_{i,j} \sum_\beta (-1)^{|i|} D_i\left(\frac{\partial^2 L}{\partial u_i^\alpha \partial u_j^\beta} D_j \varphi^\beta\right)$$

(see Remark 6.3). Hence,

$$f^*(\varphi)_\alpha = \sum_{i,j} \sum_\beta (-1)^{|i|+|i|+|j|} D_j\left(\frac{\partial^2 L}{\partial u_i^\beta \partial u_j^\alpha} D_i \varphi^\beta\right) = f_*(\varphi)_\alpha$$

for all $\alpha=1,\dots,a$ and $\varphi\in\mathcal{C}^\infty(M)^a$.

Lemma 6.16. *Let the differential operator f_* defined by a function $f\in\mathcal{C}^\infty(M)_a$ be self-adjoint, i.e., $f_*=f^*$. Then f belongs to the image of Euler-Lagrange operator, i.e., $f_\alpha=\delta L/\delta u^\alpha$ for some $L\in\mathcal{C}^\infty(M)$.*

Proof. Consider the form

$$\theta = \sum_\alpha f_\alpha \, \delta u^\alpha \wedge d^\mu x \in E_1^{1,m}(\mathcal{M}),$$

then $d_1^{1,m}\theta=[d\theta]$, where

$$d\theta = \sum_i \sum_{\alpha,\beta} \frac{\partial f_\alpha}{\partial u_i^\beta} \delta u_i^\beta \wedge \delta u^\alpha \wedge d^m x = \sum_{\alpha,\beta} F_{\alpha\beta}\left(\delta u^\alpha\right) \wedge \delta u^\beta \wedge d^m x.$$

By the hypothesis of the lemma, here

$$\sum_{\alpha} F_{\alpha\beta}\left(\varphi^{\alpha}\right) = f_{*}\left(\varphi\right)_{\beta} = f^{*}\left(\varphi\right)_{\beta} = \sum_{\alpha} F^{*}_{\alpha\beta}\left(\varphi^{\alpha}\right), \quad \beta = 1,\ldots,a,$$

for all $\varphi\in\mathcal{C}^{\infty}(M)^{a}$. Thus, $(d\theta)^{*}=d\theta\in\partial E^{2,m-1}$, by Corollary 6.14. In other words, $d_1^{1,m}\theta=0\in E_1^{2,m}(\mathcal{M})$. Now, by Theorem 6.4, $E_2^{1,m}(\mathcal{M})=0$, i.e., $\operatorname{Im}d_1^{0,m}=\operatorname{Ker}d_1^{1,m}$, so $\theta\in\operatorname{Im}d_1^{0,m}$ and f belongs to the image of Euler-Lagrange operator.

Proposition 6.17. *The image of Euler-Lagrange operator is equal to*

$$\operatorname{Im}\frac{\delta}{\delta u} = \left\{f\in\mathcal{C}^{\infty}(M)_{a} \;:\; f_{*}=f^{*}\right\}.$$

Proof. The proof follows from Lemmas 6.15 and 6.16.

The following two propositions can be useful in applications.

Proposition 6.18. *The condition $f_{*}=f^{*}$ is fulfilled iff a function $f\in\mathcal{C}^{\infty}(M)_{a}$ satisfies to the following system of partial differential equations on M:*

$$\frac{\partial f_{\alpha}}{\partial u_{i}^{\beta}} - \sum_{k\in I}(-1)^{|i+k|}\binom{i+k}{k}D_{k}\frac{\partial f_{\beta}}{\partial u_{i+k}^{\alpha}} = 0, \quad \alpha,\beta = 1,\ldots,a, \quad i\in I.$$

Proof. To get this system one should use the fact that two differential operators are equal iff their coefficients are equal.

Proposition 6.19. *Suppose that a function $f=(f_{\alpha})\in\mathcal{C}^{\infty}(M)_{a}$ satisfies the condition $f_{*}=f^{*}$, then*

$$f_{\alpha} = \frac{\delta L}{\delta u^{\alpha}}, \quad \text{where} \quad L = \sum_{\beta} u^{\beta}\int_{0}^{1} f_{\beta}(x,s\mathbf{u})ds.$$

Proof. Indeed, in this case

$$\frac{\delta L}{\delta u^\alpha} = \sum_i \sum_\beta (-1)^{|i|} D_i \frac{\partial}{\partial u_i^a}\left(u^\beta \int_0^1 f_\beta(x, s\mathbf{u})ds\right)$$

$$= \int_0^1 \left\{f_\alpha(x, s\mathbf{u}) + \sum_i \sum_\beta (-1)^{|i|} D_i\left(u^\beta \frac{\partial f_\beta(x, s\mathbf{u})}{\partial u_i^\alpha}\right)\right\}ds$$

$$= \int_0^1 \left\{f_\alpha + f^*(u)_\alpha\right\}(x, s\mathbf{u})ds = \int_0^1 \left\{f_\alpha + f_*(u)_\alpha\right\}(x, s\mathbf{u})ds$$

$$= \int_0^1 \left\{f_\alpha(x, s\mathbf{u}) + \sum_i \sum_\beta u_i^\beta \frac{\partial f_\alpha(x, s\mathbf{u})}{\partial u_i^\beta}\right\}ds$$

$$= \int_0^1 \left\{f_\alpha(x, s\mathbf{u}) + s\frac{\partial f_\alpha(x, s\mathbf{u})}{\partial s}\right\}ds = \int_0^1 \frac{\partial}{\partial s}\left(sf_\alpha(x, s\mathbf{u})\right)ds$$

$$= f_\alpha(x, \mathbf{u}).$$

Example 6.20. Consider a function

$$f = u_{tt} - g(t, u, u_t) \in \mathscr{C}^\infty(M),$$

where $m=a=1$, $M=\mathbb{R}\times\mathbb{R}_{\mathbb{N}}$ (see Example 1.5), $u=u_0$, $u_t=u_1$, $u_{tt}=u_2$ and so on. Here

$$f_* = D_t^2 - g_{u_t} D_t - g_u, \quad f^* = D_t^2 + D_t \circ g_{u_t} - g_u = D_t^2 + g_{u_t} D_t + (D_t g_{u_t}) - g_u.$$

Hence

$$f_* - f^* = -2g_{u_t} D_t - (D_t g_{u_t}),$$

so $f_*=f^*$ iff $g_{u_t}=0$ or $g=g(t,u)$. In the last case

$$f = \frac{\delta L}{\delta u} \quad \text{for} \quad L = -\left(\frac{u_t^2}{2} + G(t,u)\right), \quad \text{where} \quad G_u = g.$$

Note that Proposition 6.19 gives

$$L' = u\int_0^1 \left(su_{tt} - g(t, su) \right) ds = \frac{uu_{tt}}{2} - G(t,u) + G(t,0),$$

and

$$L' - L = D_t\left(\frac{uu_t}{2} + h(t) \right), \quad \text{where} \quad h_t = G(t,0).$$

Every function $L \in \mathcal{C}^\infty(M)$ defines the differential operator $L_*: \mathcal{C}^\infty(M)^a \to \mathcal{C}^\infty(M)$, acting by the rule

$$L_*(\varphi) = \sum_i \sum_\alpha \frac{\partial L}{\partial u_i^\alpha} D_i \varphi^\alpha, \quad \varphi \in \mathcal{C}^\infty(M)^a.$$

Its characteristic (see Appendix) $\chi(L) \in \mathcal{C}^\infty(M)_a$ has components

$$\chi(L)_\alpha = \sum_i (-1)^{|i|} D_i \frac{\partial L}{\partial u_i^\alpha} = \frac{\delta L}{\delta u^\alpha}, \quad \alpha = 1, \ldots, a.$$

Proposition 6.21 (Noether's identity). *For any functions $L \in \mathcal{C}^\infty(M)$ and $\varphi \in \mathcal{C}^\infty(M)^a$ the formula*

$$L_*(\varphi) - \sum_{\alpha=1}^{a} \frac{\delta L}{\delta u^\alpha} \varphi^\alpha = \sum_{\mu=1}^{m} D_\mu J^\mu$$

is valid, where

$$J^\mu(L,\varphi) = \sum_\alpha \sum_{i,k} (-1)^{|i|} M(i,k,\mu) \left(D_i \frac{\partial L}{\partial u^\alpha_{i+k+(\mu)}} \right) \left(D_k \varphi^\alpha \right),$$

$\mu = 1, \ldots, m$.

Proof. It is enough to use Proposition A.4 from Appendix. The factors $M(i,k,\mu)$ are defined there.

We shall discuss applications of the Lagrangian formalism to PDE in Section 15 below.

7. Evolution Equations

Evolution systems of PDE (evolution equations, for brevity) describe *dynamics,* i.e., processes developing in space and time. Hence, in this case independent variables split in several space variables and one time variable. We already studied evolution in one-dimensional space (see Example 1.4). Here we present multi-dimensional generalization of this example. Note that pseudo Cauchy-Kowalevski systems (Tsujishita [1]), normal systems (Martinez [1], Olver [1]) and regular systems (Vinogradov [4]) are reduced to evolution systems by suitable change of variables.

We shall keep the multi-index notation of the previous section. Note that in many situations it is convenient to treat the time variable as a parameter.

Let $(t,x)\in\mathbb{R}^{1+m}$ and $u=(u^1,\dots,u^a)\in\mathbb{R}^a$ be independent and dependent variables, respectively, where $t\in\mathbb{R}$ is the time variable and $x=(x^1,\dots,x^m)\in\mathbb{R}^m$ is the space variables, we assume $m\geq 1$. Suppose that an evolution system of PDE

$$u^\alpha_t = f^\alpha\left(t,x,u,\frac{\partial u}{\partial x},\dots,\frac{\partial^p u}{\partial x^p}\right), \quad \alpha = 1,\dots,a,$$

is given (we shall write $u_t=f$, and call it an evolution equation, for brevity); here $\partial^r u\big/\partial x^r$ denotes the set of all space derivatives of an order r. To construct the corresponding differential manifold let us introduce *differential variables* $\mathbf{u}=(u_i^\alpha)\in\mathbb{R}_I^a$, where $\alpha=1,\dots,m$, $i\in I=\mathbb{N}^m$. Then our manifold M will be the manifold with coordinates $t\in\mathbb{R}$, $x\in\mathbb{R}^m$, $\mathbf{u}\in\mathbb{R}_I^a$, i.e., $M=\mathbb{R}\times\mathbb{R}^m\times\mathbb{R}_I^a$. The Cartan distribution D_t, D_x on M is defined by the basis fields

$$D_\mu = \frac{\partial}{\partial x^\mu} + \sum_{\alpha=1}^{a} \sum_{i\in I} u^\alpha_{i+(\mu)} \frac{\partial}{\partial u_i^\alpha}, \quad \mu = 1,\dots,m,$$

$$D_t = \frac{\partial}{\partial t} + \sum_{\alpha=1}^{a} \sum_{i\in I} \left(D_i f^\alpha(t,x,\mathbf{u})\right) \frac{\partial}{\partial u_i^\alpha},$$

where

$$D_i = D_1^{i_1} \dots D_m^{i_m}, \quad i \in I,$$

the functions $f^\alpha\in\mathcal{C}^\infty(M)$ are defined by the given system $u_t=f$. Clear that the differentiations D_μ are pairwise commuting (see Section 6), and it easy to verify that the commutators

$$\left[D_t, D_\mu\right] = 0, \quad \mu = 1,\dots,m.$$

The basis Cartan *1*-forms are

$$\delta u_i^\alpha = du_i^\alpha - \sum_{\mu=1}^{m} u^\alpha_{i+(\mu)}\, dx^\mu - \left(D_i f^\alpha\right) dt, \quad \alpha = 1,\dots,a, \ i \in I,$$

and in this case

$$\partial\delta u_i^\alpha = - \sum_{\mu=1}^{m} \delta u^\alpha_{i+(\mu)} \wedge dx^\mu - \left(D_i f^\alpha\right)_*(\delta u)\wedge dt,$$

where

$$g_*(h) = \sum_{\beta=1}^{a} \sum_{k\in I} \frac{\partial g}{\partial u_k^\beta} D_k h^\beta, \quad \text{for } g \in \mathcal{C}^\infty(M),\ h \in \mathcal{C}^\infty(M)^a,$$

(see the analogous formulas in Section 6, t is a parameter here).

Thus, we defined the differential manifold $\mathcal{M}=(M,D_t,D_x)$. We begin our study of this manifold with the description of its integral submanifolds.

Note first that the differentiations D_μ are characterized by the following property:

$$\Big[D_\mu g(t,x,\mathbf{u})\Big]\Big|_{u=\varphi(t,x)} = \frac{\partial}{\partial x^\mu}\Big[g(t,x,\mathbf{u})\Big|_{u=\varphi(t,x)}\Big], \quad \mu = 1,\dots,m,$$

for all functions $g\in\mathcal{C}^\infty(M)$ and functions $\varphi\in\mathcal{C}^\infty(\mathbb{R}^{1+m})^a$, where

$$u=\varphi(t,x) \text{ stands for } u_i^\alpha = \left(\frac{\partial}{\partial x}\right)^i \varphi^\alpha(t,x),\ \alpha = 1,\dots,a,\ i \in I$$

(cf. Proposition 1.9, t is a parameter here). Further, the differentiation D_t is characterized by the following

Proposition 7.1. *Let a function $\varphi(t,x)\in\mathcal{C}^\infty(\mathbb{R}^{1+m})^a$. Then the equality*

$$\Big[D_t g(t,x,\mathbf{u})\Big]\Big|_{u=\varphi(t,x)} = \frac{\partial}{\partial t}\Big[g(t,x,\mathbf{u})\Big|_{u=\varphi(t,x)}\Big]$$

is valid for all functions $g\in\mathcal{C}^\infty(M)$ iff φ is a solution of the system $u_t=f$.

Proof. One can easily check that the equality is true for all $g\in\mathcal{C}^\infty(M)$ iff

$$\Big(D_i f^\alpha\Big)\Big|_{u=\varphi} = \frac{\partial}{\partial t}\left(\frac{\partial}{\partial x}\right)^i \varphi^\alpha = \left(\frac{\partial}{\partial x}\right)^i \frac{\partial}{\partial t}\varphi^\alpha,$$

i.e., iff

$$\left(\frac{\partial}{\partial x}\right)^i \left(f^\alpha\Big|_{u=\varphi} - \frac{\partial\varphi^\alpha}{\partial t}\right) = 0 \quad \text{for all } i \in I,\ \alpha = 1,\dots,a.$$

Clear, the last condition is fulfilled iff φ is a solution of the evolution system $u_t=f$.

Now, the integral submanifolds are described by

Proposition 7.2. *A (1+m)-dimensional smooth submanifold $S\subset M$ is integral iff it has a local parameterization*

$$S = \left\{t,x,\mathbf{u}\ :\ u_i^\alpha = \left(\frac{\partial}{\partial x}\right)^i \varphi^\alpha(t,x),\ \alpha = 1,\dots,a,\ i \in I\right\},$$

where a smooth function $\varphi(t,x)$ is a solution of the evolution system $u_t=f$ (cf. Example 1.4).

Proof. By Proposition 1.8, a *(1+m)*-dimensional smooth submanifold $S\subset M$ is integral iff it has a local parameterization

$$S = \left\{t,x,\mathbf{u}\ :\ u_i^\alpha = s_i^\alpha(t,x)\right\},$$

where functions s_i^α, $\alpha=1,\dots,a$, $i\in I$, satisfy the contact equations

$$\frac{\partial s_i^\alpha}{\partial x^\mu} = s_{i+(\mu)}^\alpha, \qquad \mu = 1,\dots,m;$$

$$\frac{\partial s_i^\alpha}{\partial t} = \left(D_i f^\alpha\right)\Big|_{u=s}.$$

The first subsystem easily yields

$$s_i^\alpha(t,x) = \left(\frac{\partial}{\partial x}\right)^i \varphi^\alpha(t,x), \quad \varphi \in \mathscr{C}^\infty(\mathbb{R}^{1+m})^a,$$

and then the second is satisfied iff φ is a solution of the evolution system (see Proposition 7.1).

Infinitesimal symmetries are the next step of our study.

Proposition 7.3. *A vector field*

$$\eta = \sum_{\alpha=1}^{a} \sum_{i\in I} \eta_i^\alpha(t,x,\mathbf{u}) \frac{\partial}{\partial u_i^\alpha}$$

will be Lie-Backlund iff

$$\eta_i^\alpha = D_i\phi^\alpha(t,x,\mathbf{u}), \quad \alpha = 1,\dots,a, \ i \in I,$$

where a function $\phi\in\mathcal{C}^\infty(M)^a$ *satisfies the* defining equations

$$D_t\phi^\alpha - f_*^\alpha(\phi) = 0, \quad \alpha = 1,\dots,a.$$

Explain that

$$f_*^\alpha(\phi) = \sum_{\beta=1}^{a} \sum_{i\in I} \frac{\partial f^\alpha}{\partial u_i^\beta} D_i\phi^\beta$$

(see the definition of the operator g_* above and Section 6).

Proof. By Proposition 3.1, the field η will be Lie-Backlund iff its components η_i^α, $\alpha=1,\dots,a$, $i\in I$, satisfy the system

$$D_\mu\eta_i^\alpha - \eta_{i+(\mu)}^\alpha = 0, \quad \mu = 1,\dots,m;$$

$$D_t\eta_i^\alpha - \sum_{\beta=1}^{a} \sum_{k\in I} \frac{\partial}{\partial u_k^\beta}\left(D_i f^\alpha\right) \eta_k^\beta.$$

Again, the first subsystem yields

$$\eta_i^\alpha = D_i\phi^\alpha, \quad \phi \in \mathcal{C}^\infty(M)^a,$$

and then the second reads as follows

$$D_tD_i\phi^\alpha - \left(D_if^\alpha\right)_*(\phi)$$

$$= D_tD_i\phi^\alpha - \mathrm{E}_\phi\left(D_if^\alpha\right) = D_i\left(D_t - \mathrm{E}_\phi(f^\alpha)\right)$$

$$= D_i\left(D_t - f_*^\alpha(\phi)\right) = 0,$$

where we used Remark 6.3 (here t is a parameter). Now, the last system is clearly equivalent to the defining system we claimed.

In other words, infinitesimal symmetries of evolution systems are in fact *evolution differentiations*

$$\mathrm{E}_{\varphi} = \sum_{\alpha=1}^{a} \sum_{i\in I} \left(D_i\varphi^{\alpha}(t,x,\mathbf{u})\right) \frac{\partial}{\partial u_i^{\alpha}},$$

where functions $\varphi\in\mathcal{C}^{\infty}(M)^a$ satisfy the defining equation $D_t\varphi-f_*(\varphi)=0$ (cf. Example 3.8). Again the Lie bracket of two evolution differentiations

$$\left[\mathrm{E}_{\varphi},\mathrm{E}_{\psi}\right] = \mathrm{E}_{\chi}$$

is the evolution differentiation, where

$$\chi^{\alpha} = \{\varphi,\psi\}^{\alpha} = \mathrm{E}_{\varphi}(\psi^{\alpha}) - \mathrm{E}_{\psi}(\varphi^{\alpha}), \quad \alpha = 1,\dots,a,$$

is the *Jacoby bracket* of functions φ,ψ (note that χ is a solution of the defining equation if φ and ψ are). Thus, *the Lie algebra* Sym$\mathcal{M}$ *of infinitesimal symmetries of the differential manifold* $\mathcal{M}$ *(i.e., of the evolution system* $u_t=f$*) is isomorphic to the Lie algebra of solutions of the defining equation* $D_t\varphi-f_*(\varphi)=0$. We shall identify these two algebras.

Let us turn now to the C-spectral sequence $\left\{E_r^{p,q}(\mathcal{M}),d_r^{p,q}\right\}$, associated with $\mathcal{M}$. First, note that by Proposition 5.8, terms $E_r^{p,q}(\mathcal{M})=0$ for all $p+q>0$ and $r>1$. For further study we need special notations. Let $E^{p,q}$ be the linear spaces introduced in Section 5, $p=0,1,\dots$, $q=0,\dots,m+1$. Denote by $'E^{p,q}$ the subspace of all forms from $E^{p,q}$, which don't depend on the differential dt explicitly (only via the terms δu^{α}_{i}). In more detail,

$$'E^{p,q} = \left\{\Omega = \sum_{(\alpha)_p} \sum_{(i)_p} \sum_{(\mu)_q} \Omega^{(\alpha)_p}_{(i)_p;(\mu)_q}(t,x,\mathbf{u})\left(\delta u^{i}\right)^{p}\wedge\left(dx^{\mu}\right)^{q}\right\},$$

where $p=0,1,\ldots$, $q=0,\ldots,m$, and $'E^{p,q}=0$ for $q<0$ and for $q>m$. We have the evident

Lemma 7.4. *There is the decomposition*

$$E^{p,q} = {'E^{p,q}} + {'E^{p,q-1}}\wedge dt, \quad p = 0,1,\ldots, \; q = 0,\ldots,m+1,$$

of $E^{p,q}$ into the direct sum of linear subspaces.

In other words every form $\omega\in E^{p,q}$ has the unique representation $\omega=\theta+\chi$, where $\theta\in {'E^{p,q}}$ and $\chi\in {'E^{p,q-1}}\wedge dt$. Moreover, the mappings $\partial: E^{p,q}\longrightarrow E^{p,q+1}$ split on $'E^{p,q}$ in two parts, $\partial = {'\partial} + {''\partial}$, where

$$'\partial : {'E^{p,q}} \longrightarrow {'E^{p,q+1}}; \quad ''\partial : {'E^{p,q}} \longrightarrow {'E^{p,q}}\wedge dt.$$

For example,

$$'\partial\varphi = \sum_{\mu=1}^{m} D_{\mu}\varphi \; dx^{\mu}, \quad ''\partial\varphi = D_{t}\varphi \; dt$$

for $\varphi\in {'E^{0,0}}=E^{0,0}=\mathcal{C}^{\infty}(M)$, and

$$'\partial\delta u^{\alpha}_{i} = -\sum_{\mu=1}^{m} \delta u^{\alpha}_{i+(\mu)}\wedge dx^{\mu}, \quad ''\partial\delta u^{\alpha}_{i} = -\left(D_{i}f^{\alpha}\right)_{*}\left(\delta u\right)\wedge dt$$

for $\delta u^{\alpha}_{i}\in {'E^{1,0}}$, $\alpha=1,\ldots,m$, $i\in I$ (we use the convention about the action of differential operators on differential forms, see Section 6). Note that,

$$\partial\left(\omega\wedge dt\right) = \left({'\partial\omega}\right)\wedge dt \text{ and } ''\partial\omega = \tau(\omega)\wedge dt \quad \text{for all} \quad \omega \in {'E^{p,q}},$$

where $\tau : {}'E^{p,q} \to {}'E^{p,q}$ is a well-defined linear mapping.

Lemma 7.5. *Let $'\partial$ and τ be the linear mappings defined above. Then*

(a) $'\partial \circ {}'\partial = 0$;

(b) $'\partial \circ \tau + \tau \circ {}'\partial = 0$.

Proof. Indeed,

$$0 = \partial \circ \partial = \partial \circ \left({}'\partial + {}''\partial\right) = \left({}'\partial + {}''\partial\right) \circ {}'\partial + \partial \circ {}''\partial$$

$$= {}'\partial \circ {}'\partial + \left({}''\partial \circ {}'\partial + \partial \circ {}''\partial\right),$$

hence, $'\partial \circ {}'\partial = 0$ (i.e., (a) is valid) and $''\partial \circ {}'\partial + \partial \circ {}''\partial = 0$. Now, let $\omega \in {}'E^{p,q}$, then the second equality reads

$$\tau({}'\partial\omega) \wedge dt + {}'\partial(\tau(\omega)) \wedge dt = 0,$$

and (b) is also true.

Further, let

$$E_X^{p,q} = \left\{\Omega = \sum_{(\alpha)_p} \sum_{(i)_p} \sum_{(\mu)_q} \Omega^{(\alpha)_p}_{(i)_p;(\mu)_q}(t,x,\mathbf{u}) \left(\delta_X u^i\right)^p \wedge \left(dx^\mu\right)^q\right\},$$

$p=0,1,\ldots,$ $q=0,\ldots,m,$ where,

$$\delta_X u_i^\alpha = du_i^\alpha - \sum_{\mu=1}^{m} u^\alpha_{i+(\mu)} dx^\mu, \quad \alpha = 1,\ldots,a, \; i \in I,$$

and let $\partial_X : E_X^{p,q} \to E_X^{p,q+1}$ be the respective mappings (cf. the definition of the spaces $E^{p,q}$ in Section 6, t is a parameter here). At last, denote by $\rho : {}'E^{p,q} \to E_X^{p,q}$ the linear mappings, defined by the rule: all δu_i^α are replaced by $\delta_X u_i^\alpha$, i.e.,

$$\Omega = \sum \Omega^{\cdots}_{\cdots}\left(\delta u^{\alpha}_{i}\right)^{p}\wedge\left(dx^{\mu}\right)^{q} \mapsto \rho\Omega = \sum \Omega^{\cdots}_{\cdots}\left(\delta_x u^{\alpha}_{i}\right)^{p}\wedge\left(dx^{\mu}\right)^{q}.$$

Clear, the mappings ρ are isomorphisms of the linear spaces and

$$\rho\left({}'\partial\Omega\right) = \partial_x\left(\rho\Omega\right) \quad \text{for all} \quad \Omega \in E^{p,q},$$

$p=0,1,\ldots,\ q=0,\ldots,m.$

Lemma 7.6. *Let $\omega\in{}'E^{p,q}$, where $p+q>0$ and $q<m$. Then ${}'\partial\omega=0$ iff $\omega={}'\partial\theta,\ \theta\in{}'E^{p,q-1}$.*

Proof. We only need to prove that ${}'\partial\omega=0$ implies $\omega={}'\partial\theta$. Suppose that ${}'\partial\omega=0$, so $\partial_x\rho\omega=o$, $\rho\omega\in E^{p,q}_x$, i.e., $\rho\omega\in\operatorname{Ker}\partial_x : E^{p,q}_x \longrightarrow E^{p,q+1}_x$. By Theorem 6.4, $\rho\omega\in\operatorname{Im}\partial_x : E^{p,q-1}_x \longrightarrow E^{p,q}_x$ (note that theorem 6.4 is valid here because the additional variable t can be treated as a parameter with all expressions depending smoothly on t). Thus, $\rho\omega=\partial_x\phi$, $\phi\in E^{p,q-1}_x$, and the form $\theta=\rho^{-1}\phi\in{}'E^{p,q-1}$ is the form we need.

In the same way we have

Lemma 7.7. *Let $\omega\in{}'E^{p,q}\wedge dt$, where $p+q>0$ and $q<m$. Then $\partial\omega=0$ iff $\omega=\partial\theta,\ \theta\in{}'E^{p,q-1}\wedge dt$.*

Proof. It is enough to recall that

$$\partial\left(\phi\wedge dt\right) = \left({}'\partial\phi\right)\wedge dt \quad \text{for all} \quad \phi \in {}'E^{p,q},$$

and to use the previous lemma.

After all these preparations we are ready to prove

Theorem 7.8 (cf. Tsujishita [1]). *Let $\mathcal{M}$ be the differential manifold, defined by an evolution system $u_t=f$. Then the only non-trivial terms of the C-spectral sequence $\left\{E^{p,q}_r(\mathcal{M}), d^{p,q}_r\right\}$ are:*

$$E_0^{p,q}(\mathcal{M}) = E^{p,q}, \quad p = 0,1,\ldots, \quad q = 0,\ldots,m+1$$

$$E_r^{0,0}(\mathcal{M}) = \mathbb{R}, \quad r = 1,2,\ldots;$$

$$E_1^{p,m}(\mathcal{M}) = \frac{E^{p,m}\cap\partial^{-1}\{0\}}{\partial E^{p,m-1}}, \quad p = 0,1,\ldots;$$

$$E_1^{p,1+m}(\mathcal{M}) = \frac{E^{p,1+m}}{\partial E^{p,m}}, \quad p = 0,1,\ldots.$$

Proof. In view of general results of Section 5, we need only to check that (a) $E_1^{0,0}(\mathcal{M})=\mathbb{R}$ and (b) $E_1^{p,q}(\mathcal{M})=0$ $p+q>0$, $q<m$. Consider first (a). Let $\varphi\in\mathcal{C}^\infty(\mathcal{M})$, $D_\mu\varphi=0$, $\mu=1,\ldots,m$ and $D_t\varphi=0$. One can easily find that the first m equations imply $\varphi(t,x,\mathbf{u})=\varphi(t)$, then the last equation reads $D_t\varphi=\frac{d\varphi}{dt}$ and gives $\varphi=const$. Hence $\varphi\in E_1^{00}(\mathcal{M})$ iff $\varphi=const$.

Suppose now that a form $\omega\in E^{p,q}\cap d^{-1}\{0\}$, i.e., $\omega\in E^{p,q}$, $\partial\omega = 0$. By lemma 7.4, $\omega=\theta+\chi$, where $\theta\in{}'E^{p,q}$ and $\chi\in{}'E^{p,q-1}\wedge dt$, so

$$\partial\omega = \partial\left(\theta + \chi\right) = {}'\partial\theta + {}''\partial\theta + \partial\chi = {}'\partial\theta + \left({}''\partial\theta + \partial\chi\right) = 0.$$

In particular, ${}'\partial\theta=0$, and by Lemma 7.6, $\theta={}'\partial\varphi$, $\varphi\in{}'E^{p,q-1}$. Now,

$$\partial\omega = \partial\left({}'\partial\varphi + \chi\right) = \partial\left(-{}''\partial\varphi + \chi\right) = 0,$$

where we used the following evident equality

$$\partial\partial\varphi = \partial\left({}'\partial\varphi + {}''\partial\varphi\right) = 0.$$

Hence by Lemma 7.7, $-{}''\partial\varphi+\chi=\partial\psi$, $\psi\in{}'E^{p,q-1}\wedge dt$, and

$$\omega = {}'\partial\varphi + \chi = {}'\partial\varphi + {}''\partial\varphi + \partial\psi = \partial\left(\varphi + \psi\right).$$

Thus, $\omega\in\partial E^{p,q-1}$. This implies

$$E_1^{p,q}(\mathcal{M}) = \frac{E^{p,q}\cap\partial^{-1}\{0\}}{\partial E^{p,q-1}} = 0, \quad p+q>0, q<m,$$

and the proof is finished.

In particular, the above theorem states that evolution systems have no low-dimensional conservation laws. Let us study the space

$$E_1^{0,m}(\mathcal{M}) = \frac{E^{0,m}\cap\partial^{-1}\{0\}}{\partial E^{0,m-1}}$$

of conservation laws of the maximal dimension. In this case,

$$E^{0,m} \ni \omega = (-1)^m T d^m x + \sum_{\mu=1}^{m} X^\mu d^{m-1}x_\mu \wedge dt, \quad \partial\omega = \left(D_t T + \sum_{\mu=1}^{m} D_\mu X^\mu\right) d^m x\wedge dt.$$

Here $J=(T,X)\in\mathcal{C}^\infty(M)^{m+1}$ is called the *current*, $T\in\mathcal{C}^\infty(M)$ is called the *conserved density* and $X=(x^\mu)\in\mathcal{C}^\infty(M)^m$ is called the *associated flux*. Remind that

$$d^{m-1}x_\mu = (-1)^{\mu-1}dx^1\wedge\ldots\wedge\widehat{dx^\mu}\wedge\ldots\wedge dx^m, \quad \mu = 1,\ldots,m$$

(see Section 6). Now, $\partial\omega=0$ iff the *divergency equation*

$$D_t T + \sum_{\mu=1}^{m} D_\mu X^\mu = 0$$

holds. A current J is called *trivial* if the form $\omega=\partial\theta$, where

$$\theta = \sum_{\mu} \varphi^\mu d^{m-1}x_\mu + \sum_{\mu<\nu} \psi^{\mu\nu} d^{m-2}x_{\mu\nu}\wedge dt \in E^{0,m-1},$$

$\psi^{\mu\nu}=-\psi^{\nu\mu}$, differentials

$$d^{m-2}x_{\mu\nu} = (-1)^{\mu+\nu}dx^1\wedge\ldots\wedge\widehat{dx^\mu}\ldots\wedge\widehat{dx^\nu}\ldots\wedge dx^m,$$

so

$$\partial\theta = \sum_\mu D_\mu\varphi^\mu d^m x + \sum_\mu \left((-1)^{m-1}D_t\varphi^\mu + \sum_\nu D_\nu\psi^{\nu\mu}\right)d^{m-1}x_\mu\wedge dt.$$

Hence, for trivial currents

$$T = (-1)^m\sum_\mu D_\mu\varphi^\mu \quad\text{and}\quad X^\mu = (-1)^{m-1}D_t\varphi^\mu + \sum_\nu D_\nu\psi^{\nu\mu},\ \ \mu = 1,\ldots,m.$$

To get rid of trivial currents one should have a direct description of the space $E_1^{0,m}(\mathcal{M})$ without any quotients. The C-spectral sequence presents such description. The equalities $E_2^{p,m}(\mathcal{M})=0$, $p=0,1,2,$ lead to the exact sequence of the linear spaces

$$0 \longrightarrow E_1^{0,m}(\mathcal{M}) \xrightarrow{d_1^{0,m}} E_1^{1,m}(\mathcal{M}) \xrightarrow{d_1^{1,m}} E_1^{2,m}(\mathcal{M}),$$

hence we have the isomorphism

$$d_1^{0,m} : E_1^{0,m}(\mathcal{M}) \simeq \operatorname{Ker} d_1^{1,m}.$$

Now, this isomorphism has a simple clear form, which we present below.

We begin with the spaces $E_1^{p,m}(\mathcal{M})$, $p=0,1,\ldots$. Denote by σ the projection of the space $E^{p,m}$ on the subspace $'E^{p,m}$ (see Lemma 7.4), so $\sigma(\omega)=\varphi$ for every $\omega=\varphi+\psi\in E^{p,m}$, where $\varphi\in{}'E^{p,m}$, $\psi\in{}'E^{p,m-1}\wedge dt$.

Lemma 7.9. *The projection σ defines the isomorphisms*

$$E_1^{p,m}(\mathcal{M}) \simeq \frac{\left\{\varphi \in {}'E^{p,m} : \tau(\varphi) \in {}'\partial' E^{p,m-1}\right\}}{\left\{\varphi = {}'\partial\Phi\ ;\ \Phi \in {}'E^{p,m-1}\right\}},\quad p = 0,1,\ldots.$$

Proof. Let $\omega=\varphi+\psi\wedge dt\in E^{p,m}$ then $\partial\omega=\{\tau(\varphi)+'\partial\psi\}\wedge dt$, hence, $\partial\omega=0$ implies $\tau(\varphi)=-'\partial\psi\in'\partial'E^{p,m-1}$. On the other hand, suppose $\varphi\in'E^{p,m}$, $\tau(\varphi)\in'\partial'E^{p,m-1}$, i.e., $\tau(\varphi)='\partial\chi$, $\chi\in'E^{p,m-1}$, then $\omega=\varphi-\chi\wedge dt\in E^{p,m}$ and $\partial\omega=0$. Further, if $\omega=\varphi+\psi\wedge dt=\partial\Omega$, where $\Omega=\Phi+\Psi\wedge dt\in E^{p,m-1}$, then $\varphi='\partial\Phi$. At last, if $\varphi='\partial\Phi$, $\Phi\in'E^{p,m-1}$, then $\omega=\varphi+\psi\wedge dt=\partial\Phi\in\partial E^{p,m-1}$, $\psi=\tau(\Phi)$.

Let $p=0$. Every form $\varphi\in'E^{0,m}$ is written as $\varphi=(-1)^m T d^m x$, where $T(t,x,\mathbf{u})\in\mathcal{C}^\infty(M)$, and $\tau(\varphi)=D_t T d^m x$. By Proposition 6.8, $\varphi\in'\partial'E^{0,m-1}$ iff

$$\frac{\delta T}{\delta u^\alpha}=\sum_{i\in I}(-1)^{|i|}D_i\left(\frac{\partial T}{\partial u_i^\alpha}\right)=0,\quad \alpha=1,\dots,a,$$

(note, once again, that t is a parameter here) and the same is true for $\tau(\varphi)$. Thus, we have the representation

$$E_1^{0,m}(M)=\frac{\left\{T\in\mathcal{C}^\infty(M)\ :\ \delta D_t T\big/\delta u=0\right\}}{\left\{T\in\mathcal{C}^\infty(M)\ :\ \delta T\big/\delta u=0\right\}}.$$

In other words, every conservation law is completely defined by its conserved density, but there are trivial densities.

Now, let $p=1$. In this case a form $\varphi\in'E^{1,m}$ is written as

$$\begin{aligned}\varphi&=(-1)^m\sum_{\alpha=1}^{a}\sum_{i\in I}\varphi_\alpha^i\,\delta u_i^\alpha\wedge d^m x\\ &=(-1)^m\sum_{\alpha=1}^{a}\varphi_\alpha\,\delta u^\alpha\wedge d^m x\mod '\partial'E^{1,m-1},\end{aligned}$$

where coefficients $\varphi_\alpha\in\mathcal{C}^\infty(M)$ are uniquely defined (see Proposition 6.7). Assume that the last representation is valid, then

$$\tau(\varphi) = -\sum_{\alpha=1}^{m} \left\{ D_t\varphi_\alpha \delta u^\alpha + \varphi_\alpha f_*^\alpha(\delta u) \right\} \wedge d^m x,$$

where

$$f_*^\alpha(\delta u) = \sum_{\beta,k} \frac{\partial f^\alpha}{\partial u_i^\alpha} D_t \delta u^\beta, \quad \alpha = 1,\dots,a,$$

(see Section 6). Further,

$$\tau(\varphi) = -\sum_{\alpha=1}^{m} \left\{ D_t\varphi_\alpha + f^*(\varphi)_\alpha \right\} \delta u^\alpha \wedge d^m x \mod {}'\partial' E^{1,m-1},$$

where the adjoint operator

$$f^*(\varphi)_\alpha = \sum_{\beta,k} (-1)^{|k|} D_k\left(\varphi_\beta \frac{\partial f^\beta}{\partial u_k^\alpha}\right), \quad \alpha = 1,\dots,a,$$

(see Section 6). Again coefficients

$$D_t\varphi_\alpha + f^*(\varphi)_\alpha, \quad \alpha = 1,\dots,a,$$

are defined uniquely, so $\tau(\varphi) \in {}'\partial' E^{1,m-1}$ iff these coefficients vanish. Thus,

$$E_1^{1,m}(M) = \left\{ \varphi \in \mathcal{C}^\infty(M)_a \ : \ D_t\varphi + f^*(\varphi) = 0 \right\}.$$

In the case $p=2$ every form $\varphi \in {}'E^{2,m}$ has a standard representation (see Corollary 6.10)

$$\varphi = (-1)^m \sum_{\alpha,\beta} F_{\alpha\beta}\left(\delta u^\alpha\right) \wedge \delta u^\beta \wedge d^m x \mod {}'\partial' E^{2,m-1},$$

where a differential operator

$$F_{\alpha\beta} = \sum_{i \in I} f_{\alpha\beta}^i D_i \quad \text{coefficients} \quad f_{\alpha\beta}^i \in \mathcal{C}^\infty(M).$$

By Corollary 6.14, a standard form $\varphi\in{}'\partial'E^{2,m-1}$ if it is self-adjoint, i.e., if $\varphi^*=\varphi$, where

$$\varphi^* = (-1)^m \sum_{\alpha,\beta} F^*_{\alpha\beta}\left(\delta u^\alpha\right)\wedge\delta u^\beta\wedge d^m x, \quad F^*_{\alpha\beta} = \sum_{i\in I} (-1)^{|i|} D_i \circ f^i_{\beta\alpha}.$$

Our next object is the mapping $d_1^{0,m}: E_1^{0,m}(\mathcal{M})\longrightarrow E_1^{0,m}(\mathcal{M})$. Let

$$\omega = (-1)^m T d^m x + \sum_{\mu=1}^{m} X^\mu d^{m-1}x_\mu\wedge dt \in E^{0,m}\cap\partial^{-1}\{0\},$$

then

$$\delta\omega = (-1)^m \sum_{\alpha,i} \frac{\partial T}{\partial u_i^\alpha}\, \delta u_i^\alpha\wedge d^m x + \sum_\mu \sum_{\alpha,i} \frac{\partial X^\mu}{\partial u_i^\alpha}\, \delta u_i^\alpha\wedge d^{m-1}x_\mu\wedge dt,$$

hence,

$$\begin{aligned}\sigma(\delta\omega) &= (-1)^m \sum_{\alpha,i} \frac{\partial T}{\partial u_i^\alpha}\, \delta u_i^\alpha\wedge d^m x = (-1)^m\, T_*(\delta u)\wedge d^m x \\ &= (-1)^m \sum_\alpha \frac{\delta T}{\delta u^\alpha}\, \delta u^\alpha \mod {}'d'E^{1,m-1}\end{aligned}$$

(see Lemma 6.5). Thus, the mapping $d_1^{0,m}$ is defined by the rule

$$E_1^{0,m}(\mathcal{M}) \ni [T] \mapsto \frac{\delta T}{\delta u} \in E_1^{1,m}(\mathcal{M}).$$

In particular, if T is a conserved density, i.e., if $\delta D_t T \big/ \delta u=0$, then the variational derivative $\varphi=\delta T\big/\delta u$ is a solution of the equation $D_t\varphi+f^*(\varphi)=0$. This property of conservation laws was found by Lax [1] in his studies of the KdV equation. It can be checked directly by straightforward calculations (see Zharinov [2]).

Now, using Proposition 6.17, one can easily derive that an element φ from $E_1^{1,m}(\mathcal{M})$ belongs to $\mathrm{Im}\, d_1^{0,m}=\mathrm{Ker}\, d_1^{1,m}$ iff $\varphi_*=\varphi^*$, where the differential operator $\varphi_*: \mathcal{C}^\infty(M)^a\longrightarrow\mathcal{C}^\infty(M)_a$,

$$\varphi_*(g)_\alpha = \sum_{\beta,i} \frac{\partial\varphi_\alpha}{\partial u_i^\beta} D_i g^\beta, \quad g \in \mathcal{C}^\infty(M)^a, \ \alpha = 1,\dots,a,$$

and the adjoint operator $\varphi^*:\mathcal{C}^\infty(M)^a \to \mathcal{C}^\infty(M)_a$,

$$\varphi^*(g)_\alpha = \sum_{\beta,i} (-1)^{|i|} D_i\left(g^\beta \frac{\partial\varphi_\beta}{\partial u_i^\alpha}\right), \quad g \in \mathcal{C}^\infty(M)^a, \ \alpha = 1,\dots,a.$$

Thus we proved the following

Theorem 7.10 (Zharinov [2]). *Let $\mathcal{M}$ be the differential manifold defined by an evolution system $u_t=f$. Then there exists the isomorphism*

$$\chi : \mathrm{Cons}^m\mathcal{M} \simeq \left\{\varphi \in \mathcal{C}^\infty(M)_a \ : \ D_t\varphi + f^*(\varphi)=0, \ \varphi_*=\varphi^*\right\}$$

defined by the rule

$$\mathrm{Cons}^m\mathcal{M} \ni [Td^m x + \dots] \mapsto \frac{\delta T}{\delta u} \in \mathcal{C}^\infty(M)_a.$$

The mapping χ is called *characteristic*. Note that the condition $\varphi_*=\varphi^*$ is in fact a system of linear PDE (see Proposition 6.18). Further, if a characteristic $\varphi=\chi(T)$ is known, then the density T can be recovered using Proposition 6.19.

Example 7.11. Consider the heat equation in m-dimensional space, $u_t=\Delta u$, where Δ is the Laplace operator. Here $a=1$,

$$f = \sum_{\mu=1}^m u_{2(\mu)}, \quad f_* = f^* = \sum_{\mu=1}^m D_\mu^2.$$

Let $\varphi(t,x,\mathbf{u})\in\mathcal{C}^\infty(M)$, $\mathrm{ord}\varphi\le r$ (so φ depends only on u_i, $|i|\le r$), then

$$D_t\varphi = f^*(\varphi) = \sum_{|i|=r} \sum_\mu u_{i+2(\mu)}\, \varphi_{u_i} + \dots,$$

hence, the equality

$$D_t\varphi + f^*(\varphi) = \sum_{|i|=r} D_i\left(\sum_{\mu} u_{2(\mu)}\right) \varphi_{u_i} + \ldots = 0$$

implies $\mathrm{ord}\varphi \le r-1$. By the induction, $\varphi=\varphi(t,x)$, where

$$D_t\varphi + f^*(\varphi) = \varphi_t + \Delta\varphi = 0.$$

Clear, in this case $\varphi_* = \varphi^* = 0$. Thus,

$$\mathrm{Cons}^m\{u_t=\Delta u\} \simeq \left\{\varphi(t,x) \;:\; \varphi_t+\Delta\varphi=0\right\}.$$

By Proposition 6.19,

$$T = u\int_0^1 \varphi(t,x)ds = \varphi u$$

(cf. Example 4.7(a)).

CHAPTER II. EXTERNAL GEOMETRY OF PDE

8. Differential Submanifolds

A differential manifold $\mathcal{F}=(F,CTF)$ is called a *differential submanifold* of a differential manifold $\mathcal{M}=(M,CTM)$ if the manifold F is a submanifold of M and $CTF=TF\cap CT_F M$ (i.e., the imbedding $\mathcal{F}\longrightarrow\mathcal{M}$ is a Lie-Backlund immersion, see Section 2, note that

$$CT_F M = \cup_{a\in F} CT_a M,$$

cf. Section 0.4). If the Cartan dimensions of the differential manifolds $\mathcal{F}$ and $\mathcal{M}$ coincide (i.e., $CTF=CT_F M$), then the differential submanifold $\mathcal{F}$ is called a *reduction* of the differential manifold $\mathcal{M}$. In this case every integral submanifold of $\mathcal{F}$ is an integral submanifold of the background differential manifold $\mathcal{M}$, and conversely, every integral submanifold of $\mathcal{M}$, lying in F, is an integral submanifold of $\mathcal{F}$.

Example 8.1. Let $u_t=f(t,u)$ be an ordinary differential equation (see Example 1.1) and let $\mathcal{F}=(F,\partial_t)$ be the corresponding differential manifold with the coordinates $(t,u)\in\mathbb{R}^2$ and the basis Cartan field

$$\partial_t = \frac{\partial}{\partial t} + f(t,u)\,\frac{\partial}{\partial u}.$$

The differential manifold $\mathcal{F}$ can be realized as a differential submanifold of the differential manifold $\mathcal{M}=(M,D_t)$ with the coordinates $(t,\mathbf{u})\in\mathbb{R}^{1+\mathbb{N}}$ and the basis Cartan field

$$D_t = \frac{\partial}{\partial t} + \sum_{i\in\mathbb{N}} u_{i+1}\,\frac{\partial}{\partial u_i}$$

(cf. Example 0.6.1). Indeed, if we put

$$F = \left\{t,\mathbf{u}\ :\ u_0=u,\ u_{i+1}=f_i,\ i\in\mathbb{N}\right\} \in M,$$

where

$$f_i(t,u_0,\ldots,u_i) = D_t^i f(t,u_0),\ i\in\mathbb{N},$$

are smooth functions on M, then we realize F as a subset of M. Let us introduce new coordinates $(t,\mathbf{v})\in\mathbb{R}^{1+\mathbb{N}}$ on M with

$$v_0 = u_0,\ v_{i+1} = u_{i+1} - f_i(t,u_0,\ldots,u_i),\ i\in\mathbb{N}.$$

The inverse transition formulas will be:

$$u_0 = v_0,$$
$$u_1 = v_1 + f_0(t,v_0),$$
$$u_2 = v_2 + f_1(t,v_0,v_1+f_0(t,v_0)),$$

and so on. In these new coordinates F is written as follows

$$F = \left\{t,\mathbf{v}\ :\ v_i=0,\ i=1,2,\ldots\right\},$$

thus, F is a submanifold of M. Further,

$$D_t = \frac{\partial}{\partial t} + f(t,v_0)\,\frac{\partial}{\partial v_0} + \sum_{i\in\mathbb{N}} v_{i+1}\,\frac{\partial}{\partial v_i},$$

because

$$D_t(v_0) = v_1+f(t,v_0) \quad \text{and} \quad D_t(v_i) = v_{i+1}, \quad i = 1,2,\ldots .$$

The imbedding $F\longrightarrow M$ is given here by the formulas

$$F\ni(t,u) \mapsto (t,v_0=u,v_i=0,i=1,2,\ldots)\in M,$$

so,

$$\partial_t = \frac{\partial}{\partial t} + f(t,v_0)\frac{\partial}{\partial v_0} = D_t\Big|_F .$$

Hence, we proved that $\mathcal{F}$ is a differential submanifold of $\mathcal{M}$. In fact, it is a reduction.

Example 8.2(a). Consider the heat equation in m-dimensional space, $u_t=\Delta u$, Δ is the Laplace operator, and let $\mathcal{M}=(M,D_t,D_x)$ be the corresponding differential manifold with coordinates $t\in\mathbb{R}$, $x\in\mathbb{R}^m$, $\mathbf{u}\in\mathbb{R}_I$, $I=\mathbb{N}^m$, and basis fields

$$D_\mu = \frac{\partial}{\partial x^\mu} + \sum_{i\in I} u_{i+(\mu)}\frac{\partial}{\partial u_i}, \quad \mu = 1,\ldots,m,$$

$$D_t = \frac{\partial}{\partial t} + \sum_{i\in I} f_i\frac{\partial}{\partial u_i}, \quad f_i(\mathbf{u}) = D_i f, \quad f = \sum_{\mu=1}^{m} u_{2(\mu)},$$

(see Example 7.11, note that $\operatorname{ord} f=2$, $\operatorname{ord} f_i=|i|+2$). Static solutions of the heat equation are described by the Laplace equation $\Delta u=0$. This leads to the realization of the Laplace equation as a reduction of the heat equation. Namely, let

$$F = \Big\{t,x,\mathbf{u} : f_i(\mathbf{u})=0, \ i\in I\Big\}\subset M,$$

so

$$D_t\Big|_F = \frac{\partial}{\partial t}.$$

Let us show that F is a submanifold of M. We split a multi-index $i\in I=\mathbb{N}^m$ in two parts, $i=(r,i')$, $r=i_1\in\mathbb{N}$, $i'=(i_2,\dots,i_m)\in I'=\mathbb{N}^{m-1}$ and introduce new coordinates on M:

$$t,\ x,\ u_{(0,i')},\ u_{(1,i')},\ v_i=f_i(\mathbf{u}),\quad i'\in I',\ i\in I.$$

The inverse transition formulas are easily calculated. For instance,

$$u_{(2,i')} = D_{(0,i')}u_{(2,0')} = D_{(0,i')}\left(f - \sum_{\mu=2}^{m} u_{2(\mu)}\right)$$

$$= v_{(0,i')} - \sum_{\mu=2}^{m} u_{(0,i'+2(\mu))},$$

$$u_{(3,i')} = D_{(1,i')}u_{(2,0')} = D_{(1,i')}\left(f - \sum_{\mu=2}^{m} u_{2(\mu)}\right)$$

$$= v_{(1,i')} - \sum_{\mu=2}^{m} u_{(1,i'+2(\mu))},$$

$$u_{(4,i')} = D_{(2,i')}\left(f - \sum_{\mu=2}^{m} u_{2(\mu)}\right) = v_{(2,i')} - \sum_{\mu=2}^{m} u_{(2,i'+2(\mu))}$$

$$= v_{(2,i')} - \sum_{\mu=2}^{m} D_{(0,i'+2(\mu))}\left(f - \sum_{\nu=2}^{m} u_{2(\nu)}\right)$$

$$= v_{(2,i')} - \sum_{\mu=2}^{m} v_{(0,i'+2(\mu))} + \sum_{\mu,\nu=2}^{m} u_{(0,i'+2(\mu)+2(\nu))},$$

and so on. In these new coordinates

$$F = \left\{t,x,u_{(0,i')},u_{(1,i')},v_i\ :\ v_i = 0,\ i\in I\right\},$$

and we proved that F is a submanifold of M. Now, let us define basis Cartan fields ∂_t, ∂_μ on F as the restrictions of the basis fields D_t, D_μ, $\mu=1,\dots,m$:

$$\partial_t = D_t\big|_F = \frac{\partial}{\partial t} \text{ (we already noted the last equality)},$$

$$\partial_\mu = D_\mu\big|_F, \quad \mu = 1,\dots,m.$$

To prove that this definition is correct we should show that these restrictions are tangent to F, i.e., $\partial_t,\partial_\mu \in TF$, and that commutators $[\partial_t,\partial_\mu]=[\partial_\mu,\partial_\nu]=0$ for all $\mu,\nu=1,\dots,m$. First, let us write the expressions for D_μ, D_t in the new coordinates. Easy calculations yield:

$$D_1 = \frac{\partial}{\partial x^1} + \sum_{i'\in I'} \left\{ u_{(1,i')} \frac{\partial}{\partial u_{(0,i')}} - \sum_{\nu=2}^{m} u_{(0,i'+2(\nu))} \frac{\partial}{\partial u_{(1,i')}} \right\}$$

$$+ \sum_{i'\in I'} v_{(0,i')} \frac{\partial}{\partial u_{(1,i')}} + \sum_{i\in I} v_{i+(1)} \frac{\partial}{\partial v_i},$$

$$D_\mu = \frac{\partial}{\partial x^\mu} + \sum_{i'\in I'} \left\{ u_{(0,i'+(\mu))} \frac{\partial}{\partial u_{(0,i')}} + u_{(1,i'+(\mu))} \frac{\partial}{\partial u_{(1,i')}} \right\}$$

$$+ \sum_{i\in I} v_{i+(\mu)} \frac{\partial}{\partial v_i}, \quad \mu = 2,\dots,m,$$

$$D_t = \frac{\partial}{\partial t} + \sum_{i'\in I'} \left\{ v_{(0,i')} \frac{\partial}{\partial u_{(0,i')}} + v_{(1,i')} \frac{\partial}{\partial u_{(1,i')}} \right\}$$

$$+ \sum_{i\in I} \sum_{\nu=2}^{m} v_{i+2(\nu)} \frac{\partial}{\partial v_i},$$

Hence,

$$\partial_1 = \frac{\partial}{\partial x^1} + \sum_{i'\in I'} \left\{ u_{(1,i')} \frac{\partial}{\partial u_{(0,i')}} - \sum_{\nu=2}^{m} u_{(0,i'+2(\nu))} \frac{\partial}{\partial u_{(1,i')}} \right\},$$

$$\partial_\mu = \frac{\partial}{\partial x^\mu} + \sum_{i'\in I'} \left\{ u_{(0,i'+(\mu))} \frac{\partial}{\partial u_{(0,i')}} + u_{(1,i'+(\mu))} \frac{\partial}{\partial u_{(1,i')}} \right\},$$

$$\mu = 2,\dots,m,$$

$$\partial_t = \frac{\partial}{\partial t},$$

and clear that the tangency condition is fulfilled. After that the commutation condition follows from the general properties of the Lie brackets. It can be also checked by the direct computations. Thus, we constructed the differential submanifold $\mathcal{F}=(F,\partial_t,\partial_x)$ of the manifold $\mathcal{M}=(M,D_t,D_x)$. Note that standard coordinates on $\mathcal{F}$ are $t,x,u_{(0,i')},u_{(1,i')}$, $i'\in I'$, and basis Cartan fields are ∂_t,∂_μ, $\mu=1,\dots,m$. Let us find integral submanifolds of $\mathcal{F}$. By Proposition 1.8, a $(m+1)$-dimensional submanifold $S\subset F$ will be integral iff

$$S = \left\{ u_{(0,i')}=\varphi_{(0,i')}(t,x),\ u_{(1,i')}=\varphi_{(1,i')}(t,x),\ i'\in I' \right\},$$

where smooth functions $\varphi_{(0,i')}$, $\varphi_{(1,i')}$ satisfy the contact equations

$$\frac{\partial\varphi_{(0,i')}}{\partial x^1} = \varphi_{(1,i')}, \qquad \frac{\partial\varphi_{(1,i')}}{\partial x^1} = -\sum_{\nu=2}^{m} \varphi_{(0,i'+2(\nu))},$$

$$\frac{\partial\varphi_{(0,i')}}{\partial x^\mu} = \varphi_{(0,i'+(\mu))}, \qquad \frac{\partial\varphi_{(1,i')}}{\partial x^\mu} = \varphi_{(1,i'+(\mu))},$$

$$\frac{\partial\varphi_{(0,i')}}{\partial t} = 0, \qquad \frac{\partial\varphi_{(1,i')}}{\partial t} = 0,$$

where $i'\in I'$, $\mu=2,\dots,m$. A simple study shows that the general solution of this linear system of PDE is

$$\varphi_{(0,i')}(t,x) = \left(\frac{\partial}{\partial x}\right)^{(0,i')}\varphi(x), \quad \varphi_{(1,i')}(t,x) = \left(\frac{\partial}{\partial x}\right)^{(1,i')}\varphi(x),$$

where a smooth function $\varphi(x)$ is a solution of the Laplace equation $\Delta\varphi=0$. So, the differential manifold $\mathcal{F}$ corresponds to the Laplace equation, and we in fact realized *the Laplace equation as a reduction of the heat equation, corresponding to the static solutions of the latter*. Note that the crucial part in our construction played three facts: (a) the function f (i.e., the coefficients of the heat equation) doesn't depend on t, (b) it was possible to choose the functions f_i as a part of the new coordinates, $i\in I$, and

(c) $D_\mu(f_i)=f_{i+(\mu)}$, $D_t(f_i)=\Sigma_{\nu=2}^m f_{i+2(\nu)}$, $\mu=1,\ldots,m$, $i\in I$.

Example 8.2(b). This time we shall realize the Laplace equation as a differential submanifold of the heat equation, which is not a reduction. Such realization corresponds to the fact, that in the case of the Laplace equation there are m independent variables while there are $m+1$ independent variables in the case of the heat equation. So, let $\mathcal{M}=(M,D_t,D_x)$ be the differential manifold, defined by the heat equation (we are using the notation of Example 8.2(a)), with the standard coordinates t, x, $u_{(0,i')}$, $u_{(1,i')}$, $v_i=f_i(\mathbf{u})$, $i'\in I'$, $i\in I$, and the basis Cartan fields D_t, D_μ, $\mu=1,\ldots,m$. Further, let $\mathcal{F}=(F,\partial_x)$ be the differential manifold with the coordinates x, $u_{(0,i')}$, $u_{(1,i')}$, $i'\in I'$, and the basis Cartan fields ∂_μ, $\mu=1,\ldots,m$. Like above, one can easily check that integral submanifolds of $\mathcal{F}$ are defined by solutions of Laplace equation, so $\mathcal{F}$ again corresponds to the Laplace equation. Now, let us define the imbedding $\mathcal{F}\longrightarrow\mathcal{M}$ by the rule:

$$F \ni \left(x,u_{(0,i')},u_{(1,i')}\right) \mapsto \left(t=0,x,u_{(0,i')},u_{(1,i')},v_i=0\right) \in M,$$

then we get the desired realization.

Proposition 8.3. *Let $\mathcal{F}=(F,CTF)$ be a differential submanifold of a differential manifold $\mathcal{M}=(M,CTM)$. Then for every point $a\in F$ there exists a standard chart $(V,\psi,\mathbb{R}^{m+n+I+J},D_x,D_y)$ on $\mathcal{M}$, $a\in V$, with coordinates x^μ,u^i,y^ν,v^j, where $\mu=1,\ldots,m$, $i\in I$, $\nu=1,\ldots,n$, $j\in J$, and basis Cartan fields*

$$D_{x^\mu} = \frac{\partial}{\partial x^\mu} + \sum_{i\in I} A^i_\mu(x,u,y,v)\frac{\partial}{\partial u^i} + \sum_{j\in J} B^j_\mu(x,u,y,v)\frac{\partial}{\partial v^j}, \quad \mu=1,\ldots,m,$$

$$D_{y^\nu} = \frac{\partial}{\partial y^\nu} + \sum_{i\in I} C^i_\nu(x,u,y,v)\frac{\partial}{\partial u^i} + \sum_{j\in J} D^j_\nu(x,u,y,v)\frac{\partial}{\partial v^j}, \quad \nu=1,\ldots,n,$$

such that:

$$F\cap V = \left\{x,u,y,v \;:\; y^\nu=0,\ v^j=0,\ \nu=1,\ldots,n,\ j\in J\right\},$$

$(U,\varphi,\mathbb{R}^{m+I},\partial_x)$ is a standard chart on F, $U=F\cap V$, $\varphi=\psi\big|_F$, with coordinates x^μ, u^i, $\mu=1,\ldots,m$, $i\in I$, and basis Cartan fields

$$\partial_{x^\mu} = D_{x^\mu}\big|_F = \frac{\partial}{\partial x^\mu} + \sum_{i\in I} a^i_\mu(x,u)\frac{\partial}{\partial u^i}, \quad \mu=1,\ldots,m,$$

(i.e., $A^i_\mu(x,u,0,0)=a^i_\mu(x,u)$ and $B^j_\mu(x,u,0,0)=0$).

Proof. It is enough to use Proposition 2.10. Note that here m and $m+n$ are the Cartan dimensions of $\mathcal{F}$ and $\mathcal{M}$, respectively.

Let $\mathcal{M}=(M,CTM)$ be a differential manifold and F be a submanifold of M. When F defines a differential submanifold of $\mathcal{M}$? The natural sufficient condition is the following:

the vector subbundle

$$CTF = TF\cap CT_F M$$

is an involutive distribution (see Section 0.5) on F. For example, this condition is fulfilled if $CT_F M\subset TF$. We used these conditions in our previous examples. Further, let $\mathcal{I}_F=I\mathcal{C}^\infty(M)$ be the *defining ideal* of F, i.e., the set of all functions f from the algebra $C^\infty(M)$ vanishing on the submanifold F ($f\big|_F=0$, see Section 0.4).

Proposition 8.4. *A vector field $\zeta\in\mathcal{T}(M)$ is tangent to the submanifold F, i.e., $\zeta\big|_F\in\mathcal{T}(F)$, iff $\zeta:\mathcal{I}_F\to\mathcal{I}_F$, i.e.,*

$$\zeta(f)\big|_F = 0 \quad \text{if } f\big|_F = 0.$$

Proof. Clear that the statement is local. So, let $(W,\chi,\mathbb{R}^{I+J})$ be a defining chart on M (see Section 0.4) with coordinates u^i,v^j, $i\in I, j\in J$, such that $F\cap W=\{u,v: v=0\}$. Let

$$\zeta = \sum_{i\in I} \xi^i(u,v)\,\frac{\partial}{\partial u^i} + \sum_{j\in J} \vartheta^j(u,v)\,\frac{\partial}{\partial v^j} \in \mathcal{T}(M),$$

then $\zeta\big|_F\in\mathcal{T}(F)$ iff $\vartheta^j(u,0)=0$, $j\in J$. In the same way $f(u,v)\in\mathcal{C}^\infty(M)$, $f\big|_F=0$, iff $f(u,0)=0$. Now,

$$\zeta(f)(u,0) = \sum_{i\in I} \xi^i(u,0)\,\frac{\partial f(u,0)}{\partial u^i} + \sum_{j\in J} \vartheta^j(u,0)\left(\frac{\partial f(u,v)}{\partial v^j}\right)\bigg|_{v=0},$$

hence, $\zeta(f)(u,0)=0$ for all $f\in\mathcal{I}_F$ iff $\zeta\big|_F\in\mathcal{T}(F)$.

Corollary 8.5. *Let $\mathcal{M}=(M,CTM)$ be a differential manifold and F be a submanifold of M. Then $\mathcal{F}=(F,CT_F M)$ will be a reduction of $\mathcal{M}$ iff $\zeta:\mathcal{I}_F\to\mathcal{I}_F$ for every Cartan field $\zeta\in C\mathcal{T}(M)$.*

Corollary 8.6. *Let $\mathcal{F}$ be a reduction of a differential manifold $\mathcal{M}$, and let $(V,\psi,\mathbb{R}^{m+I+J},D_x)$ and $(U,\varphi,\mathbb{R}^{m+I},\partial_x)$ be standard defining charts on $\mathcal{M}$ and $\mathcal{F}$, respectively* (see Proposition 8.3). *Then*

$$\partial_{x^\mu}\phi = \left(D_{x^\mu}\Phi\right)\Big|_U, \quad \mu = 1,\dots,m,$$

where $\Phi\in\mathcal{C}^\infty(V)$ is a continuation of $\phi\in\mathcal{C}^\infty(U)$.

In a typical situation there are a differential manifold $\mathcal{M}=(M,CTM)$ (usually it is trivial, see Section 6) and an initial "seed" subset F_0 of M. Using these, one wants to construct a reduction $\mathcal{F}=(F,CTF)$ of $\mathcal{M}$, $F\subset F_0$, maximal in some sense. The standard procedure is the following. Let $\mathcal{I}_{F_0}$ be the defining ideal of the subset F_0 (i.e., the set of all functions $f\in\mathcal{C}^\infty(M)$ vanishing on F_0). Denote by $\mathcal{I}^c_{F_0}$ the *differential completion of* $\mathcal{I}_{F_0}$, i.e., the least differential ideal containing $\mathcal{I}_{F_0}$ (an ideal $\mathcal{Y}$ of the algebra $\mathcal{C}^\infty(M)$ is called *differential* if it is closed with respect to Cartan differentiations, i.e., if $\zeta(f)\in\mathcal{Y}$ for all $\zeta\in C\mathcal{T}(M)$ and $f\in\mathcal{Y}$, cf. Appendix). Let F be the common zero set of $\mathcal{I}^c_{F_0}$,

$$F = \left\{z \in F_0 \ : \ f(z)=0 \text{ for all } f \in \mathcal{I}^c_{F_0}\right\},$$

and suppose that:

(a) F has a structure of a smooth submanifold of M,

(b) the ideal $\mathcal{I}^c_{F_0} = \mathcal{I}_F$ is defining for F,

($\mathcal{I}_F$ is set of all functions $f\in\mathcal{C}^\infty(M)$ vanishing on F). Then, by Corollary 8.5, $\mathcal{F}=(F,CT_FM)$ will be a reduction of $\mathcal{M}$. Note that just described process of construction of the submanifold F, proceeding from an initial subset F_0, is usually called the *prolongation*, and the submanifold F is called the *infinite prolongation* of F_0 (see, for example, Ibragimov [1], Olver [1], Ovsyannikov [1] and Vinogradov, Krasil'shchik, Lychagin [1]). In particular, let the initial subset F_0 be defined as the solution set of a finite system of equations,

$$F_0 = \left\{ z \in M \; : \; F^\sigma(z) = 0, \; \sigma = 1,\dots,s \right\},$$

where $F^1,\dots,F^s$ are some smooth "seed" functions on M. In nondegenerated case, when the system $F^1,\dots,F^s$ has the maximal rank ($=s$) on F_0, the ideal $\mathcal{I}_{F_0}$ is generated by this system, i.e.,

$$\mathcal{I}_{F_0} = \left\{ f = \sum_{\sigma=1}^{s} f_\sigma F^\sigma ; \; f_\sigma \in \mathcal{C}^\infty(M) \right\}.$$

Suppose that this is the case. Further, suppose that the Cartan distribution CTM is defined by basis fields $D_1,\dots,D_m$ (m is the Cartan dimension of $\mathcal{M}$), then the differential completion

$$\mathcal{I}^c_{F_0} = \left\{ f = \sum_{\sigma=1}^{s} \sum_{i\in I} f^i_\sigma D_i F^\sigma \; ; \; f^i_\sigma \in \mathcal{C}^\infty(M) \right\}$$

$$= \left\{ f = P(F) \; ; \; P \in \mathcal{P}_s \right\},$$

(we use the notation of Appendix, here $\mathcal{A}=\mathcal{C}^\infty(M)$, $\mathcal{P}=\mathcal{C}^\infty(M)[D]$,

$F=(F^{\sigma})\in\mathcal{A}^{s}$). If the functions F^{σ}, $\sigma=1,\dots,s$, satisfy some further nondegeneracy conditions (cf. Olver [1]), then $\mathcal{I}^{c}_{F_0}=\mathcal{I}_{F}$, and the above conditions (a) and (b) will be fulfilled. Thus, if seed functions $F^{1},\dots,F^{s}$ are sufficiently good, the prolongation process leads to a reduction of the ambient differential manifold $\mathcal{M}$.

9. Normal Projection. External Fields and Forms

Let $\mathcal{F}=(F,CTF)$ be a reduction of a differential manifold $\mathcal{M}=(M,CTM)$. We shall assume that $\mathcal{M}$ has a simple structure, and $\mathcal{F}$ is a more complicated differential manifold we want to study. From this point of view the existence of a background manifold $\mathcal{M}$ provides $\mathcal{F}$ with an additional external geometry together with its own internal geometry, studied in Chapter I. By the same reason, surfaces in $\mathbb{R}^n$ have additional properties in comparison with abstract manifolds.

As it was noted above (see Section 8), the natural imbedding $i:F\longrightarrow M$ is a Lie-Backlund immersion. In particular, the tangent mapping $i_*:TF\longrightarrow TM$ is a monomorphism of vector bundles. Moreover, it is an isomorphism on the fibers of the Cartan subbundle, i.e., $i_*:CT_aF\simeq CT_{i(a)}M$, $a\in F$. Further, according to Section 0.4, we have the exact sequence

$$0 \longrightarrow TF \xrightarrow{j} T_FM \xrightarrow{\nu} NF \longrightarrow 0$$

of the vector bundles over F, where TF is the tangent bundle on F,

$$T_FM = \bigcup_{a\in F} T_aM,$$

the bundle monomorphism j is defined by the decomposition $i_*=\rho\circ j$, $\rho:T_FM\to TM$ is the natural imbedding, and ν is the normal projection of the bundle T_FM on the quotient bundle $NF=T_FM\big/j(TF)$. In our case, this sequence has the exact subsequence $0\to CTF\xrightarrow{j}CT_FM\to 0$.

Let $\mathcal{T}(F)$, $\mathcal{V}(F)$ and $\mathcal{N}(F)$ be the linear spaces of tangent, vector and normal fields on F, respectively (see Section 0.4). Denote by $\rho^*:\mathcal{T}(M)\to\mathcal{V}(F)$ the linear mapping consisting in the restriction of the vector fields from M to the submanifold F, and let $F\mathcal{T}(M)=(\rho^*)^{-1}(\mathcal{T}(F))$ be the linear space of all vector fields $\zeta\in\mathcal{T}(M)$ tangent to F, i.e., such that $\zeta\big|_F=\rho^*(\zeta)\in\mathcal{T}(F)\subset\mathcal{V}(F)$.

Proposition 9.1. *The following locally exact commutative diagram of the linear spaces is defined:*

$$\begin{array}{ccccccccc}
 & & 0 & & 0 & & 0 & & \\
 & & \uparrow & & \uparrow & & \uparrow & & \\
0 & \to & \mathcal{T}(F) & \xrightarrow{j} & \mathcal{V}(F) & \xrightarrow{\nu} & \mathcal{N}(F) & \to & 0 \\
 & & \uparrow & & \uparrow\rho^* & & \uparrow & & \\
0 & \to & F\mathcal{T}(M) & \longrightarrow & \mathcal{T}(M) & \longrightarrow & \dfrac{\mathcal{T}(M)}{F\mathcal{T}(M)} & \to & 0 \\
 & & \uparrow & & \uparrow & & \uparrow & & \\
0 & \to & R\mathcal{T}(M) & \longrightarrow & R\mathcal{T}(M) & \longrightarrow & 0 & & \\
 & & \uparrow & & \uparrow & & & & \\
 & & 0 & & 0 & & & &
\end{array}$$

Note that here and below we say that a statement is valid *locally* if it is valid in every star-shaped chart.

Proof. We only need to explain that here $R\mathcal{T}(M)$ is the kernel of the linear mapping ρ^*, it consists of all vector fields $\zeta\in\mathcal{T}(M)$

vanishing on F, i.e., $\zeta\big|_F=0$.

Clear that if ζ is a Lie-Backlund field on M (i.e., $\zeta\in LB\mathcal{T}(M)$) and it is tangent to F (i.e., $\zeta\in F\mathcal{T}(M)$), then the restriction $\zeta\big|_F$ is a Lie-Backlund field on F (i.e., $\zeta\big|_F\in LB\mathcal{T}(F)$). In other words, the Lie algebra

$$LB\mathcal{T}_{ext}(F) = \rho^*\Big(LB\mathcal{T}(M)\cap F\mathcal{T}(M)\Big) \subset LB\mathcal{T}(F)$$

of *exterior Lie-Backlund fields* on F is defined, and it is a Lie subalgebra of the Lie algebra $LB\mathcal{T}(F)$. The question of the coincidence of the algebras $LB\mathcal{T}_{ext}(F)$ and $LB\mathcal{T}(F)$ is not clear even locally in the general case, although in examples occurring in practice they do coincide.

Let us define $LBV(F)=\rho^*(LB\mathcal{T}(M))$, $LBF\mathcal{T}(M)=LB\mathcal{T}(M)\cap F\mathcal{T}(M)$ and $RLB\mathcal{T}(M)=R\mathcal{T}(M)\cap LB\mathcal{T}(M)$.

Proposition 9.2. *The following exact commutative diagram of the linear spaces is defined:*

$$\begin{array}{ccccccccc}
 & & 0 & & 0 & & 0 & & \\
 & & \uparrow & & \uparrow & & \uparrow & & \\
0 & \longrightarrow & LB\mathcal{T}_{ext}(F) & \xrightarrow{\ j\ } & LBV(F) & \xrightarrow{\ \nu\ } & LB\mathcal{N}(F) & \longrightarrow & 0 \\
 & & \uparrow & & \uparrow{\scriptstyle \rho^*} & & \uparrow & & \\
0 & \longrightarrow & LBF\mathcal{T}(M) & \longrightarrow & LB\mathcal{T}(M) & \longrightarrow & \dfrac{LB\mathcal{T}(M)}{LBF\mathcal{T}(M)} & \longrightarrow & 0 \\
 & & \uparrow & & \uparrow & & \uparrow & & \\
0 & \longrightarrow & RLB\mathcal{T}(M) & \longrightarrow & RLB\mathcal{T}(M) & \longrightarrow & 0 & & \\
 & & \uparrow & & \uparrow & & & & \\
 & & 0 & & 0 & & & &
\end{array}$$

Proof. The only thing we need to explain is that, by the definition, $LB\mathcal{N}(F)=LB\mathcal{V}(F)\big/LB\mathcal{T}_{ext}(F)$.

Let us define the quotient space $\mathrm{Sym}\mathcal{M}_F=LB\mathcal{V}(F)\big/C\mathcal{T}(F)$.

Proposition 9.3. *The following exact commutative diagram of the linear spaces is defined:*

$$
\begin{array}{ccccccccc}
 & & 0 & & 0 & & 0 & & \\
 & & \uparrow & & \uparrow & & \uparrow & & \\
0 & \longrightarrow & C\mathcal{T}(F) & \xrightarrow{\ j\ } & LB\mathcal{V}(F) & \xrightarrow{\ \nu\ } & \mathrm{Sym}\mathcal{M}_F & \longrightarrow & 0 \\
 & & \uparrow & & \uparrow{\scriptstyle \rho^*} & & \uparrow & & \\
0 & \longrightarrow & C\mathcal{T}(M) & \longrightarrow & LB\mathcal{T}(M) & \longrightarrow & \mathrm{Sym}\mathcal{M} & \longrightarrow & 0 \\
 & & \uparrow & & \uparrow & & \uparrow & & \\
0 & \longrightarrow & RC\mathcal{T}(M) & \longrightarrow & RLB\mathcal{T}(M) & \longrightarrow & \dfrac{RLB\mathcal{T}(M)}{RC\mathcal{T}(M)} & \longrightarrow & 0 \\
 & & \uparrow & & \uparrow & & \uparrow & & \\
 & & 0 & & 0 & & 0 & &
\end{array}
$$

Proof. It is enough to note that $RC\mathcal{T}(M)=R\mathcal{T}(M)\cap C\mathcal{T}(M)$.

Corollary 9.4. *We have*

$$\mathrm{Sym}\ \mathcal{M}_F = \frac{LB\mathcal{T}(M)}{RLB\mathcal{T}(M)}\Big/\frac{C\mathcal{T}(M)}{RC\mathcal{T}(M)} = \frac{LB\mathcal{T}(M)}{RLB\mathcal{T}(M)+C\mathcal{T}(M)}.$$

Let us define the quotient space

$$\mathrm{Sym}_{ext}\mathcal{F} = \frac{LB\mathcal{F}_{ext}(F)}{C\mathcal{T}(F)} \subset \mathrm{Sym}\ \mathcal{F}.$$

Clear that $\mathrm{Sym}_{ext}\mathcal{F}$ is a Lie subalgebra of the Lie algebra $\mathrm{Sym}\mathcal{F}$, because $C\mathcal{T}(F)$ is an ideal of $LB\mathcal{T}_{ext}(F)$.

Corollary 9.5. *We have*

$$\mathrm{Sym}_{ext}\mathcal{F} = \frac{LBF\mathcal{T}(M)}{RLB\mathcal{T}(M)}\Big/\frac{C\mathcal{T}(M)}{RC\mathcal{T}(M)} = \frac{LBF\mathcal{T}(M)}{RLB\mathcal{T}(M)+C\mathcal{T}(M)}.$$

Proof. It is enough to use Propositions 9.2 and 9.3.

We introduce also the quotient space $\mathrm{Sym}\mathcal{N}\mathcal{F}=\mathrm{Sym}\mathcal{M}_F\big/\mathrm{Sym}_{ext}\mathcal{F}$.

Corollary 9.6. *We have*

$$\mathrm{Sym}\ \mathcal{N}\mathcal{F} = \frac{LB\mathcal{T}(M)}{RLB\mathcal{T}(M)+C\mathcal{T}(M)}\bigg/\frac{LBF\mathcal{T}(M)}{RLB\mathcal{T}(M)+C\mathcal{T}(M)} = \frac{LB\mathcal{T}(M)}{LBF\mathcal{T}(M)} = LB\mathcal{N}(F).$$

Proof. It is enough to use Corollaries 9.4, 9.5 and Proposition 9.2.

Theorem 9.7. *The following exact sequence of the linear spaces is defined:*

$$0 \longrightarrow \mathrm{Sym}_{ext}\mathcal{F} \xrightarrow{\ j\ } \mathrm{Sym}\ \mathcal{M}_F \xrightarrow{\ \nu\ } \mathrm{Sym}\ \mathcal{N}\mathcal{F} \longrightarrow 0,$$

where

$$\mathrm{Sym}_{ext}\mathcal{F} = \frac{LBF\mathcal{T}(M)}{RLB\mathcal{T}(M)+C\mathcal{T}(M)},\quad \mathrm{Sym}\ \mathcal{M}_F = \frac{LB\mathcal{T}(M)}{RLB\mathcal{T}(M)+C\mathcal{T}(M)},\quad \mathrm{Sym}\ \mathcal{N}\mathcal{F} = \frac{LB\mathcal{T}(M)}{LBF\mathcal{T}(M)}$$

Proof. It is enough to combine Corollaries 9.4, 9.5, 9.6.

Corollary 9.8. *We have the equality*

$$\mathrm{Sym}_{ext}\mathcal{F} = \mathrm{Ker}\ \nu\ :\ \mathrm{Sym}\ \mathcal{M}_F \longrightarrow \mathrm{Sym}\ \mathcal{N}\mathcal{F}.$$

Suppose now that the differential manifold $\mathcal{M}$ and its reduction $\mathcal{F}$ are covered by a single chart (see Proposition 8.3) with coordinates x^μ, u^i, v^j, $\mu=1,\dots,m$, $i\in I$, $j\in J$, and basis Cartan fields

$$D_\mu = \frac{\partial}{\partial x^\mu} + \sum_{i\in I} A^i_\mu(x,u,v)\frac{\partial}{\partial u^i} + \sum_{j\in J} B^j_\mu(x,u,v)\frac{\partial}{\partial v^j},\quad \mu=1,\dots,m,$$

such that:

$$F = \left\{x,u,v\ :\ v^j=0,\ \ j\in J\right\};$$

x^μ, u^i, $\mu=1,\dots,m$, $i\in I$, are coordinates on F;

$$\partial_\mu = D_\mu\Big|_F = \frac{\partial}{\partial x^\mu} + \sum_{i\in I} a^i_\mu(x,u)\frac{\partial}{\partial u^i}, \quad \mu=1,\dots,m,$$

are basis Cartan fields on $\mathcal{F}$, i.e., $A^i_\mu(x,u,0)=a^i_\mu(x,u)$, $B^j_\mu(x,u,0)=0$ (we shall call such chart a *standard defining chart*). Then a vector field $\zeta\in LB\mathcal{T}(F)$ iff

$$\xi = \sum_{i\in I} \xi^i(x,u)\ \frac{\partial}{\partial u^i} \quad \text{mod } C\mathcal{T}(F),$$

where smooth functions ξ^i satisfy the defining equations (see Proposition 3.1). In other words,

$$\text{Sym }\mathcal{F} = \left\{\xi=(\xi^i)\ :\ \partial_\mu\xi^i - \sum_{k\in I} \frac{\partial a^i_\mu}{\partial u^k}\,\xi^k = 0\right\}.$$

By the same reasons, a vector field $Z\in LB\mathcal{T}(M)$ iff

$$Z = \sum_{i\in I} \Xi^i(x,u,v)\ \frac{\partial}{\partial u^i} + \sum_{j\in J} \Theta^j(x,u,v)\ \frac{\partial}{\partial v^j} \quad \text{mod } C\mathcal{T}(M),$$

where smooth functions Ξ^i,Θ^j satisfy the defining equations. Thus,

$$\text{Sym }\mathcal{M} = \left\{Z=(\Xi^i,\Theta^j)\ :\ D_\mu\Xi^i - \sum_{k\in I} \frac{\partial A^i_\mu}{\partial u^k}\,\Xi^k - \sum_{l\in J}\frac{\partial A^i_\mu}{\partial v^l}\,\Theta^l = 0,\right.$$

$$\left. D_\mu\Theta^j - \sum_{k\in I} \frac{\partial B^j_\mu}{\partial u^k}\,\Xi^k - \sum_{l\in J}\frac{\partial B^j_\mu}{\partial v^l}\,\Theta^l = 0\right\}.$$

Further,

$$\text{Sym }\mathcal{M}_F = \left\{\zeta=(\xi^i,\theta^j)\ :\ \xi^i=\Xi^i\Big|_F,\ \theta^j=\Theta^j\Big|_F,\ Z=(\Xi^i,\Theta^j)\in\text{Sym}\mathcal{M}\right\},$$

where, for example, $\Xi^i\Big|_F(x,u)=\Xi^i(x,u,0)$,

$$\mathrm{Sym}_{ext}\mathcal{F} = \left\{\xi=(\xi^i) \ : \ \xi^i=\Xi^i\Big|_F, \ \Theta^j\Big|_F=0, \ Z=(\Xi^i,\Theta^j)\in\mathrm{Sym}\mathcal{M}\right\},$$

$$\mathrm{Sym}\ N\mathcal{F} = \left\{\theta=(\theta^j) \ : \ \theta^j=\Theta^j\Big|_F, \ Z=(\Xi^i,\Theta^j)\in\mathrm{Sym}\mathcal{M}\right\}.$$

Remark 9.9. Note that

$$\mathrm{Sym}\mathcal{F} = \mathrm{Sym}_{ext}\mathcal{F} \quad \text{if} \quad A^i_\mu = A^i_\mu(x,u) \quad \text{and} \quad B^j_\mu = B^j_\mu(x,v),$$

for all $\mu=1,\dots,m$, $i\in I$, $j\in J$. Indeed, here $Z=(\Xi^i=\xi^i,\Theta^j=0)\in\mathrm{Sym}\mathcal{M}$ if $\xi=(\xi^i)\in\mathrm{Sym}\mathcal{F}$.

Let us turn now to forms. We start with the dual exact sequence of covector bundles over F:

$$0 \longleftarrow T^*F \xleftarrow{\ j^*\ } T^*_FM \xleftarrow{\ \nu^*\ } N^*F \longleftarrow 0,$$

where T^*F is the cotangent bundle on F,

$$T^*_FM = (T_FM)^* = \bigcup_{a\in F} T^*_aM$$

is the restriction to F of the cotangent bundle T^*M,

$$N^*F = (NF)^* = \bigcup_{a\in F} N^*_aF$$

is the conormal bundle on F, j^* and ν^* are the dual linear mappings.

Remind that $\Lambda(F)$, $\Omega(F)$ and $N\Omega(F)$ are the linear spaces of of tangent 1-forms, 1-forms and normal 1-forms on F, respectively (see Section 0.4).

Proposition 9.10. *The following locally exact commutative diagram of linear spaces is defined:*

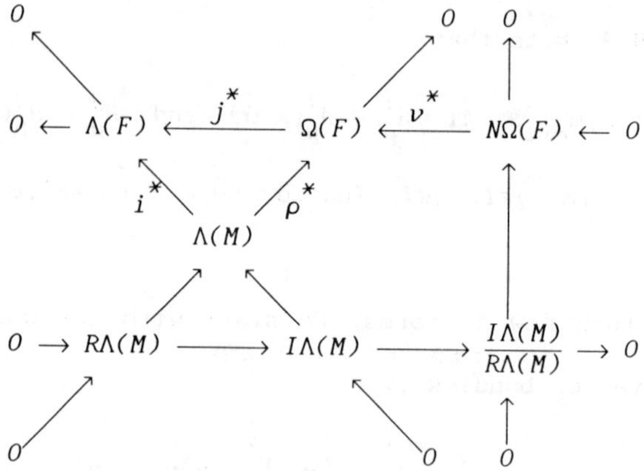

Proof. It is enough to explain, that $R\Lambda(M)$ and $I\Lambda(M)$ are the kernels of the linear mappings ρ^* and i^*, respectively.

Consider the linear space $\Omega^q(F)$ of q-forms on F, $q=0,1,\ldots$ (see Section 0.4), and let us call a q-form $\omega\in\Omega^q(F)$ *p-Cartan* if $\omega(\zeta_1,\ldots,\zeta_q)=0$ when at least $q-p+1$ of the fields $\zeta_1,\ldots,\zeta_q\in V(F)$ are Cartan, $p=0,1,\ldots$. We denote by $C^p\Omega^q(F)$ the linear subspace of all p-Cartan q-forms on F. Elements of the space $C^p\Lambda^q(F)$ (see Section 5) we shall call now *p-Cartan tangent q-forms*.

Proposition 9.11. *The following locally exact commutative diagram of linear spaces is defined:*

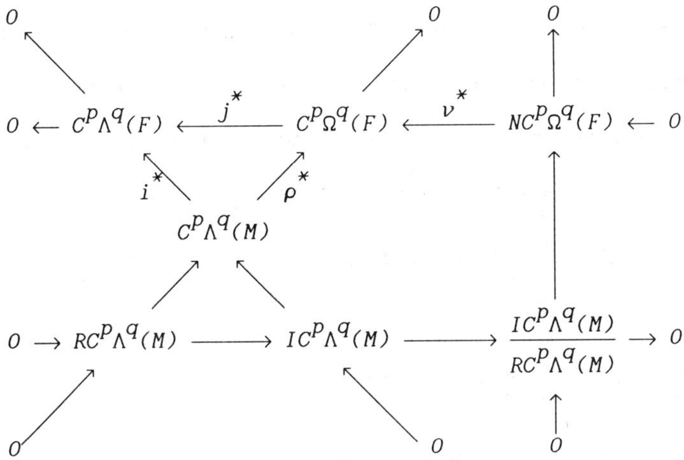

Proof. It enough to explain that here, by the definition, $NC^p\Omega^q(F)$ is the kernel of the linear mapping j^*, and so in this case ν^* is the natural imbedding. Of course, $RC^p\Lambda^q(M)$ and $IC^p\Lambda^q(M)$ are the kernels of the linear mappings ρ^* and i^*, respectively.

Lemma 9.12. *The equality*

$$I\Lambda^q(M) = R\Lambda^q(M) + IC\Lambda^q(M), \quad q = 0,1,\dots,$$

is valid locally (note that this statement is nontrivial only for $0\le q\le m$).

Proof. Clear, $R\Lambda^q(M)+IC\Lambda^q(M)\subset I\Lambda^q(M)$. Hence, we only need to prove the inclusion $I\Lambda^q(M)\subset R\Lambda^q(M)+IC\Lambda^q(M)$. So, let $\omega\in I\Lambda^q(M)$, i.e., $\omega\in\Lambda^q(M)$ and $i^*\omega=0$. In particular, $\rho^*\omega=0$ on $TF\subset T_FM$, and thus $\rho^*\omega=0$ on $CT_FM=CTF\subset TF$. Consequently, $\rho^*\omega\in C\Omega^q(F)$. It follows from the surjectivity of ρ^* (see Proposition 9.11) that there exists an $\omega_1\in C\Lambda^q(M)$ such that $\rho^*\omega_1=\rho^*\omega$. What is more, $\omega_1\in IC\Lambda^q(M)$, since $i^*\omega_1=j^*\circ\rho^*\omega_1=j^*\circ\rho^*\omega=i^*\omega=0$. Let $\omega=\omega_1+\omega_2$. Then $\omega_2=\omega-\omega_1\in\Lambda^q(M)$ and

$\rho^*\omega_2=\rho^*\omega-\rho^*\omega_1=0$, i.e., $\omega_2\in R\Lambda^q(M)$, and we proved the inclusion.

Suppose now that the differential manifold M and its reduction $\mathcal{F}$ are covered by a single standard defining chart. In this chart the basis Cartan 1-forms on M have the form

$$\delta u^i = du^i - \sum_{\mu=1}^{m} A^i_\mu(x,u,v)\, dx^\mu, \quad i\in I,$$

$$\delta v^j = dv^j - \sum_{\mu=1}^{m} B^j_\mu(x,u,v)\, dx^\mu, \quad j\in J.$$

Their restrictions on T_FM are equal to

$$\rho^*(\delta u^i) = du^i - \sum_{\mu=1}^{m} a^i_\mu(x,u)\, dx^\mu, \quad i\in I,$$

$$\rho^*(\delta v^j) = dv^j, \qquad j\in J,$$

and form a basis of Cartan 1-forms on T_FM. Further, the restrictions of the 1-forms δu^i and δv^j on TF are

$$i^*(\delta u^i) = du^i - \sum_{\mu=1}^{m} a^i_\mu(x,u)\, dx^\mu, \quad i\in I,$$

$$i^*(\delta v^j) = 0, \qquad j\in J.$$

In particular, the restrictions $i^*(\delta u^i)$ coincide with the basis Cartan 1-forms on F.

Remind that $dC^p\Lambda^q(M)\subset C^p\Lambda^{q+1}(M)$. Also $dIC^p\Lambda^q(M)\subset IC^p\Lambda^{q+1}(M)$, since $d\circ i^*=i^*\circ d$.

Lemma 9.13. *The inclusion*

$$dRC^p\Lambda^q(M)\subset RC^p\Lambda^{q+1}(M)+IC^{p+1}\Lambda^{q+1}(M)$$

is valid locally, $p,q=0,1,\ldots$.

Proof. Let M be covered by a standard defining chart, and let $\omega \in RC^p\Lambda^q(M)$. Clear that ω has the representation $\omega=\omega_0+\omega_1$, where $\omega_0 \in RC^{p+1}\Lambda^q(M)$ (so $d\omega_0 \in IC^{p+1}\Lambda^{q+1}(M)$, since $RC^{p+1}\Lambda^q \subset IC^{p+1}\Lambda^q(M)$), and

$$\omega_1 = \sum_{k+r=p} \sum_{(i)} \sum_{(j)} \sum_{(\mu)} \omega_{(i);(j);(\mu)} \left(\delta u^i\right)^k \wedge \left(\delta v^j\right)^r \wedge \left(dx^\mu\right)^{q-p},$$

where $(i)=(i)_k$, $(j)=(j)_r$, $(\mu)=(\mu)_{q-p}$ (cf. the notation of Proposition 4.2, note that the statement is nontrivial only for $p \le q$), and $\omega_{(i);(j);(\mu)}(x,u,0)=0$, since $\omega_1 \in RC^p\Lambda^q(M)$. The exterior derivative is $d\omega_1=\omega_2+\omega_3$, where

$$\begin{aligned}\omega_3 &= \sum \omega_{(i);(j);(\mu)} d\left[\left(\delta u^i\right)^k \wedge \left(\delta v^j\right)^r\right] \wedge \left(dx^\mu\right)^{q-p} \\ &\in RC^p\Lambda^{q+1}(M),\end{aligned}$$

since $\omega_{(i);(j);(\mu)}(x,u,0)=0$, and

$$\begin{aligned}\omega_2 &= \sum d\omega_{(i);(j);(\mu)} \wedge \left(\delta u^i\right)^k \wedge \left(\delta v^j\right)^r \wedge \left(dx^\mu\right)^{q-p} \\ &= \omega_4 + \omega_5,\end{aligned}$$

where

$$\begin{aligned}\omega_4 &= \sum \sum_{\mu=1}^{m} \left(D_\mu \omega_{(i);(j);(\mu)}\right) dx^\mu \wedge \left(\delta u^i\right)^k \wedge \left(\delta v^j\right)^r \wedge \left(dx^\mu\right)^{q-p} \\ &\in RC^p\Lambda^{q+1}(M),\end{aligned}$$

since

$$\left(D_\mu \omega_{(i);(j);(\mu)}\right)(x,u,0) = \partial_\mu\left(\omega_{(i);(j);(\mu)}(x,u,0)\right) = 0,$$

because the fields D_μ are tangent to the submanifold F, and

$$w_5 = \sum_{i\in I} \sum \frac{\partial\omega_{(i);(j);(\mu)}}{\partial u^i} \delta u^i \wedge \left(\delta u^i\right)^k \wedge \left(\delta v^j\right)^r \wedge \left(dx^\mu\right)^{q-p}$$

$$+ \sum_{j\in J} \sum \frac{\partial\omega_{(i);(j);(\mu)}}{\partial v^j} \delta v^j \wedge \left(\delta u^i\right)^k \wedge \left(\delta v^j\right)^r \wedge \left(dx^\mu\right)^{q-p}$$

$$\in RC^{p+1}\Lambda^{q+1}(M) + IC^{p+1}\Lambda^{q+1}(M)$$

(note that the fields ∂ / du^i are tangent to the submanifold F). Thus, $d\omega = d\omega_0 + \omega_3 + \omega_4 + \omega_5 \in RC^p\Lambda^{q+1}(M) + IC^{p+1}\Lambda^{q+1}(M)$, and the proof is finished.

According to Section 5, the C-spectral sequences $E_r^{p,q}(\mathcal{F})$ and $E_r^{p,q}(\mathcal{M})$ are defined, and the imbedding $i: F \longrightarrow M$ induces the linear mapping

$$i^* : E_r^{p,q}(\mathcal{M}) \longrightarrow E_r^{p,q}(\mathcal{F}),$$

for all $r, p, q = 0, 1, \ldots$. Further, we have the filtration

$$IC^0\Lambda^q(M) \supset IC^1\Lambda^q(M) \supset IC^2\Lambda^q(M) \supset \ldots \supset IC^q\Lambda^q(M) \supset 0,$$

compatible with the exterior differentiation

$$d: IC^p\Lambda^q(M) \longrightarrow IC^p\Lambda^{q+1}(M),$$

as we noted above. Hence, the spectral subsequence of $E_r^{p,q}(\mathcal{M})$ is defined, and we denote this subsequence by $E_r^{p,q}(I\mathcal{M})$. On the other hand, the filtrations

$$RC^0\Lambda^q(M) \supset RC^1\Lambda^q(M) \supset RC^2\Lambda^q(M) \supset \ldots \supset RC^q\Lambda^q(M) \supset 0,$$

$$C^0\Omega^q(M) \supset C^1\Omega^q(M) \supset C^2\Omega^q(M) \supset \ldots \supset C^q\Omega^q(M) \supset 0,$$

$$NC^0\Omega^q(M) \supset NC^1\Omega^q(M) \supset NC^2\Omega^q(M) \supset \ldots \supset NC^q\Omega^q(M) \supset 0$$

are not connected with the exterior differentiation (cf. Lemma 9.13), and there is no natural way to define the corresponding spectral sequences. Nevertheless, together with the linear spaces

$$E_0^{p,q}(\mathcal{F}) = \frac{C^p\Lambda^{p+q}(F)}{C^{p+1}\Lambda^{p+q}(F)}, \quad E_0^{p,q}(\mathcal{M}) = \frac{C^p\Lambda^{p+q}(M)}{C^{p+1}\Lambda^{p+q}(M)}$$

$$\text{and } E_0^{p,q}(I\mathcal{M}) = \frac{IC^p\Lambda^{p+q}(M)}{IC^{p+1}\Lambda^{p+q}(M)},$$

the linear spaces

$$E_0^{p,q}(R\mathcal{M}) = \frac{IC^p\Lambda^{p+q}(M)}{IC^{p+1}\Lambda^{p+q}(M)}, \quad E_0^{p,q}(\mathcal{M}_F) = \frac{C^p\Omega^{p+q}(F)}{C^{p+1}\Omega^{p+q}(F)}$$

$$\text{and } E_0^{p,q}(N\mathcal{F}) = \frac{NC^p\Omega^{p+q}(F)}{NC^{p+1}\Omega^{p+q}(F)}$$

are defined.

Proposition 9.14. *The following equalities (more precisely, natural isomorphisms)*

$$E_0^{p,q}(\mathcal{F}) = \frac{C^p\Lambda^{p+q}(M)}{C^{p+1}\Lambda^{p+q}(M)+IC^p\Lambda^{p+q}(M)},$$

$$E_0^{p,q}(\mathcal{M}_F) = \frac{C^p\Lambda^{p+q}(M)}{C^{p+1}\Lambda^{p+q}(M)+RC^p\Lambda^{p+q}(M)},$$

$$E_0^{p,q}(N\mathcal{F}) = \frac{IC^p\Lambda^{p+q}(M)}{IC^{p+1}\Lambda^{p+q}(M)+RC^p\Lambda^{p+q}(M)}$$

are valid locally.

Note that these equalities can be regarded as external representations of the corresponding linear spaces.

Proof. Indeed,

$$E_0^{p,q}(\mathcal{F}) = \frac{C^p\Lambda^{p+q}(F)}{C^{p+1}\Lambda^{p+q}(F)} = \frac{C^p\Lambda^{p+q}(M)}{IC^p\Lambda^{p+q}(M)} \Big/ \frac{C^{p+1}\Lambda^{p+q}(M)}{IC^{p+1}\Lambda^{p+q}(M)}$$

$$= \frac{C^p\Lambda^{p+q}(M)}{C^{p+1}\Lambda^{p+q}(M)+IC^p\Lambda^{p+q}(M)},$$

where we used the exactness of the i^*-diagonal of the diagram in Proposition 9.11. The second and the third equalities are proved similarly with the exactness of the ρ^*-diagonal and the vertical of the same diagram taken into account.

Theorem 9.15. *The following locally exact commutative diagram of linear spaces is defined:*

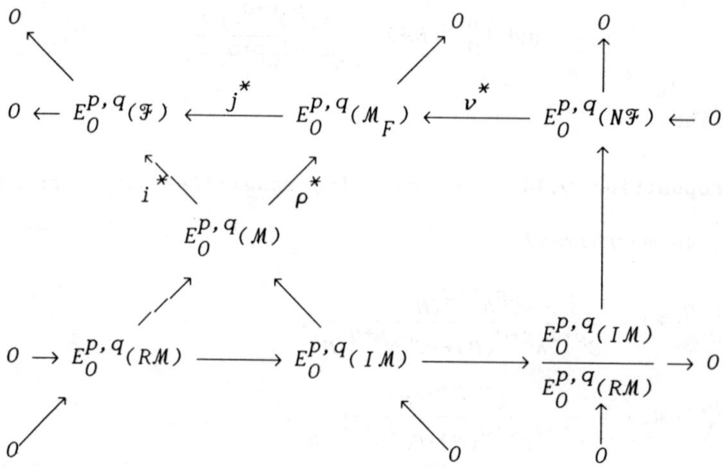

for $p,q=0,1,\ldots.$

Proof. The proof is by the direct verification, with the use of Proposition 9.14.

We single out the case $p=0$.

Corollary 9.16. *The equalities*

$$E_0^{0,q}(\mathcal{F}) = E_0^{0,q}(\mathcal{M}_F) \quad \text{and} \quad E_0^{0,q}(R\mathcal{M}) = E_0^{0,q}(I\mathcal{M})$$

are valid locally for $q=0,1,\ldots.$

Proof. It is enough to note that by Lemma 9.12,

$$E_0^{0,q}(N\mathcal{F}) = \frac{I\Lambda^q(M)}{IC\Lambda^q(M)+R\Lambda^q(M)} = 0,$$

and to use Theorem 9.15.

Let us write now the first-order terms of the spectral sequences defined above. We have

$$E_1^{p,q}(\mathcal{F}) = \frac{C^p\Lambda^{p+q}(F)\cap d^{-1}C^{p+1}\Lambda^{p+q+1}(F)}{C^{p+1}\Lambda^{p+q}(F)+dC^p\Lambda^{p+q-1}(F)},$$

$$E_1^{p,q}(\mathcal{M}) = \frac{C^p\Lambda^{p+q}(M)\cap d^{-1}C^{p+1}\Lambda^{p+q+1}(M)}{C^{p+1}\Lambda^{p+q}(M)+dC^p\Lambda^{p+q-1}(M)},$$

$$E_1^{p,q}(I\mathcal{M}) = \frac{IC^p\Lambda^{p+q}(M)\cap d^{-1}IC^{p+1}\Lambda^{p+q+1}(M)}{IC^{p+1}\Lambda^{p+q}(M)+dIC^p\Lambda^{p+q-1}(M)},$$

where $p,q=0,1,\ldots.$

We temporarily denote the right-hand sides of the equalities in Proposition 9.14 by $E_0^{p,q}(\mathcal{F})_{ext}$, $E_0^{p,q}(\mathcal{M}_F)_{ext}$ and $E_0^{p,q}(N\mathcal{F})_{ext}$, respectively.

Proposition 9.17. *The exterior differentiation induces the liner mappings*

$$d_{0,ext}^{p,q} : E_0^{p,q}(\mathcal{F})_{ext} \longrightarrow E_0^{p,q+1}(\mathcal{F})_{ext},$$

$$d_{0,ext}^{p,q} : E_0^{p,q}(\mathcal{M}_F)_{ext} \longrightarrow E_0^{p,q+1}(\mathcal{M}_F)_{ext},$$

$$d^{p,q}_{0,ext} : E^{p,q}_0(N\mathcal{F})_{ext} \to E^{p,q+1}_0(N\mathcal{F})_{ext},$$

for $p,q=0,1,\ldots.$

Proof. It is enough to use the inclusions

$$dC^p\Lambda^{p+q}(M) \subset C^p\Lambda^{p+q+1}(M), \quad dIC^p\Lambda^{p+q}(M) \subset IC^p\Lambda^{p+q+1}(M),$$

$$dRC^p\Lambda^{p+q}(M) \subset RC^p\Lambda^{p+q+1}(M) + IC^{p+1}\Lambda^{p+q+1}(M)$$

$$\subset RC^p\Lambda^{p+q+1}(M) + C^{p+1}\Lambda^{p+q+1}(M)$$

(see Lemma 9.13).

Corollary 9.18. *There is the local external representation*

$$E^{p,q}_1(\mathcal{F}) = \frac{C^p\Lambda^{p+q}(M)\cap d^{-1}(C^{p+1}\Lambda^{p+q+1}(M)+IC^p\Lambda^{p+q+1}(M))}{C^{p+1}\Lambda^{p+q}(M)+dC^p\Lambda^{p+q-1}(M)+IC^p\Lambda^{p+q}(M)}.$$

Proof. To prove the representation one should use the equality $i^*\circ d=d\circ i^*$ and to verify that the linear mapping $d^{p,q}_0$ passes into $d^{p,q}_{0,ext}$ when the spaces $E^{p,q}_0(\mathcal{F})$ and $E^{p,q}_0(\mathcal{F})_{ext}$ identified.

For the spaces $E^{p,q}_0(\mathcal{M}_F)$ and $E^{p,q}_0(N\mathcal{F})$ there is no alternative mapping and we put by definition

$$E^{p,q}_1(\mathcal{M}_F) = \frac{\mathrm{Ker}d^{p,q}_{0,ext}}{\mathrm{Im}d^{p,q-1}_{0,ext}}$$

$$= \frac{C^p\Lambda^{p+q}(M)\cap d^{-1}(C^{p+1}\Lambda^{p+q+1}(M)+RC^p\Lambda^{p+q+1}(M))}{C^{p+1}\Lambda^{p+q}(M)+dC^p\Lambda^{p+q-1}(M)+RC^p\Lambda^{p+q}(M)},$$

$$E^{p,q}_1(N\mathcal{F}) = \frac{\mathrm{Ker}d^{p,q}_{0,ext}}{\mathrm{Im}d^{p,q-1}_{0,ext}}$$

$$= \frac{IC^p\Lambda^{p+q}(M)\cap d^{-1}(IC^{p+1}\Lambda^{p+q+1}(M)+RC^p\Lambda^{p+q+1}(M))}{IC^{p+1}\Lambda^{p+q}(M)+dIC^p\Lambda^{p+q-1}(M)+RC^p\Lambda^{p+q}(M)}.$$

It is easy to check that the mappings j^* and ν^*, defined on the zero-order terms, induce corresponding mappings on the first-order terms.

Theorem 9.19. *The following locally exact sequence of the linear spaces is defined:*

$$E_1^{p,q}(\mathcal{F}) \xleftarrow{\ j^*\ } E_1^{p,q}(\mathcal{M}_F) \xleftarrow{\ \nu^*\ } E_1^{p,q}(N\mathcal{F}) \longleftarrow NE_1^{p,q}(N\mathcal{F}) \longleftarrow 0,$$

where

$$NE_1^{p,q}(N\mathcal{F}) = \frac{IC^p\Lambda^{p+q}(M)\cap(C^{p+1}\Lambda^{p+q}(M)+dC^p\Lambda^{p+q-1}(M))}{IC^{p+1}\Lambda^{p+q}(M)+dIC^p\Lambda^{p+q-1}(M)+RC^p\Lambda^{p+q}(M)},$$

$p,q=0,1,\ldots.$

Proof. The proof can be done by the easy direct computations of $\mathrm{Ker}j^*$, $\mathrm{Im}\nu^*$ and $\mathrm{Ker}\nu^*$. Note that $\mathrm{Ker}\nu^*$ is trivial in the zero-order.

10. Trivial Ambient Differential Manifolds

A standard procedure allows us to realize a given system of nonlinear PDE as a reduction of an appropriate trivial differential manifold from the class studied in Section 6, in the same way as usual manifolds are often realized as surfaces in Euclidean spaces $\mathbb{R}^n$.

Let $\mathcal{M}=(M,D)$ be a trivial differential manifold, where $M=\mathbb{R}^m\times\mathbb{R}^a_K$ is the linear space with global coordinates $x=(x^\mu)$, $\mathbf{w}=(w^\alpha_k)$, $K=\mathbb{N}^m$, $\mu=1,\ldots,m$, $\alpha=1,\ldots,a$, $k\in K$, and the Cartan distribution D is determined by basis fields

$$D_\mu = \frac{\partial}{\partial x^\mu} + \sum_{\alpha=1}^{a} \sum_{k\in K} w^\alpha_{k+(\mu)} \frac{\partial}{\partial w^\alpha_k}, \quad \mu = 1,\ldots,m,$$

(cf. Section 6, we use the standard multi-index notation).

Suppose that we are given a system of nonlinear partial differential equations (PDE)

$$F^\sigma\left(x, w, \frac{\partial w}{\partial x}, \ldots, \frac{\partial^r w}{\partial w^r}\right) = 0, \quad \sigma = 1,\ldots,s,$$

or $\{F=0\}$, for brevity. Here:

$x = (x^1,\ldots,x^m)$ are independent variables;

$w = (w^1,\ldots,w^a)$ are dependent variables;

$\dfrac{\partial^p w}{\partial x^p} = \left(\dfrac{\partial^p w^\alpha}{(\partial x^1)^{k_1}\dots(\partial x^m)^{k_m}},\ |k|=p\right)$ are partial derivatives of the order $p=0,\dots,r$;

$F = (F^1,\dots,F^s)$, where $F^\sigma(x,\mathbf{w})$, $\sigma=1,\dots,s$, are smooth functions on M determined by the correspondence

$$w_k^\alpha \longleftrightarrow \frac{\partial^p w^\alpha}{(\partial x^1)^{k_1}\dots(\partial x^m)^{k_m}};$$

m is a number of independent variables;

a is a number of dependent variables;

s is a number of equations;

r is an order of the system.

Remind that a smooth function $w=\varphi(x)\in\mathcal{C}^\infty(\mathcal{O})^a$, $\mathcal{O}\subset\mathbb{R}^m$, is called a local solution of the system $\{F=0\}$ if

$$F^\sigma(x,\mathbf{w})\Big|_{w=\varphi(x)} = 0, \quad x\in\mathcal{O},\ \sigma = 1,\dots,s,$$

where

$w=\varphi(x)$ stands for $w_k^\alpha = \left(\dfrac{\partial}{\partial x}\right)^k \varphi^\alpha(x)$, $\alpha = 1,\dots,a$, $k\in K$.

We denote by $\mathrm{Sol}\{F=0\}$ the set of all local solutions of the system $\{F=0\}$.

Consider a variety

$$\mathcal{F} = \left\{x,\mathbf{w} \,:\, D_k F^\sigma(x,\mathbf{w}) = 0,\ k\in K,\ \sigma = 1,\dots,s\right\}$$

of M, defined as a solution set of infinite system of algebraic (i.e., not differential) equations $D_k F^\sigma(x,\mathbf{w})=0$ (each equation is called a *prolongation of the original system* $F^\sigma(x,\mathbf{w})=0$, and the whole infinite system is called *the infinitely prolongated system*).

Remind that every smooth function $w=\varphi(x)\in\mathcal{C}^\infty(O)^a$, $O\subset\mathbb{R}^m$, defines an integral submanifold

$$S(\varphi) = \left\{x,\mathbf{w} \ :\ w_k^\alpha = \left(\frac{\partial}{\partial x}\right)^k \varphi^\alpha(x),\ \alpha=1,\dots,a,\ k\in K\right\},$$

of $\mathcal{M}$ (see Proposition 6.1). The variety F has the remarkable property: *a smooth function $w=\varphi(x)$ is a local solution of the system $\{F=0\}$* (i.e., $\varphi\in\mathrm{Sol}\{F=0\}$) *iff the corresponding integral submanifold $S(\varphi)$ lies in F* (i.e., $S(\varphi)\subset F$). Because of this property the variety F is often called *the solution variety of the system $\{F=0\}$*. Natural questions arising here are the following: when for every point $z\in F$ there exists a local solution $\varphi\in\mathrm{Sol}\{F=0\}$ such that the integral submanifold $S(\varphi)$ passes through the point z (*the existence theorem*), and when such solution is unique (*the uniqueness theorem*)? As far as I know, these problems are not solved yet in any general setting. Note that close problems are studied in Olver [1,Ch.2] in the real-analytic category (when functions F^σ and solutions φ are real-analytic, i.e., locally represented by convergent power series). On the other hand, the formal theory of PDE is concerned with formal power series solutions, see Pommaret [1]. Further, in the general case, when no restrictions are imposed on the system $\{F=0\}$, we have no reasons to expect that the variety F will have a structure of a submanifold of the manifold M (cf. the discussion in Section 0.4). Note that F can be even empty if the original system $\{F=0\}$ is not compatible.

To get a better understanding of the situation, let us treat a pair of examples from Olver [1,Ch.2] from our point of view.

Example 10.1 (Olver [1,(2.117)). Consider the over-determined system

$$\begin{cases} F^1 = w_x - yw = 0, \\ F^2 = w_y = 0. \end{cases}$$

Here M has coordinates $x^1=x$, $x^2=y$, w_{ij}, $i,j\in\mathbb{N}$,

$$w_{ij} \longleftrightarrow \frac{\partial^{i+j} w}{\partial x^i \partial y^j}.$$

Cross-differentiation shows

$$0 = w_{xy} = (yw)_y = w + yw_y = w,$$

and hence,

$$F = \left\{x, y, w_{ij} : w_{ij}=0, \quad i,j \in \mathbb{N}\right\}.$$

Thus, from one side the system has only the trivial solution $w=\varphi(x,y)\equiv 0$, i.e., it is obviously poor from the traditional point of view. On the other side, the corresponding variety F is a 2-dimensional plane in M, and through every point of F there passes one and only one integral submanifold, i.e., F itself!

Example 10.2 (Olver [1,(2.119)). Consider the evolution system

$$\begin{cases} F^1 = u_t - h(t,x)v_x = 0, \\ F^2 = v_t + h(t,x)u_x = 0. \end{cases}$$

Here M has coordinates t, x, u_{ij}, v_{ij}, $i, j \in \mathbb{N}$,

$$u_{ij} \longleftrightarrow \frac{\partial^{i+j}u}{\partial t^i \partial x^j}, \quad v_{ij} \longleftrightarrow \frac{\partial^{i+j}v}{\partial t^i \partial x^j}.$$

In this case

$$F = \Big\{ t, x, u_{ij}, v_{ij} \ : \ u_{i+1,j} = D_t^i D_x^j\big(hv_{0,1}\big),$$
$$v_{i+1,j} = D_t^i D_x^j\big(-hu_{0,1}\big), \quad i, j \in \mathbb{N} \Big\}$$

is a smooth submanifold of M with global coordinates t, x, u_{0j}, v_{0j}, $i\in\mathbb{N}$, (cf. Section 7), for any smooth function $h(t,x)$, and the situation seems to be perfect. On the other side, Nirenberg [1] have constructed a smooth function $h(t,x)$ such that the corresponding evolution system has only constant solutions in a neighborhood of the origin. In our setting this means that for such function h through each point

$$t=0, x=0, u_{00}=a, v_{00}=b, u_{0j}=0, v_{0j}=0, \ \ j=1,2,\ldots,$$

passes one and only one integral submanifold

$$\Big\{ u_{00}=a, v_{00}=b, u_{0j}=0, v_{0j}=0, \ \ j=1,2,\ldots, \Big\},$$

and no integral submanifold passes through any point

$$t=0, x=0, u_{0j}, v_{0j}, \ \ j \in \mathbb{N},$$

where at least one of the coordinates u_{0j}, v_{0j} differs from zero for $j=1,2,\ldots$.

In spite of these and other similar examples there are sound reasons to expect that if a given system of PDE $\{F=0\}$ describes a well-posed physical, geometrical or some other problem then the corresponding variety F will have a structure of a smooth submanifold of M, and through every point of F will pass an integral submanifold.

In a standard situation, to which we confine ourselves here, coordinates w^α_k, $(\alpha,k)\in a\times K$, can be split in two parts:

$$w^\alpha_k,\ (\alpha,k)\in I, \quad \text{and } w^\alpha_k,\ (\alpha,k)\in I', \quad I\cup I'=a\times K,\ I\cap I'=\varnothing,$$

such that the first group can be considered *free* and the second *restricted*, i.e., coordinates from the first group can be chosen arbitrary and then coordinates from the second group can be determined from the infinitely prolongated system $D_kF^\sigma=0$, $(\sigma,k)\in s\times K$. Further, we suppose that the free coordinates and a part of the functions D_kF^σ, $(\sigma,k)\in J\subset s\times K$, can be chosen as coordinates for a standard defining chart on M, i.e., that there exist transition formulas $w^\alpha_k=W^\alpha_k(x,u,v)$, $(\alpha,k)\in I'$, where $\mathbf{u}=(u^i=w^\alpha_k)$, $i=(\alpha,k)\in I$, $\mathbf{v}=(v^j=D_kF^\sigma)$, $j=(\sigma,k)\in J$, W^α_k are smooth functions (cf. Section 0.4). At last, we suppose that all depended variables $w^\alpha=w^\alpha_0$ are free (i.e., $(\alpha,0)\in I$, $\alpha=1,\dots,a$), and all functions F^σ are chosen as part of v-coordinates (i.e., $(\sigma,0)\in J$, $\sigma=1,\dots,s$). Thus, we come to new coordinates

$$x^\mu,u^i,v^j,\quad \mu = 1,\dots,m,\ i = (\alpha,k) \in I,\ j = (\sigma,k) \in J,$$

on M, which we shall call *F-coordinates.* In F-coordinates

$$F = \left\{x^\mu,u^i,v^j\ :\ v^j=0,\ j \in J\right\},$$

variables $x=(x^\mu)$, $\mathbf{u}=(u^i)$, $\mu=1,\ldots,m$, $i\in I$ are coordinates on the submanifold F, and the basis Cartan fields

$$D_\mu = \frac{\partial}{\partial x^\mu} + \sum_{i\in I} A^i_\mu(x,\mathbf{u},\mathbf{v}) \frac{\partial}{\partial u^i} + \sum_{j\in J} B^j_\mu(x,\mathbf{u},\mathbf{v}) \frac{\partial}{\partial v^j}, \quad \mu = 1,\ldots,m,$$

where

$$A^i_\mu = D_\mu u^i = w^\alpha_{k+(\mu)}, \qquad \mu = 1,\ldots,m, \; i = (\alpha,k) \in I,$$

$$B^j_\mu = D_\mu v^j = D_{k+(\mu)} F^\sigma, \quad \mu = 1,\ldots,m, \; j = (\sigma,k) \in J.$$

In particular,

$$B^j_\mu \Big|_F = B^j_\mu(x,\mathbf{u},0) = 0, \quad \mu = 1,\ldots,m, \; j \in J,$$

by the definition of F. Hence, the restrictions

$$\partial_\mu = D_\mu \Big|_F = \frac{\partial}{\partial x^\mu} + \sum_{i\in I} a^i_\mu(x,\mathbf{u}) \frac{\partial}{\partial u^i}, \quad \mu = 1,\ldots,m,$$

where $a^i_\mu(x,\mathbf{u})=A^i_\mu(x,\mathbf{u},0)$, are tangent to F and determine a Cartan distribution ∂ on F (clear that the fields ∂_μ are pairwise commuting). By this procedure, *the differential manifold* $\mathcal{F}=(F,\partial)$ *is defined and it is a reduction of the trivial differential manifold* $\mathcal{M}=(M,D)$. In the described situation we shall say that the system $\{F=0\}$ has *a standard realization.*

Remind that infinitesimal symmetries of $\mathcal{M}$ are evolution differentiations,

$$\text{Sym}\,\mathcal{M} = \left\{ \mathrm{E}_\Phi = \sum_{\alpha=1}^{a} \sum_{k\in K} \left(D_k \Phi^\alpha(x,\mathbf{w}) \right) \frac{\partial}{\partial w^\alpha_k};\ \Phi \in \mathcal{C}^\infty(M)^a \right\}$$

$$= \left\{ \Phi \in \mathcal{C}^\infty(M)^a \right\}$$

(see Section 6). In the coordinates x,u,v a representing field $E_\Phi \in \mathrm{Sym}\mathcal{M}$ has the form

$$E_\Phi = \sum_{i\in I} U^i \frac{\partial}{\partial u^i} + \sum_{j\in J} V^j \frac{\partial}{\partial v^j},$$

where

$$U^i = E_\Phi(u^i) = D_k \Phi^\alpha, \quad i = (\alpha,k) \in I,$$

and

$$V^j = E_\Phi(v^j) = E_\Phi\left(D_k F^\sigma\right) = D_k\left(E_\Phi(F^\sigma)\right)$$

$$= D_k F_*(\Phi)^\sigma, \qquad j = (\sigma,k) \in J,$$

(remind that $E_\Phi D_\mu = D_\mu E_\Phi$, see Remark 6.3),

$$F_*(\Phi)^\sigma = \sum_{\alpha=1}^{a} \sum_{k\in K} \frac{\partial F^\sigma}{\partial w_k^\alpha} D_k \Phi^\alpha, \quad \sigma = 1,\dots,s,$$

(cf. Section 6). This implies that

$$\mathrm{Sym}\,\mathcal{M}_F = \left\{\varepsilon_\varphi = E_\Phi\Big|_F = \sum_{\alpha=1}^{a} \sum_{k\in K} \left(\partial_k \varphi^\alpha(x,\mathbf{u})\right) \frac{\partial}{\partial w_k^\alpha} \; ; \; \varphi \in \mathcal{C}^\infty(F)^a\right\}$$

$$= \left\{\varepsilon_\varphi = \sum_{i\in I} \bar{U}^i \frac{\partial}{\partial u^i} + \sum_{j\in J} \bar{V}^j \frac{\partial}{\partial v^j}\right\}$$

$$= \left\{\varphi \in \mathcal{C}^\infty(F)^a\right\},$$

where

$$\bar{U}^i = U^i\Big|_F = \partial_k \varphi^\alpha, \qquad i = (\alpha,k) \in I,$$

and

$$\bar{V}^j = V^j\Big|_F = \partial_k \overline{F_*}(\varphi)^\sigma, \quad j = (\sigma,k) \in J,$$

with

$$\overline{F_*}(\varphi)^\sigma = \sum_{\alpha=1}^{a} \sum_{k\in K} \left(\frac{\partial F^\sigma}{\partial w_k^\alpha}\right)\Bigg|_F \partial_k\varphi^\alpha, \quad \varphi \in \mathcal{C}^\infty(F)^a, \ \sigma = 1,\ldots,s.$$

A vector field $\varepsilon_\varphi \in \mathrm{Sym}\mathcal{M}_F$ will be tangent to F (i.e., will belong to $\mathcal{T}(F)$) iff $\overline{F_*}(\varphi)=0$, and we get the representations (cf. Section 9)

$$\mathrm{Sym}_{ext}\mathcal{F} = \left\{\varphi \in \mathcal{C}^\infty(F)^a \ : \ \overline{F_*}(\varphi)=0\right\},$$

$$\mathrm{Sym}\ N\mathcal{F} = \left\{\psi=\overline{F_*}(\varphi) \in \mathcal{C}^\infty(F)^s \ ; \ \varphi \in \mathcal{C}^\infty(F)^a\right\}.$$

It is worth to recall that

$$\mathrm{Sym}\ \mathcal{F} = \left\{\xi \in \mathcal{C}^\infty(F)^I \ : \ \partial_\mu\xi^i - \sum_{r\in I} \frac{\partial a_\mu^i}{\partial u^r}\xi^r = 0, \ \mu=1,\ldots,m, \ i\in I\right\}$$

(cf. Proposition 3.1).

The normal projection $\nu:Sym\mathcal{M}_F \longrightarrow SymN\mathcal{F}$ is given here by the differential operator $\nu=\overline{F_*}:\mathcal{C}^\infty(F)^a \longrightarrow \mathcal{C}^\infty(F)^s$ acting according to the rule

$$\mathcal{C}^\infty(F)^a \ni \varphi \mapsto \overline{F_*}(\varphi) \in \mathcal{C}^\infty(F)^s.$$

The expressions for the basis Cartan 1-forms δu^i and δv^j and their restrictions on F have nothing special in this case with respect to the general standard defining chart (see Section 9).

Example 10.3 (Duzhin, Tsujishita [1]). Consider the BBM equation

$$w_t + w_x + ww_x - w_{txx} = 0.$$

Here the ambient manifold M has coordinates t,x,w_k, $k=(p,q)\in K=\mathbb{N}^2$,

$$w_{pq} \longleftrightarrow \frac{\partial^{p+q} w}{\partial t^p \partial x^q}, \quad p,q \in \mathbb{N},$$

the basis Cartan fields D_t and D_x are standard. Further,

$$F = \left\{ t,x,w_k \; : \; D_t^p D_x^q (w_{12}-f)=0, \;\; p,q \in \mathbb{N} \right\},$$

where $f=w_{10}+w_{01}+w_{00}w_{01}$. We choose coordinates $u_i=w_{pq}$, $i=(p,q)\in I$, to be free, where

$$I = \{p=0; q=0,1\} \cup \{p=1,2,\ldots; q=0,1\} \cup \{p=0; q=2,3,,\ldots\},$$

and the other coordinates w_k, $k\in I'=K\setminus I$, to be restricted. Let us put

$$v_j = D_t^p D_x^q (w_{12}-f) = w_{p+1,q+2} - D_t^p D_x^q f$$

$$= w_{p+1,q+2} - f_{pq}, \quad j=(p,q) \in J=K=\mathbb{N}^2,$$

where $f_{pq}(\ldots,w_{p+1,q},w_{p,q+1})$, dots denote the junior arguments. Then, using the recursion formula

$$w_{pq} = v_{p-1,q-2} + f_{p-1,q-2}, \quad k=(p,q) \in I',$$

we can express $w_k=W_k(u,v)$, $k\in K$, (for $k\in I$ we simply have $w_k=u_k$). Thus, the variables $t,x,u=(u_i; i\in I), v=(v_j; j\in J)$ are F-coordinates on M, such that

$$F = \left\{ t,x,u,v \; : \; v_j=0, \quad j \in J \right\},$$

and so F is a submanifold of M. Further, in these coordinates

$$D_t = \frac{\partial}{\partial t} + \sum_{p=0}^{\infty} \sum_{q=0}^{1} u_{p+1,q} \frac{\partial}{\partial_{u_{pq}}} + \sum_{q=2}^{\infty} W_{1q}(u,v) \frac{\partial}{\partial_{u_{0q}}}$$

$$+ \sum_{p,q=0}^{\infty} v_{p+1,q} \frac{\partial}{\partial v_{pq}},$$

$$D_x = \frac{\partial}{\partial x} + \sum_{q=0}^{\infty} u_{0,q+1} \frac{\partial}{\partial u_{0q}} + \sum_{p=1}^{\infty} u_{p1} \frac{\partial}{\partial u_{p0}} + \sum_{p=1}^{\infty} W_{p2}(u,v) \frac{\partial}{\partial u_{p1}}$$

$$+ \sum_{p,q=0}^{\infty} v_{p,q+1} \frac{\partial}{\partial v_{pq}},$$

and their restrictions on F are

$$\partial_t = D_t\Big|_F = \frac{\partial}{\partial t} + \sum_{p=0}^{\infty} \sum_{q=0}^{1} u_{p+1,q} \frac{\partial}{\partial_{u_{pq}}} + \sum_{q=2}^{\infty} U_{1q}(u) \frac{\partial}{\partial_{u_{0q}}},$$

$$\partial_x = D_x\Big|_F = \frac{\partial}{\partial x} + \sum_{q=0}^{\infty} u_{0,q+1} \frac{\partial}{\partial u_{0q}} + \sum_{p=1}^{\infty} u_{p1} \frac{\partial}{\partial u_{p0}} + \sum_{p=1}^{\infty} U_{p2}(u) \frac{\partial}{\partial u_{p1}},$$

where

$$U_{1q}(u) = W_{1q}(u,0) \text{ and } U_{p2}(u) = W_{p2}(u,0).$$

This completes the realization of BBM equation as the reduction $\mathcal{F}=(F,\partial_t,\partial_x)$ of the trivial differential manifold $\mathcal{M}=(M,D_t,D_x)$.

Example 10.4 (Evolution equations). Consider an evolution system of PDE

$$F^\alpha = w_t^\alpha - f^\alpha\left(t,x,w,\frac{\partial w}{\partial x},\ldots,\frac{\partial^r w}{\partial x^r}\right) = 0, \quad \alpha = 1,\ldots,a,$$

(cf. Section 7). Here we take $\mathcal{M}$ to be a manifold with coordinates

$t,x=(x^\mu),\mathbf{w}=(w_k^\alpha),\ \mu=1,\ldots,m,\ \alpha=1,\ldots,a,\ k\in\mathbb{N}^{1+m},$

$$w_k^\alpha \longleftrightarrow \frac{\partial^{p+|i|}w^\alpha}{\partial t^p(\partial x)^i}, \quad \alpha = 1,\dots,a, \ k=(p,i), \ p\in\mathbb{N}, \ i\in\mathbb{N}^m.$$

The basis Cartan fields D_t, D_μ, $\mu=1,\dots,m$, have the standard form.

Now,

$$F = \left\{t,x,\mathbf{w} \ : \ D_t^p D_i F^\alpha(t,x,\mathbf{w})=0, \ p\in\mathbb{N}, \ i\in\mathbb{N}^m, \ \alpha=1,\dots,a\right\},$$

and we choose the coordinates

$$u_i^\alpha = w_{0i}^\alpha, \quad \alpha = 1,\dots,a, \ i \in \mathbb{N}^m,$$

to be free, and all the other will be restricted:

$$w_{p+1,i}^\alpha = D_t^p D_i f^\alpha, \quad \alpha = 1,\dots,a, \ p \in \mathbb{N}, \ i \in \mathbb{N}^m.$$

Let us put

$$v_k^\alpha = D_t^p D_i F^\alpha = w_{p+1,i}^\alpha - D_t^p D_i f^\alpha, \ \alpha = 1,\dots,a, \ k=(p,i) \in \mathbb{N}^{1+m},$$

then, using the recursion formula

$$w_{p+1,i}^\alpha = v_{pi}^\alpha + D_t^p D_i f^\alpha, \ \alpha = 1,\dots,a, \ (p,i) \in \mathbb{N}^{1+m},$$

we can express $w_k^\alpha = W_k^\alpha(t,x,\mathbf{u},\mathbf{v})$, $\alpha=1,\dots,a$, $k\in\mathbb{N}^{1+m}$, (for $k=(0,i)$ we simply have $w_{0i}^\alpha=u_i^\alpha$). Thus, we get F-coordinates on M: t, $x=(x^\mu)$, $\mathbf{u}=(u_i^\alpha)$, $\mathbf{v}=(v_k^\alpha)$, $\mu=1,\dots,m$, $\alpha=1,\dots,a$, $i\in I=\mathbb{N}^m$, $k\in K=\mathbb{N}^{1+m}$, such that

$$F = \left\{t,x,\mathbf{u},\mathbf{v} \ : \ v_k^\alpha=0, \ \alpha=1,\dots,a, \ k\in K\right\},$$

and so F is a submanifold of M. Further, in these coordinates the basis Cartan fields on M have the form

$$D_\mu = \frac{\partial}{\partial x^\mu} + \sum_{\alpha,i} u_{i+(\mu)}^\alpha \frac{\partial}{\partial u_i^\alpha} + \sum_{\alpha,p,i} v_{p,i+(\mu)}^\alpha \frac{\partial}{\partial v_{pi}^\alpha}, \quad \mu = 1,\dots,m,$$

$$D_t = \frac{\partial}{\partial t} + \sum_{\alpha, i} \left(v^\alpha_{0i}+D_i f^\alpha\right) \frac{\partial}{\partial u^\alpha_i} + \sum_{\alpha, p, i} v^\alpha_{p+1, i} \frac{\partial}{\partial v^\alpha_{pi}},$$

and their restrictions on F are

$$\partial_\mu = D_\mu\Big|_F = \frac{\partial}{\partial x^\mu} + \sum_{\alpha, i} u^\alpha_{i+(\mu)} \frac{\partial}{\partial u^\alpha_i}, \quad \mu = 1,\dots,m,$$

$$\partial_t = D_t\Big|_F = \frac{\partial}{\partial t} + \sum_{\alpha, i} \partial_i f^\alpha \frac{\partial}{\partial u^\alpha_i},$$

where $\partial_i f^\alpha = f^\alpha_i(t,x,\mathbf{u})$ are smooth functions on F (cf. Section 7). This completes the realization of the given evolution system of PDE as the reduction $\mathcal{F}=(F,\partial_t,\partial_x)$ of the trivial differential manifold $\mathcal{M}=(M,D_t,D_x)$.

Note that in this case

$$\begin{aligned}\overline{F_*}(\varphi)^\alpha &= \partial_t\varphi^\alpha - \sum_{\beta=1}^{a} \sum_{i\in I} \frac{\partial f^\alpha}{\partial u^\beta_i} \partial_i \varphi^\beta \\ &= \partial_t\varphi^\alpha - f_*(\varphi)^\alpha, \quad \alpha = 1,\dots,a,\end{aligned}$$

(cf. Proposition 7.3), hence, for evolution systems

$$\operatorname{Sym}\mathcal{F} = \operatorname{Sym}_{ext}\mathcal{F} = \left\{\varphi \in \mathcal{C}^\infty(F)^a \ : \ \overline{F_*}(\varphi)=0\right\}.$$

Let us return to the general case. Let $\mathcal{M}=(M,D)$ be a trivial differential manifold and $\{F=0\}$ be a system of nonlinear PDE defined above. Remind that two systems of equations (not necessary differential) are considered equivalent if they have the same set of solutions. In our situation we call two systems of PDE *equivalent* if they have the same solution variety (clear that in this case they have the same set of local solutions). Thus, the system from Example 10.1 is equivalent to the single algebraic (i.e., not differential) equation $F=w=0$. In the geometrical setting all equivalent systems have the same description, *only the solution variety is essential*. But for actual calculations the explicit form the system under investigation is also important.

Let us introduce some characteristics of the given system $\{F=0\}$. Denote by DO the set of all linear differential operators on the algebra $\mathcal{C}^\infty(M)$, i.e., the set of all polynomials

$$P = \sum_{k\in K} a^k(x,\mathbf{w})D_k$$

in D's with coefficients $a^k\in\mathcal{C}^\infty(M)$, where $k\in K=\mathbb{N}^m$ (see Appendix), the action $P:\mathcal{C}^\infty(M)\longrightarrow\mathcal{C}^\infty(M)$ is defined as follows

$$\mathcal{C}^\infty(M) \ni \Psi \mapsto P(\Psi) = \sum_{k\in K} a^k D_k\Psi \in \mathcal{C}^\infty(M).$$

Clear, DO is a noncommutative algebra with the multiplication given by the operator composition rule. Further, denote by DO_S the left DO-module of all linear differential operators $P=(P_\sigma)$ acting from $\mathcal{C}^\infty(M)^S$ into $\mathcal{C}^\infty(M)$,

$$\mathcal{C}^\infty(M)^S \ni \Psi=(\Psi^\sigma) \mapsto P(\Psi)=P_\sigma(\Psi^\sigma) \in \mathcal{C}^\infty(M).$$

Now, a differential operator $P\in\mathrm{DO}_S$ is called *a differential identity of the system $\{F=0\}$* if $P(F)\equiv 0$ on M, where $F=(F^\sigma)$ are defining functions of the system. The left DO-module of all differential identities of the system $\{F=0\}$ we denote by DI_F, i.e.,

$$\mathrm{DI}_F = \left\{ P \in \mathrm{DO}_s \ : \ P(F)=0 \text{ on } M \right\}.$$

Every system $\{F=0\}$, $s>1$, has a lot of differential identities. For example, any system $\{F^1=0, F^2=0\}$ has the evident differential identity $P=(P_1=F^2, P_2=-F^1)$, because in this case

$$P_1(F^1)+P_2(F^2) = F^2F^1-F^1F^2 \equiv 0$$

(here $s=2$, the operators P_1, P_2 have the zero order). Thus, there is a question of nontriviality of differential identities. We shall call a differential identity

$$P = \left(\sum_k a^k_\sigma D_k \right) \in \mathrm{DI}_F$$

essential if the coefficients a^k_σ don't vanish on the solution variety F simultaneously, i.e., if

$$\sum_{\sigma,k} \left(a^k_\sigma(x,\mathbf{w}) \right)^2 > 0 \quad \text{for every point } (x,\mathbf{w}) \in F.$$

Note that the differential identity from the above example is not essential, since there

$$\left(F^2 \right)^2 + \left(-F^1 \right)^2 \equiv 0 \text{ on } F.$$

We shall call the system $\{F=0\}$ *well-determined* if it has no essential differential identities (cf. definitions of over- and under-determined systems in Olver [1, Definition 2.86]).

Let

$$\mathcal{I}_F = \left\{ \Phi \in \mathcal{C}^\infty(M) \ : \ \Phi=0 \text{ on } F \right\}$$

be the defining ideal of the solution variety F (see Sections 0.4 and 8), and denote by $\mathrm{DO}_s(F)$ the differential ideal of $\mathcal{C}^\infty(M)$, generated by $F=(F^\sigma)\in\mathcal{C}^\infty(M)^s$, i.e.,

$$\mathrm{DO}_s(F) = \left\{\Phi=P(F) \in \mathcal{C}^\infty(M) \ ; \ P \in \mathrm{DO}_s\right\}.$$

Clear that $\mathrm{DO}_s(F)\subset\mathcal{I}_F$. We shall call the system $\{F=0\}$ *generating* if the functions F^σ, $\sigma=1,\dots,s$, generate the defining ideal of the solution variety F, i.e., if

$$\mathrm{DO}_s(F) = \mathcal{I}_F.$$

Lemma 10.5. *Let $f(u,v)\in\mathcal{C}^\infty(U\times V)$, where $U\subset\mathbb{R}^q$ is an open set and $0\in V\subset\mathbb{R}$ is an open interval. Suppose that $f(u,0)=0$ then*

$$f(u,v)=g(u,v)v, \quad g(u,v)\in\mathcal{C}^\infty(U\times V).$$

Proof. It suffices to show that the function

$$g(u,v) = \frac{f(u,v)}{v}, \ v\neq 0, \quad g(u,0) = 0,$$

belongs to $\mathcal{C}^\infty(U\times V)$, and this can be easily done.

Lemma 10.6. *Let $f(u,v)\in\mathcal{C}^\infty(U\times V)$, where $U\subset\mathbb{R}^q$ and $0\in V\subset\mathbb{R}^r$ are open sets. Suppose that $f(u,0)=0$ then*

$$f(u,v) = \sum_{\rho=1}^{r} g_\rho(u,v)v^\rho, \quad g_\rho(u,v) \in \mathcal{C}^\infty(U\times V), \ \rho = 1,\dots,r.$$

Proof. The proof can be done, using the induction in r, the case $r=1$ is already treated in Lemma 10.5.

Lemma 10.7. *Let $f(u,v)\in\mathcal{C}^\infty(U\times V)$, where $U\subset\mathbb{R}^I$ and $0\in V\subset\mathbb{R}^J$ are open sets, the index sets I, J can be finite or countable* (see

Section 0.1). *Suppose that* $f(u,0)=0$ *then*

$$f(u,v) = \sum_{j\in J} g_j(u,v)v^j, \quad g_j(u,v) \in \mathcal{C}^\infty(U\times V), \ j \in J.$$

Proof. It is enough to note that, by the definition, every smooth function $f(u,v)\in\mathcal{C}^\infty(U\times V)$ actually depends on a finite number of the variables u^i, v^j, $i\in I, j\in J$, and use Lemma 10.6.

Proposition 10.8. *If a system* $\{F=0\}$ *has a standard realization then it is well-determined iff* $J=s\times K$.

Proof. Indeed, suppose that $s\times K\backslash J\neq\emptyset$, and let $(\sigma',k')\in s\times K\backslash J$. Then $D_{k'}F^{\sigma'}=f(x,\mathbf{u},\mathbf{v})$ and $f(x,\mathbf{u},0)=0$ by the definition of the solution variety F. Hence, by Lemma 10.7,

$$f(x,\mathbf{u},\mathbf{v}) = \sum_{j\in J} f_j(x,\mathbf{u},\mathbf{v})v^j,$$

and we have the essential differential identity

$$D_{k'}F^{\sigma'} - \sum_{(\sigma,k)\in J} f^k_\sigma(x,\mathbf{w})D_kF^\sigma.$$

On the other hand, suppose that there exists a differential identity

$$\sum_{\sigma,k} a^k_\sigma(x,\mathbf{w})D_kF^\sigma(x,\mathbf{w}) \equiv 0 \text{ on } M,$$

where a coefficient $a^{\sigma'}_{k'}\neq 0$ in a neighborhood of a point $(x_0,\mathbf{w}_0)\in F$. Then

$$D_{k'}F^{\sigma'} = \sum_{(\sigma,k)\neq(\sigma',k')} \left(-\frac{a^k_\sigma}{a^{k'}_{\sigma'}}\right) D_kF^\sigma,$$

that is,

$$D_{k'}F^{\sigma'}(x,\mathbf{u},\mathbf{v}) = \sum_{j\neq j'} f_j(x,\mathbf{u},\mathbf{v})v^j,$$

and clear that the index $j'=(\sigma',k')$ cannot belong to the index set J. This ends the proof.

Remark 10.9. Thus, in a standard realization of a well-determined system $\{F=0\}$ the index set $J=s\times K$, i.e., $j=(\sigma,k)\in s\times K$, $v^j=v^\sigma_k$, and

$$B^\sigma_{k,\mu}=D_\mu v^\sigma_k=v^\sigma_{k+(\mu)}, \quad \sigma = 1,\dots,s,\ k \in K,\ \mu = 1,\dots,m.$$

Proposition 10.10. *If a system $\{F=0\}$ has a standard realization then it is generating.*

Proof. Indeed, let a smooth function $\Phi\in\mathcal{I}_F$, i.e., $\Phi(x,\mathbf{u},0)=0$. Then, by Lemma 10.7,

$$\Phi(x,\mathbf{u},\mathbf{v}) = \sum_{j\in J} a_j(x,\mathbf{u},\mathbf{v})v^j,$$

or in the old coordinates

$$\Phi(x,\mathbf{w}) = \sum_{(\sigma,k)\in J} a^k_\sigma(x,\mathbf{w})D_kF^\sigma(x,\mathbf{w}),$$

and hence, $\Phi\in\mathrm{DO}_s(F)$.

Remark 10.11. Note that every evolution system is both generating and well-determined.

Lemma 10.12. *Let a system $\{F=0\}$ be generating. Suppose that $\mathrm{E}_\Phi(F^\sigma)\in\mathcal{I}_F$, $\sigma=1,\dots,s$, for some $\Phi\in\mathcal{C}^\infty(M)^a$, then $\mathrm{E}_\Phi:\mathcal{I}_F\to\mathcal{I}_F$.*

Proof. Remind that evolution differentiations are characterized by the property: $D_\mu\mathrm{E}_\Phi=\mathrm{E}_\Phi D_\mu$, $\mu=1,\dots,m$ (see Remark 6.3), so in our case $E_{\Phi}(D_k F^{\sigma}) \in \mathcal{I}_F$ for all $k \in K$, $\sigma = 1,\dots,s$. Now, let a function $\chi \in \mathcal{I}_F$, then

$$\chi = \sum_{k,\sigma} a^k_{\sigma}(x,\mathbf{w}) D_k F^{\sigma}, \quad a^k_{\sigma} \in \mathcal{C}^{\infty}(M),$$

because the system $\{F=0\}$ is generating. Hence

$$E_{\Phi}(\chi) = \sum_{k,\sigma} \left(E_{\Phi}\left(a^k_{\sigma}\right) D_k F^{\sigma} + a^k_{\sigma} E_{\Phi}\left(D_k F^{\sigma}\right)\right) \in \mathcal{I}_F,$$

and the proof is completed.

Proposition 10.13. *Suppose that a system $\{F=0\}$ has a standard realization. Let $\phi \in \mathrm{Sym}_{ext}\mathcal{F}$, then $E_{\Phi} : \mathcal{I}_F \longrightarrow \mathcal{I}_F$ for every $\Phi \in \mathcal{C}^{\infty}(M)^a$ such that $\Phi\Big|_F = \phi$.*

Proof. Indeed, let $\Phi \in \mathcal{C}^{\infty}(M)$ and $\Phi\Big|_F = \phi$. Then, by definition,

$$E_{\Phi}(F^{\sigma})\Big|_F = F_*(\Phi)^{\sigma}\Big|_F = \overline{F_*}(\phi)^{\sigma} = 0,$$

i.e., $E_{\Phi}(F^{\sigma}) \in \mathcal{I}_F$ for all $\sigma = 1,\dots,s$. To finish the proof now it is enough to use Proposition 10.10 and Lemma 10.12.

11. The Characteristic Mapping

Let the differential manifold $\mathcal{F}=(F,CTF)$ under study be a reduction of a ambient differential manifold $\mathcal{M}=(M,CTM)$. We shall suppose that *the manifolds $\mathcal{F}$ and $\mathcal{M}$ are covered by a single standard defining chart* (see Section 9). Consider the linear mappings induced in the linear spaces $E_1^{p,q}$ by the exterior differentiation. The following mappings are defined according to the general rules for constructing spectral sequences (see Section 5):

$$d_1^{p,q} : E_1^{p,q}(\mathcal{F}) \longrightarrow E_1^{p+1,q}(\mathcal{F}),$$
$$d_1^{p,q} : E_1^{p,q}(\mathcal{M}) \longrightarrow E_1^{p+1,q}(\mathcal{M}),$$
$$d_1^{p,q} : E_1^{p,q}(I\mathcal{M}) \longrightarrow E_1^{p+1,q}(\mathcal{M}),$$

where $p,q=0,1,\ldots$.

The external representations of the spaces $E_1^{p,q}(\mathcal{F})$ (see Corollary 9.18) enable us to define additional mappings on $E_1^{p,q}(\mathcal{F})$. For the time being, we denote the right-hand side of the corresponding equality in Corollary 9.18 by $E_1^{p,q}(\mathcal{F})_{ext}$. Suppose that an equivalence class $[\omega]\in E_1^{p,q}(\mathcal{F})_{ext}$, where a form $\omega\in C^p\Lambda^{p+q}(M)$, the exterior derivative $d\omega\in C^{p+1}\Lambda^{p+q+1}(M)+IC^p\Lambda^{p+q+1}(M)$. The last inclusion leads to the splitting $d\omega=d\omega+d\omega$, where the forms

$d_t\omega\in C^{p+1}\Lambda^{p+q+1}(M)$ and $d_n\omega\in IC^{p}\Lambda^{p+q+1}(M)$ are defined modulo the intersection $C^{p+1}\Lambda^{p+q+1}(M)\cap IC^{p}\Lambda^{p+q+1}(M)=IC^{p+1}\Lambda^{p+q+1}(M)$. Further, $dd_t\omega=-dd_n\omega\in C^{p+1}\Lambda^{p+q+2}(M)\cap IC^{p}\Lambda^{p+q+2}(M)=IC^{p+1}\Lambda^{p+q+2}(M)$, since $dd\omega=0$. It is easy to verify that the rule $d_t[\omega]=[d_t\omega]$, where the equivalence class on the right-hand side is taken in $E_1^{p+1,q}(\mathcal{F})_{ext}$, determines a mapping

$$d_t^{p,q} : E_1^{p,q}(\mathcal{F})_{ext} \to E_1^{p+1,q}(\mathcal{F})_{ext}.$$

Moreover, the mapping $d_t^{p,q}$ passes into the mapping $d_1^{p,q}$ when the spaces $E_1^{p,q}(\mathcal{F})_{ext}$ and $E_1^{p,q}(\mathcal{F})$ are identified. Similarly, the rule $d_n[\omega]=[d_n\omega]$, where the equivalence class in the right-hand side is taken in $E_1^{p,q+1}(I\mathcal{M})$, determines a mapping

$$d_n^{p,q} : E_1^{p,q}(\mathcal{F}) \to E_1^{p,q+1}(I\mathcal{M})$$

(we have identified the spaces $E_1^{p,q}(\mathcal{F})_{ext}$ and $E_1^{p,q}(\mathcal{F})$).

Proposition 11.1. *There is the equality*

$$d_1^{p,q+1}\circ d_n^{p,q} + d_n^{p+1,q} = 0,$$

where $p,q=0,1,\ldots$.

Proof. It is enough to use the familiar property $d\circ d=0$ of the exterior differentiation.

Denote by $k^{p,q}:E_1^{p,q}(I\mathcal{M})\to E_1^{p,q}(N\mathcal{F})$ the linear mapping induced by the natural imbedding, and let us call the composition

$$\chi = \chi^{p,q} = k^{p+1,q+1}\circ d_1^{p,q+1}\circ d_n^{p,q} : E_1^{p,q}(\mathcal{F}) \to E_1^{p+1,q+1}(N\mathcal{F})$$

the *characteristic mapping*, $p,q=0,1,\ldots$. Note that χ acts according to the rule

$$E_1^{p,q}(\mathcal{F}) \ni [\omega] \longmapsto [dd_n\omega] \in E_1^{p+1,q+1}(N\mathcal{F}).$$

Theorem 11.2. *The following imbedding is valid:*

$$\mathrm{Im}\ \chi^{p,q} \subset NE_1^{p+1,q+1}(N\mathcal{F}),$$

i.e., the image of the characteristic mapping χ is contained in the kernel of the mapping ν^* (see Theorem 9.19).

Proof. Indeed, let $[\omega]\in E_1^{p,q}(\mathcal{F})$. Then in terms of the external representation of Corollary 9.18 we can assume that $\omega\in C^p\Lambda^{p+q}(M)$, and $d_n\omega\in IC^p\Lambda^{p+q+1}(M)$ modulo $IC^{p+1}\Lambda^{p+q+1}(M)$ (see above). Further, $dd_n\omega\in IC^{p+1}\Lambda^{p+q+2}(M)\cap dC^{p+1}\Lambda^{p+q+1}(M)$, and hence the equivalence class $[dd_n\omega]$, computed in the sense of the space $E_1^{p+1,q+1}(N\mathcal{F})$, actually lies in the kernel $NE_1^{p+1,q+1}(N\mathcal{F})$ (see Theorem 9.19), which is what was to be proved.

In the particular case when $p=0$ and $q=m-1$ (m is the Cartan dimension of the differential manifolds $\mathcal{F}$ and $\mathcal{M}$) the linear space

$$E_1^{0,m-1}(\mathcal{F}) = \frac{\Lambda^{m-1}(M)\cap d^{-1}(C\Lambda^m(M)+I\Lambda^m(M))}{C\Lambda^{m-1}(M)+d\Lambda^{m-2}(M)+I\Lambda^{m-1}(M)}$$

(we use the external representation) is the space of conventional (i.e., $(m-1)$-dimensional) conservation laws of the differential manifold $\mathcal{F}$ (see Section 4), and the kernel

$$NE_1^{1,m}(N\mathcal{F}) = \frac{I\Lambda^{m+1}(M)\cap(C^2\Lambda^{m+1}(M)+dC\Lambda^m(M))}{IC^2\Lambda^{m+1}(M)+dIC\Lambda^m(M)+R\Lambda^{m+1}(M)}$$

(recall that $\Lambda^{m+1}(M)=C\Lambda^{m+1}(M)$, see Section 5) admits an effective description in many practically interesting cases. This indicates the importance of the characteristic mapping for the study of conservation laws.

Special cases of such a mapping have been studied (in another language) by Lax [1] (see also Ibragimov [1], Zharinov [2] and Section 7) for evolution equations, by Vinogradov [4] and Olver [1] for special classes of PDE, imbedded in trivial differential manifolds. In particular, in Olver [1] the image $\chi[\omega]$ of a conservation law $[\omega]$ is called the *characteristic* of this law.

In standard coordinates x^{μ},u^{i},v^{j} (remind that we assume that $\mathcal{F}$ and $\mathcal{M}$ are covered by a single standard defining chart) each conservation law $[\omega]\in E_1^{o,m-1}(\mathcal{F})$ has an external representation

$$\omega = \sum_{\mu=1}^{m} J^{\mu}(x,u,v)\, d^{m-1}x_{\mu}, \quad \left(\sum_{\mu=1}^{m} D_{\mu}J^{\mu}\right)(x,u,0) = 0$$

(cf. an internal representation in Section 4; remind that

$$d^{m-1}x_{\mu} = (-1)^{\mu-1}dx^{1}\wedge\ldots\wedge\widehat{dx^{\mu}}\wedge\ldots\wedge dx^{m}, \quad \mu = 1,\ldots,m$$

see Section 6). It is easy to verify that $d\omega=\left(\Sigma D_{\mu}J^{\mu}\right)d^{m}x$ modulo $C\Lambda^{m}(M)$, so $d_{n}\omega=\Sigma D_{\mu}J^{\mu}d^{m}x$ and

$$dd_{n}\omega = \sum_{i\in I} \frac{\partial\Sigma D_{\mu}J^{\mu}}{\partial u^{i}}\, du^{i}\wedge d^{m}x + \sum_{j\in J} \frac{\partial\Sigma D_{\mu}J^{\mu}}{\partial v^{j}}\, dv^{j}\wedge d^{m}x = \omega_{1} + \omega_{2}$$

(note that $du^{i}\wedge d^{m}x=\delta u^{i}\wedge d^{m}x$ and $dv^{j}\wedge d^{m}x=\delta v^{j}\wedge d^{m}x$). Moreover, in this case $\omega_{1}\in R\Lambda^{m+1}(M)$, because

$$\left(\frac{\partial\Sigma D_{\mu}J^{\mu}}{\partial u^{i}}\right)(x,u,0) = \frac{\partial}{\partial u^{i}}\left[\left(\Sigma D_{\mu}J^{\mu}\right)(x,u,o)\right] = 0,$$

and hence $dd_n\omega = \omega_2$ modulo $R\Lambda^{m+1}(M)$. Thus,

$$\chi[\omega] = \Big[\sum_{j\in J} \frac{\partial \Sigma D_\mu J^\mu}{\partial v^j}\, \delta v^j \wedge d^m x \Big],$$

where the equivalence class on the right-hand side is taken in the space $E_1^{1,m}(N\mathcal{F})$. What is most important for applications is that this class can be computed explicitly in the case of a trivial ambient deferential manifold $\mathcal{M}$.

Indeed, suppose that a nonlinear system of PDE $\{F=0\}$ is realized as a reduction $\mathcal{F}=(F,\partial)$ of a trivial differential manifold $\mathcal{M}=(M,D)$ (see Section 10, we shall use the notation introduced there). Moreover, suppose that the system $\{F=0\}$ is well-determined (see Remark 10.9). Remind also that

$$E_1^{1,m}(N\mathcal{F}) = \frac{I\Lambda^{m+1}(M)}{IC^2\Lambda^{m+1}(M)+dIC\Lambda^m(M)+R\Lambda^{m+1}(M)}$$

(see Section 9). In F-coordinates $x,\mathbf{u},\mathbf{v}$ every form $\omega\in\Lambda^{m+1}(M)$ can be written as follows

$$\omega = \sum_i \Phi_i \delta u^i \wedge d^m x + \sum_{\sigma,k} \Psi_\sigma^k \delta v_k^\sigma \wedge d^m x \mod C^2\Lambda^{m+1}(M),$$

where $\Phi_i, \Psi_\sigma^k(x,\mathbf{u},\mathbf{v})\in\mathcal{C}^\infty(M)$, $i\in I$, $\sigma=1,\dots,s$, $k\in K$. But if $\omega\in I\Lambda^{m+1}(M)$, then the restrictions $\Phi_i\big|_F=0$, and hence

$$\sum_i \Phi_i \delta u^i \wedge d^m x \in R\Lambda^{m+1}(M).$$

So, it can be proved that every $\omega\in I\Lambda^{m+1}(M)$ has the form

$$\omega = \sum_{\sigma,k} \Psi_\sigma^k \delta v_k^\sigma \wedge d^m x \mod (IC^2\Lambda^{m+1}(M)+R\Lambda^{m+1}(M)).$$

In our case (see Remark 10.9),

$$\delta v_k^\sigma = dv_k^\sigma - \sum_{\mu=1}^{m} v_{k+(\mu)}^\sigma dx^\mu,$$

and

$$d\delta v_k^\sigma = - \sum_{\mu=1}^{m} \delta v_{k+(\mu)}^\sigma \wedge dx^\mu.$$

Hence,

$$d\left(\Psi\delta v_k^\sigma \wedge d^{m-1}x_\mu\right) = -\left((D_\mu\Psi)\delta v_k^\sigma + \Psi\delta v_{k+(\mu)}^\sigma\right)\wedge d^m x \mod IC^2\Lambda^{m+1}(M)$$

for all $\Psi\in\mathscr{C}^\infty(M)$, $\sigma=1,\dots,s$, $k\in K$, $\mu=1,\dots,m$. Using this equality, we further reduce ω to the form

$$\omega = \sum_\sigma \Psi_\sigma\Big|_F \delta v^\sigma\wedge d^m x \mod (IC^2\Lambda(M)+dIC\Lambda^m(M)+R\Lambda^{m+1}(M)),$$

where

$$\Psi_\sigma = \sum_{k\in K} (-1)^{|k|} D_k \Psi_\sigma^k.$$

Thus, we have proved the representation

$$E_1^{1,m}(N\mathcal{F}) = \left\{\omega = \sum_\sigma \Psi_\sigma\Big|_F \delta v^\sigma \wedge d^m x \; ; \; \Psi_\sigma \in \mathscr{C}^\infty(M)\right\}$$

$$= \left\{\psi \in \mathscr{C}^\infty(F)_s\right\}.$$

Proposition 11.3. *Suppose that a well-determined system of PDE $\{F=0\}$ has a standard realization. Then the characteristic mapping χ carries an equivalence class*

$$\left[\sum_\mu J^\mu d^{m-1}x_\mu\right] \in E_1^{0,m-1}(\mathcal{F})$$

into the smooth function

$$\psi = \left(\psi_\sigma = \sum_k (-1)^{|k|}\left(D_k \frac{\partial}{\partial v_k^\sigma}\left(\sum_\mu D_\mu J^\mu\right)\right)\Big|_F,\ \sigma = 1,\dots,s\right) \in E_1^{1,m}(N\mathcal{F}).$$

Proof. The statement follows from the above discussion.

Let us turn now to the linear mapping $\nu^*: E_1^{1,m}(N\mathcal{F}) \longrightarrow E_1^{1,m}(\mathcal{M}_F)$. According to Section 9

$$E_1^{1,m}(\mathcal{M}_F) = \frac{\Lambda^{m+1}(M)}{C^2\Lambda^{m+1}(M)+dC\Lambda^m(M)+R\Lambda^{m+1}(M)}.$$

Every form $\omega\in\Lambda^{m+1}(M)$ has the representation

$$\omega = \sum_{\alpha=1}^{a} \sum_{k\in K} \Omega_\alpha^k(x,\mathbf{w})\delta w_k^\alpha \wedge d^m x \mod (C^2\Lambda^{m+1}(M)+dC\Lambda^m(M))$$

(note that $\delta w_k^\alpha \wedge d^m x = dw_k^\alpha \wedge d^m x$), where Ω_α^k are smooth functions on M. With the help of the equality

$$d\left(\Phi\delta w_k^\alpha \wedge d^{m-1}x_\mu\right) = -\left((D_\mu\Phi)\delta w_k^\alpha + \Phi\delta w_{k+(\mu)}^\alpha\right)\wedge d^m x \mod C^2\Lambda^{m+1}(M)$$

(cf. the similar equality above), ω can be reduced to the form

$$\omega = \sum_\alpha \Phi_\alpha \delta w^\alpha \wedge d^m x \mod (C^2\Lambda^{m+1}(M)+dC\Lambda^m(M)),$$

where

$$\Phi_\alpha = \sum_k (-1)^{|k|} D_k \Phi_\alpha^k.$$

Such a form satisfies $\omega\in R\Lambda^{m+1}(M)$ if the restrictions $\Phi_\alpha\big|_F = 0$. Thus we have obtained the representation

$$E_1^{1,m}(\mathcal{M}_F) = \left\{\omega = \sum_\alpha \Phi_\alpha\Big|_F \delta w^\alpha \wedge d^m x\ ;\ \Phi_\alpha \in \mathcal{C}^\infty(M)\right\}$$
$$= \left\{\phi \in \mathcal{C}^\infty(F)_a\right\}.$$

By the definition, the mapping ν^* must associate with each representing form $\omega \in E_1^{1,m}(N\mathcal{F})$ a form $\nu^*\omega$ representing ω in the space $E_1^{1,m}(M_F)$. To find this form $\nu^*\omega$ we consider the following chain of equalities

$$\omega = \sum_\sigma \psi_\sigma \delta v^\sigma \wedge d^m x = \sum_\sigma \Psi_\sigma(x,\mathbf{w}) dF^\sigma \wedge d^m x$$

$$= \sum_\sigma \Psi_\sigma \sum_{\alpha,k} \frac{\partial F^\sigma}{\partial w_k^\alpha} dw_k^\alpha \wedge d^m x = \sum_\alpha \left[\sum_{\sigma,k} (-1)^{|k|} D_k \left(\Psi_\sigma \frac{\partial F^\sigma}{\partial w_k^\alpha} \right) \delta w^\alpha \wedge d^m x \right.$$

$$= \sum_\alpha \overline{F^*}(\psi)_\alpha \delta w^\alpha \wedge d^m x \mod (C^2\Lambda^{m+1}(M) + dC\Lambda^m(M) + R\Lambda^{m+1}(M)),$$

where Ψ_σ are continuations of the functions ψ_σ from F to M, the differential operator $\overline{F^*}: \mathcal{C}^\infty(F)_s \rightarrow \mathcal{C}^\infty(F)_a$ is defined by the rule

$$\overline{F^*}(\psi)_\alpha = \sum_{\sigma=1}^{s} \sum_{k \in K} (-1)^{|k|} \partial_k \left(\psi_\sigma \left(\frac{\partial F^\sigma}{\partial w_k^\alpha} \bigg|_F \right) \right), \quad \alpha = 1,\ldots,a, \ \psi \in \mathcal{C}^\infty(F)_s,$$

and is adjoint to the operator $\overline{F_*}: \mathcal{C}^\infty(F)^a \rightarrow \mathcal{C}^\infty(F)^s$ (see Section 10, cf. Section 6). Thus, here $\nu^* = \overline{F^*}: \mathcal{C}^\infty(F)_s \rightarrow \mathcal{C}^\infty(F)_a$. In particular, the linear mapping ν^* is adjoint to ν (cf. Theorem 12.2 below). Recalling that by the definition the linear space $NE_1^{1,m}(N\mathcal{F})$ is the kernel of ν^*, we arrive at the representation

$$NE_1^{1,m}(N\mathcal{F}) = \left\{ \omega = \sum_\sigma \psi_\sigma \delta v^\sigma \wedge d^m x \ : \ \overline{F^*}(\psi) = 0 \right\} = \operatorname{Ker} \overline{F^*}.$$

Proposition 11.4. *Suppose that a well-determined system of PDE $\{F=0\}$ has a standard realization. Let*

$$[\omega]=\Big[\sum_{\mu} J^{\mu}d^{m-1}x_{\mu}\Big]\in E_1^{0,m-1}(\mathcal{F}),$$

then the characteristic $\chi[\omega]\in\mathrm{Ker}\overline{F^*}$.

Proof. It is enough to use the above discussion, Proposition 11.3 and Theorem 11.2.

Corollary 11.5. *If a well-determined system of PDE* $\{F=0\}$ *has a standard realization then the linear mapping*

$$\chi\ :\ \mathrm{Cons}^{m-1}\mathcal{F}\longrightarrow \mathrm{Ker}\ \overline{F^*}$$

is determined.

For evolution systems of PDE we have two definitions of the characteristic mapping, i.e., the present external definition (we denote it by χ_{ext} here) and the internal definition (we denote it by χ_{int} here) given in Section 7.

Proposition 11.6. *For any evolution system* $\{F=w_t-f=0\}$ *the linear mappings* χ_{int} *and* χ_{ext} *coincide.*

Proof. Let a conservation law $[\omega]\in E_1^{0,m}(\mathcal{F})$ (we shall use the notation of Example 10.4, note that here the Cartan dimension of $\mathcal{F}$ is equal to $m+1$) be represented by a current

$$\omega = T(t,x,\mathbf{u})d^mx - \sum_{\mu=1}^{m} X^{\mu}(t,x,\mathbf{u})dt\wedge d^{m-1}x_{\mu} \in \Lambda^m(F).$$

Then according to Theorem 7.10,

$$\chi_{int}[\omega] = \frac{\delta T}{\delta u} \in \mathcal{C}^{\infty}(F)_a,$$

where

$$\frac{\delta T}{\delta u^\alpha} = \sum_{i\in I} (-1)^{|i|}\partial_i \frac{\partial T}{\partial u_i^\alpha}, \quad \alpha = 1,\dots,a,$$

$I=\mathbb{N}^m$. On the other hand, in order to calculate $\chi_{ext}[\omega]$ let us continue the coefficients T, X^μ from F to M by the formula

$$T, X^\mu(t,x,\mathbf{u},\mathbf{v}) = T, X^\mu(t,x,\mathbf{u}), \quad \mu = 1,\dots,m$$

(note that the result should not depend on the choice of the continuation). Then

$$D_t T + \sum_\mu D_\mu X^\mu = \sum_{\alpha=1}^{a} \sum_{i\in I} v_{0i}^\alpha \frac{\partial T}{\partial u_i^\alpha} + \dots,$$

where dotes denote terms, which don't depend on the variables v_k^α. Hence by Proposition 11.3,

$$\chi_{ext}[\omega]_\alpha = \sum_{i\in I} (-1)^{|i|}\left(D_i \frac{\partial T}{\partial u_i^\alpha}\right)\Big|_F = \sum_{i\in I} (-1)^{|i|}\partial_i \frac{\partial T}{\partial u_i^\alpha}$$

$$= \frac{\delta T}{\delta u^\alpha}, \quad \alpha = 1,\dots,a,$$

(note that here $s=a$). This ends the proof.

Corollary 11.7. *Suppose that a given system $\{F=0\}$ can be transformed into an evolution system by a suitable change of independent and dependent variables. Then the characteristic mapping $\chi\colon \mathrm{Cons}^{m-1}\mathcal{F} \longrightarrow \mathrm{Ker}\overline{F^*}$ is injective* (i.e., $\mathrm{Ker}\chi=0$).

Proof. It is enough to use the coordinate-free definition of characteristic mapping, given in this section, Proposition 11.6 and Theorem 7.10.

12. THE GREEN'S FORMULA

Suppose again, that the differential manifold $\mathcal{F}=(F,CTF)$ under study is a reduction of a differential manifold $\mathcal{M}=(M,CTM)$, and that both are covered by a single standard defining chart (see Section 9).

Remind (see Proposition 5.4), that the rule

$$[\zeta],[\omega] \mapsto <[\zeta],[\omega]> = [\zeta]\rfloor_1 [\omega] = [\zeta\rfloor\omega],$$

where the equivalence class on the right-hand side is taken in the linear space $E_1^{p-1,q}(\mathcal{F})$, $p\geq 1$, determines a pairing

$$<,> : \text{Sym } \mathcal{F}\times E_1^{p,q}(\mathcal{F}) \longrightarrow E_1^{p-1,q}(\mathcal{F}),$$

and the same is true for $\mathcal{M}$.

Proposition 12.1. *The rule*

$$[\zeta],[\omega] \mapsto <[\zeta],[\omega]> = [\zeta\rfloor\omega],$$

where the equivalence class on the right-hand side is taken in the linear space $E_1^{p-1,q}(\mathcal{F})$ *determines pairings*

(a) $<,> : \text{Sym } \mathcal{M}_F \times E_1^{p,q}(\mathcal{M}_F) \longrightarrow E_1^{p-1,q}(\mathcal{F})$,

(b) $<,> : \text{Sym } N\mathcal{F} \times E_1^{p,q}(N\mathcal{F}) \longrightarrow E_1^{p-1,q}(\mathcal{F})$,

for all $p=1,2,\ldots,$ $q=0,1,\ldots.$

Proof. Suppose first that $[\zeta]\in \mathrm{Sym}\mathcal{M}_F$ and $[\omega]\in E_1^{p,q}(\mathcal{M}_F)$. In this case (see Corollary 9.4 and the definition of the space $E^{p,q}(\mathcal{M}\ )$ in Section 9) $\zeta\rfloor\omega\in C^{p-1}\Lambda^{p+q-1}(M)$,

$$d(\zeta\rfloor\omega) = L_\zeta\omega - \zeta\rfloor d\omega \in C^p\Lambda^{p+q}(M) + IC^{p-1}\Lambda^{p+q}(M),$$

and

$$\begin{aligned}[\zeta]\rfloor[\omega] &= (\zeta+C\mathcal{T}(M)+RLB\mathcal{T}(M)) \\ &\qquad \rfloor(\omega+C^{p+1}\Lambda^{p+q}(M)+dC^p\Lambda^{p+q-1}(M)+RC^p\Lambda^{p+q}(M)) \\ &= \zeta\rfloor\omega + C\mathcal{T}(M)\rfloor\omega + (\zeta+C\mathcal{T}(M))\rfloor C^{p+1}\Lambda^{p+q}(M) \\ &\quad + (\zeta+C\mathcal{T}(M))\rfloor dC^p\Lambda^{p+q-1}(M) + (\zeta+C\mathcal{T}(M))\rfloor RC^p\Lambda^{p+q}(M) \\ &\quad + RLB\mathcal{T}(M)\rfloor(\omega+C^{p+1}\Lambda^{p+q}(M)+dC^p\Lambda^{p+q-1}(M)+RC^p\Lambda^{p+q}(M)) \\ &\subset \zeta\rfloor\omega + C^p\Lambda^{p+q-1}(M) + dC^{p-1}\Lambda^{p+q-2}(M) + IC^{p-1}\Lambda^{p+q-1}(M) \\ &= [\zeta\rfloor\omega],\end{aligned}$$

where we have repeatedly used the formulas in Proposition 5.2.

Similarly, the proof of (b) amounts to verifying the inclusion (see Corollary 9.6 and the definition of the space $E_1^{p,q}(N\mathcal{F})$ in Section 9):

$$\begin{aligned}&(\zeta+LBF\mathcal{T}(M))\rfloor(\omega+IC^{p+1}\Lambda^{p+q}(M)+dIC^p\Lambda^{p+q-1}(M)+RC^p\Lambda^{p+q}(M)) \\ &\qquad \subset \zeta\rfloor\omega + C^p\Lambda^{p+q-1}(M) + dC^{p-1}\Lambda^{p+q-2}(M) + IC^{p-1}\Lambda^{p+q-1}(M)\end{aligned}$$

and is implemented with the help of the same formulas in Proposition 5.2.

Theorem 12.2 (Green's formula). *There is the duality*

$$\langle \nu[\zeta],[\omega]\rangle = \langle [\zeta],\nu^{*}[\omega]\rangle$$

for all $[\zeta]\in \mathrm{Sym}\mathcal{M}_F$ *and* $[\omega]\in E_1^{p,q}(N\mathcal{F})$, $p\geq 1$.

In other words, the linear mapping $\nu^{*}:E_1^{p,q}(N\mathcal{F})\to E_1^{p,q}(\mathcal{M}_F)$ (see Theorem 9.19) is dual to the linear mapping $\nu:\mathrm{Sym}\mathcal{M}_F\to \mathrm{Sym}N\mathcal{F}$ (see Theorem 9.7) with respect to the paring, we have introduced.

Proof. Let $[\zeta]\in \mathrm{Sym}\mathcal{M}_F$ and $[\omega]\in E_1^{p,q}(N\mathcal{F})$. The linear mapping ν takes the equivalence class $[\zeta]=\zeta+C\mathcal{T}(M)+RLB\mathcal{T}(M)$ into the equivalence class $\nu[\zeta]=\zeta+LBF\mathcal{T}(M)$, and ν^{*} takes the class

$$[\omega] = \omega+IC^{p+1}\Lambda^{p+q}(M)+dIC^{p}\Lambda^{p+q-1}(M)+RC^{p}\Lambda^{p+q}(M)$$

into the class

$$\nu^{*}[\omega] = \omega+C^{p+1}\Lambda^{p+q}(M)+dC^{p}\Lambda^{p+q-1}(M)+RC^{p}\Lambda^{p+q}(M).$$

As in the proof of Proposition 12.1(b), it can be checked that

$$\nu[\zeta]\rfloor[\zeta] \subset [\zeta\rfloor\omega],$$

where the equivalence class on the right-hand side is taken in the sense of the linear space $E_1^{p-1,q}(\mathcal{F})$, which implies that

$$\langle \nu[\zeta],[\omega]\rangle = [\zeta\rfloor\omega].$$

Similarly, $[\zeta]\rfloor\nu^{*}[\omega]\subset[\zeta\rfloor\omega]$, where the equivalence class on the right-hand side is taken again in the sense of the linear space $E_1^{p-1,q}(\mathcal{F})$, so that also

$$\langle [\zeta],\nu^{*}[\omega]\rangle = [\zeta\rfloor\omega],$$

and the required equality is proved.

It follows from the proof that the left-hand and the right-hand sides of the Green's formula are actually the equivalence class of the form $\zeta\rfloor\omega$, computed in the linear space $E_1^{p-1,q}(\mathcal{F})$ in two different ways. This raises doubts about the meaningfulness of the formula. However, if in each of the five quotient spaces involved in it a unique "canonical" way of choosing the representatives of the equivalence classes is fixed, then the linear mappings ν and ν^* become differential operators acting on these representatives, and the Green's formula becomes a nontrivial relation between them. One should also remember that the kernel of the linear mapping ν describes the symmetries (see Corollary 9.8), while the kernel of ν^* in the case $p=1$, $q=m$ (m is the Cartan dimension of $\mathcal{F}$ and $\mathcal{M}$) is connected by the characteristic mapping with the conservation laws (see Theorem 11.2) of the differential manifold $\mathcal{F}$ under study.

Suppose that a nonlinear system of PDE $\{F=0\}$ is realized as a reduction $\mathcal{F}=(F,\partial)$ of a trivial differential manifold $\mathcal{M}=(M,D)$ (see Section 10, we shall use the notation introduced there). Moreover, let us suppose that the system $\{F=0\}$ is well-determined (see Remark 10.9). Then

$$\operatorname{Sym} \mathcal{M}_F = \left\{\zeta = \sum_{\alpha,k} \left(\partial_k\varphi^\alpha\right)\frac{\partial}{\partial w_k^\alpha} \ ; \ \varphi \in \mathcal{C}^\infty(F)^a\right\},$$

$$\operatorname{Sym} N\mathcal{F} = \left\{\zeta = \sum_{\sigma,k} \left(\partial_k\psi^\sigma\right)\frac{\partial}{\partial v_k^\sigma} \ ; \ \psi=\overline{F_*}(\varphi) \in \mathcal{C}^\infty(F)^s, \ \varphi \in \mathcal{C}^\infty(F)^a\right\}$$

(see Section 10),

$$E_1^{1,m}(N\mathcal{F}) = \left\{\omega = \sum_{\sigma} \psi_{\sigma}\delta v^{\sigma}\wedge d^m x \ ; \ \psi \in \mathcal{C}^{\infty}(F)_s\right\},$$

$$E_1^{1,m}(\mathcal{M}_F) = \left\{\omega = \sum_{\alpha} \varphi_{\alpha}\delta w^{\alpha}\wedge d^m x \ ; \ \varphi \in \mathcal{C}^{\infty}(M)_a\right\}$$

(see Section 11). The mapping ν is represented by the differential operator $\overline{F_*}:\mathcal{C}^{\infty}(F)^a \rightarrow \mathcal{C}^{\infty}(F)^s$, where

$$\overline{F_*}(\varphi)^{\sigma} = \sum_{\alpha=1}^{a} \sum_{k\in K} \left(\frac{\partial F^{\sigma}}{\partial w_k^{\alpha}}\right)\bigg|_F \partial_k \varphi^{\alpha}, \quad \varphi \in \mathcal{C}^{\infty}(F)^a, \ \sigma = 1,\ldots,s,$$

(see Section 10), and the mapping ν^* is represented by the adjoint operator $\overline{F^*}:\mathcal{C}^{\infty}(F)_s \rightarrow \mathcal{C}^{\infty}(F)_a$, where

$$\overline{F^*}(\psi)_{\alpha} = \sum_{\sigma=1}^{s} \sum_{k\in K} (-1)^{|k|} \partial_k\left(\psi_{\sigma}\left(\frac{\partial F^{\sigma}}{\partial w_k^{\alpha}}\bigg|_F\right)\right), \quad \alpha = 1,\ldots,a, \ \psi \in \mathcal{C}^{\infty}(F)_s,$$

(see Section 11). We also need the space

$$E_1^{0,m}(\mathcal{F}) = \frac{\left\{\omega = g d^m x \ ; \ g \in \mathcal{C}^{\infty}(F)\right\}}{\left\{\omega = \left(\sum_{\mu} \partial_{\mu} h^{\mu}\right) d^m x \ ; \ h \in \mathcal{C}^{\infty}(F)^m\right\}}$$

(see Section 5, this is the internal representation, cf. Corollary 9.13).

Theorem 12.3 (Green's formula in coordinates). *Suppose that a nonlinear well-determined system of PDE $\{F=0\}$ is realized as a reduction of a trivial differential manifold. Then for any smooth functions $\varphi\in\mathcal{C}^{\infty}(F)^a$ and $\psi\in\mathcal{C}^{\infty}(F)_s$ there exists a smooth function $h\in\mathcal{C}^{\infty}(F)^m$ such that the equality*

$$\sum_{\sigma=1}^{s} \overline{F_*}(\varphi)^\sigma \psi_\sigma - \sum_{\alpha=1}^{a} \varphi^\alpha \overline{F^*}(\psi)_\alpha = \sum_{\mu=1}^{m} \partial_\mu h^\mu$$

holds. For example, one can take

$$h^\mu = \sum_{\alpha,\sigma,i,k} (-1)^{|k|} M(i,k,\mu) \partial_i \left(\varphi^\alpha\right) \partial_k \left(\psi_\sigma \left(\frac{\partial F^\sigma}{\partial w^\alpha_{i+k+(\mu)}}\right)\bigg|_F\right),$$

where $\mu=1,\dots,m$, *factors*

$$M(i,k,\mu) = \frac{C(i)C(k)}{C(i+k+(\mu))},$$

polynomial coefficients $C(i)=\frac{|i|!}{i!}$ (cf. Proposition A.4).

Proof. Indeed, let $\varphi\in\mathscr{C}^\infty(F)^a$ and $\psi\in\mathscr{C}^\infty(F)_s$. Consider

$$\zeta = \sum_{\alpha,k} \left(\partial_k \varphi^\alpha\right) \frac{\partial}{\partial w^\alpha_k} \in \text{Sym } \mathscr{M}_F \text{ and } \omega = \sum_\sigma \psi_\sigma \delta v^\sigma \wedge d^m x \in E_1^{1,m}(N\mathscr{F}),$$

then

$$\langle \nu[\zeta],[\omega]\rangle = \left[\sum_{\sigma=1}^{s} \overline{F_*}(\varphi)^\sigma \psi_\sigma\right] \text{ and } \langle [\zeta],\nu^*[\omega]\rangle = \left[\sum_{\alpha=1}^{a} \varphi^\alpha \overline{F^*}(\psi)_\alpha\right],$$

where the equivalence classes are taken in $E_1^{0,m}(\mathscr{F})$ (we omit the factor $d^m x$). Now, by Theorem 12.2,

$$\left[\sum_{\sigma=1}^{s} \overline{F_*}(\varphi)^\sigma \psi_\sigma - \sum_{\alpha=1}^{a} \varphi^\alpha \overline{F^*}(\psi)_\alpha\right] = 0 \text{ in } E_1^{0,m}(\mathscr{F}).$$

Hence, there exists a function $h\in\mathscr{C}^\infty(F)^m$ such that the equality we claimed holds. To check that the function h proposed above is valid one should use Proposition A.2.

Example 12.4. Consider the Laplace equation $\Delta w=0$ in $\mathbb{R}^m$. Here $a=s=1$, $x=(x^\mu)\in\mathbb{R}^m$, $\mathbf{w}=(w_k)\in\mathbb{R}_K$, $K=\mathbb{N}^m$, and

$$F = \sum_{\mu=1}^{m} w_{2(\mu)}$$

(cf. Example 8.2(b)). Simple calculations show that

$$\overline{F_*}(\varphi) = \sum_{\mu} \partial_{\mu}^2\varphi, \quad \overline{F^*}(\psi) = \sum_{\mu} \partial_{\mu}^2\psi, \quad \varphi,\psi \in \mathscr{C}^{\infty}(F),$$

$$h^{\mu} = (\partial_{\mu}\varphi)\psi - \varphi(\partial_{\mu}\psi), \quad \mu = 1,\ldots,m,$$

and we get the easy equality

$$\left(\sum_{\mu} \partial_{\mu}^2\varphi\right)\psi - \varphi\left(\sum_{\mu} \partial_{\mu}^2\psi\right) = \sum_{\mu} \partial_{\mu}\left((\partial_{\mu}\varphi)\psi - \varphi(\partial_{\mu}\psi)\right).$$

Further, let us take $\varphi=u(x),\psi=v(x)\in\mathscr{C}^{\infty}(\mathbb{R}^m)$, then we come to the classical Green's formula

$$(\Delta u)v - u(\Delta v) = \nabla((\nabla u) - u(\nabla v)).$$

Remark 12.5. If φ is a symmetry, i.e. $\overline{F_*}(\varphi)=0$, and ψ is the characteristic of a conservation law, i.e., $\overline{F^*}(\psi)=0$, then $h=h(\varphi,\psi)$ is a conserved current (maybe trivial) of the system $\{F=0\}$, i.e., $\sum_{\mu}\partial_{\mu}h^{\mu}=0$. Thus Green's formula, like Noether's identity for Lagrangian systems (see Proposition 6.21), ties together symmetries and conservation laws of the system under study. One can use it to construct new conservation laws from known ones and symmetries.

Example 12.6. Let us see what Green's formula yields if the original system $\{F=0\}$ is linear. Namely, let

$$F^{\sigma}(x,\mathbf{w}) = \sum_{\alpha,k} a_{\alpha}^{\sigma,k}(x)w_{k}^{\alpha}, \quad \sigma = 1,\ldots,s.$$

Then

$$\overline{F_*}(\varphi)^\sigma = \sum_{\alpha,k} a_\alpha^{\sigma,k}(x)\partial_k\varphi^\alpha, \quad \sigma = 1,\dots,s,$$

$$\overline{F^*}(\psi)_\alpha = \sum_{\sigma,k} (-1)^{|k|}\partial_k\left(\psi_\sigma a_\alpha^{\sigma,k}(x)\right), \quad \alpha = 1,\dots,a,$$

$$h^\mu = \sum_{\alpha,\sigma,i,k} (-1)^{|k|} M(i,k,\mu)\partial_i\left(\varphi^\alpha\right)\partial_k\left(\psi_\sigma a_\alpha^{\sigma,i+k+(\mu)}(x)\right),$$

$$\mu = 1,\dots,m,$$

where $\varphi\in\mathscr{C}^\infty(F)^a$ and $\psi\in\mathscr{C}^\infty(F)_s$. If now we restrict ourselves to φ and ψ that depend only on x, then our Green's formula reduces to the usual Green's formula for linear systems (see, for example, Vladimirov, Zharinov [1]), since in this case the differentiations ∂_μ reduce to the partial derivatives $\partial/\partial x^\mu$. In the familiar notation the latter formula is written as follows

$$(\psi, F(x,\partial_x)\varphi)_s - (F^*(x,\partial_x)\psi,\varphi)_a = \nabla h,$$

where $F(x,\partial_x)$ is a linear operator in partial derivatives with smooth matrix coefficients acting from $\mathscr{C}^\infty(\mathbb{R}^m)^a$ into $\mathscr{C}^\infty(\mathbb{R}^m)^s$, $F^*(x,\partial_x)$ is the formally adjoint operator acting from $\mathscr{C}^\infty(\mathbb{R}^m)_s$ into $\mathscr{C}^\infty(\mathbb{R}^m)_a$, $(\cdot,\cdot)_s$ and $(\cdot,\cdot)_a$ are inner products in the Euclidean spaces $\mathbb{R}^s$ and $\mathbb{R}^a$, respectively, the vector $h=h(F,\psi,\varphi)$ is defined by the above formula. Recalling that every solution φ of the linear system $F(x,\partial_x)w=0$ establishes a symmetry of this system (corresponding to the one-parameter group of transformations $w\mapsto w+t\varphi$), we conclude that our formula (Theorem 12.3) also generalizes the usual Green's formula in the linear case, indicating the close connection of the latter with the symmetries of the original system of PDE.

13. Low-dimensional Conservation Laws

Let $\mathcal{M}=(M,D)$ be a trivial differential manifold and $\{F=0\}$ be a nonlinear system of PDE (see Section 10, we shall use the notation introduced there). Let

$$F = \left\{ x,\mathbf{w} \; : \; D_k F^{\sigma}(x,\mathbf{w}) = 0, \;\; k \in K, \;\; \sigma = 1,\dots,s \right\}$$

be the solution variety of the system $\{F=0\}$ and

$$\mathcal{I}_F = \left\{ \varphi \in \mathcal{C}^{\infty}(M) \; : \; \varphi=0 \text{ on } F \right\}$$

be its defining ideal. Consider the quotient algebra

$$\mathcal{C}^{\infty}_{ext}(F) = \frac{\mathcal{C}^{\infty}(M)}{\mathcal{I}_F}.$$

If the system $\{F=0\}$ has a standard realization as a reduction of the differential manifold $\mathcal{M}$, then $\mathcal{C}^{\infty}_{ext}(F)=\mathcal{C}^{\infty}(F)$. In the general case the algebra $\mathcal{C}^{\infty}_{ext}(F)$ is the natural substitute of the latter, and its elements can be considered as smooth functions on the solution variety F (even when F has no structure of the smooth submanifold of M). By the definition, for every function $\Phi\in\mathcal{C}^{\infty}(M)$ its "restriction" $\phi\in\mathcal{C}^{\infty}_{ext}(M)$ is determined by the rule

$$\phi = \Phi\Big|_F = [\Phi] = \Phi+\mathcal{I}_F.$$

The algebra $\mathcal{C}^{\infty}(M)$ together with the total differentiations D_{μ}, $\mu=1,\dots,m$, determines a differential algebra (see Appendix). We shall assume below that *the system $\{F=0\}$ is generating*, i.e., $\mathcal{I}_F=\mathrm{DO}_s(F)$ (see Section 10). In this case the ideal $\mathcal{I}_F$ is differential, i.e., if $\Phi\in\mathcal{I}_F$ then $D_{\mu}\Phi\in\mathcal{I}_F$, $\mu=1,\dots,m$, hence the differentiations $\partial_{\mu}=D_{\mu}\big|_F$ of the quotient algebra $\mathcal{C}^{\infty}_{ext}(F)$ are induced,

$$\mathcal{C}^{\infty}_{ext}(F) \ni \phi=[\Phi] \mapsto \partial_{\mu}[\phi]=[D_{\mu}\Phi] \in \mathcal{C}^{\infty}_{ext}(F), \quad \mu = 1,\dots,m,$$

and thus $\mathcal{C}^{\infty}_{ext}(F)$ together with these differentiations is a differential algebra (see Appendix).

In our study of low-dimensional conservation laws of the system $\{F=0\}$ we shall use the following construction.

For a linear space $\mathcal{L}$ with pairwise commuting linear mappings $D_{\mu}:\mathcal{L}\longrightarrow\mathcal{L}$, $[D_{\mu},D_{\nu}]=0$, $\mu,\nu=1,\dots,m$, (cf. Appendix) let us denote by $\Lambda^q\mathcal{L}$, $q=0,1,\dots$, the linear spaces of all q-forms

$$\Omega = \sum_{\mu_1,\dots,\mu_q} \Omega_{\mu_1\dots\mu_q} dx^{\mu_1}\wedge\dots\wedge dx^{\mu_q}$$

over $\mathcal{L}$, where coefficients $\Omega_{\mu_1\dots\mu_q}\in\mathcal{L}$ are alternating in indices $\mu_1,\dots,\mu_q=1,\dots,m$, (differentials dx^{μ} are formal here, we write them only to make formulas more readable). Clear that $\Lambda^q\mathcal{L}=0$ for $q>m$. Further, let the exterior differentiations $D:\Lambda^q\mathcal{L}\longrightarrow\Lambda^{q+1}\mathcal{L}$ be determined by the rule

$$D\Omega = \sum_{\mu_0,\dots,\mu_q} D_{[\mu_0}\Omega_{\mu_1\dots\mu_q]} dx^{\mu_0}\wedge\dots\wedge dx^{\mu_q} \in \Lambda^{q+1}\mathcal{L}$$

for a q-form $\Omega\in\Lambda^q\mathscr{L}$, where the brackets [...] denote the skew-symmetrization in indices $\mu_0,\dots,\mu_q=1,\dots,m$. Clear that the composition $D\circ D=0$, hence the complex

$$0 \longrightarrow \mathbb{K} \longrightarrow \Lambda^0\mathscr{L} \xrightarrow{D} \Lambda^1\mathscr{L} \xrightarrow{D} \dots \xrightarrow{D} \Lambda^{m-1}\mathscr{L} \xrightarrow{D} \Lambda^m\mathscr{L} \longrightarrow 0$$

is determined, where

$$\mathbb{K} = \left\{\Phi \in \mathscr{L} \ : \ D_\mu\Phi=0, \quad \mu = 1,\dots,m\right\}$$

is *the subspace of constants* of the space $\mathscr{L}$, (cf. Appendix). The cohomology spaces of this complex we denote by $\mathscr{H}^q(\mathscr{L})$, i.e.,

$$\mathscr{H}^q(\mathscr{L}) = \frac{\left\{\Omega \in \Lambda^q\mathscr{L} \ : \ D\Omega=0\right\}}{\left\{\Omega=D\Gamma \ ; \ \Gamma \in \Lambda^{q-1}\mathscr{L}\right\}}, \quad q = 0,\dots,m.$$

Remark 13.1. In Appendix (see Remark A.7) the linear spaces $\mathscr{A}^{\wedge p}$ and the linear mappings $D:\mathscr{A}^{\wedge p}\longrightarrow\mathscr{A}^{\wedge(p-1)}$, $p=1,\dots,m$, were defined. Consider the linear mappings $*:\Lambda^q\mathscr{L}\longrightarrow\mathscr{A}^{\wedge p}$, $p+q=m$, defined by the rule

$$(*\Omega)^{\nu_1\dots\nu_p} = \sum_{\mu_1,\dots,\mu_q} \mathrm{sign}(\nu_1\dots\nu_p\mu_1\dots\mu_q)\Omega_{\mu_1\dots\mu_q},$$

where $\nu_1,\dots,\nu_p=1,\dots,m$, $\mathrm{sign}(\lambda_1\dots\lambda_m)$ is the sign of a permutation $\{1\dots m\}\mapsto\{\lambda_1\dots\lambda_m\}$. It is easy to check that the mappings $*$ are in fact isomorphisms, and

$$*(D\Omega) = (-1)^{p-1}D(*\Omega), \quad \Omega \in \Lambda^q\mathscr{L}, \ p+q = m.$$

Example 13.2. Let $\mathscr{L}=\mathscr{C}^\infty(\mathbb{R}^m)$ be the algebra of all smooth functions on $\mathbb{R}^m$ and $D_\mu=\partial/\partial x^\mu$, $\mu=1,\dots,m$, be the partial derivatives.

Then $\Lambda^q\mathscr{C}^\infty(\mathbb{R}^m)=\Lambda^q(\mathbb{R}^m)$ are the spaces of usual differential q-forms on $\mathbb{R}^m$, dx^μ are standard basis differentials, $D=d$ is the usual exterior differentiation, $\mathbb{K}=\mathbb{R}$, and $\mathscr{H}^q(\mathscr{C}^\infty(\mathbb{R}^m))$ are de Rham cohomology spaces, which are trivial by Poincare lemma, i.e.,

$$\mathscr{H}^q(\mathscr{C}^\infty(\mathbb{R}^m))=0, \quad q=0,\dots,m.$$

Example 13.3. Let $\mathcal{M}=(M,D)$ be the above trivial differential manifold. Let $\mathscr{L}=\mathscr{C}^\infty(M)$ be the differential algebra of all smooth functions on the manifold M with the total differentiations D_μ, $\mu=1,\dots,m$. Then $\Lambda^q\mathscr{C}^\infty(M)=E^{0,q}$ and $D=\partial:E^{0,q}\to E^{0,q+1}$ (see Sections 5 and 6). Again $\mathbb{K}=\mathbb{R}$, and

$$\mathscr{H}^q(\mathscr{C}^\infty(M)) = E_1^{0,q}(\mathcal{M}) = 0, \quad q=0,\dots,m-1,$$

by Theorem 6.4.

Remind (see Sections 6 and 10) that for every $\Psi\in\mathscr{C}^\infty(M)$ and $\varphi\in\mathscr{C}^\infty(\mathbb{R}^m)^a$ the substitution $\Psi\Big|_{w=\varphi}\in\mathscr{C}^\infty(\mathbb{R}^m)$ is defined, and

$$\Big[D_\mu\Psi(x,\mathbf{w})\Big]\Big|_{w=\varphi(x)} = \frac{\partial}{\partial x^\mu}\Big[\Psi(x,\mathbf{w})\Big|_{w=\varphi(x)}\Big], \quad \mu = 1,\dots,m,$$

where

$$w=\varphi(x) \text{ stands for } w_k^\alpha = \left(\frac{\partial}{\partial x}\right)^k\varphi^\alpha(x), \ \alpha = 1,\dots,a, \ k\in K.$$

Hence, for every q-form $\Omega\in\Lambda^q\mathscr{C}^\infty(M)$ and $\varphi\in\mathscr{C}^\infty(\mathbb{R}^m)^a$ the substitution $\Omega\Big|_{w=\varphi}\in\Lambda^q(\mathbb{R}^m)$ is defined by the rule

$$\Omega\Big|_{w=\varphi} = \sum_{\mu_1,\dots,\mu_q}\Omega_{\mu_1\dots\mu_q}\Big|_{w=\varphi}dx^{\mu_1}\wedge\dots\wedge dx^{\mu_q}.$$

Moreover,

$$\left(D\Omega\right)\Big|_{w=\varphi} = d\left(\Omega\Big|_{w=\varphi}\right).$$

Example 13.4. Let $\mathcal{L}=\mathcal{I}_F$ be a differential subalgebra of all smooth functions $\Phi\in\mathcal{C}^\infty(M)$ vanishing on F, i.e., $\Phi\big|_F=0$ (remind that we assume that the system $\{F=0\}$ is generating). Then $\Lambda^q\mathcal{I}_F\subset E^{0,q}$ and $D=\partial:\Lambda^q\mathcal{I}_F\longrightarrow\Lambda^{q+1}\mathcal{I}_F$. Clear that $\mathbb{K}=0$ if $F\neq\varnothing$, i.e., the subalgebra of constants of $\mathcal{I}_F$ is trivial under the natural assumption that *the solution variety F is not empty*. If the system $\{F=0\}$ has a standard realization as a reduction of $\mathcal{M}$ then

$$\Lambda^q\mathcal{I}_F = RE^{0,q} = R\Lambda^q(M)\cap E^{0,q}, \quad q = 0,1,\ldots,$$

(see Sections 6 and 9). We have no reasons to expect the cohomolgy spaces $\mathcal{H}^q(\mathcal{I}_F), q=0,\ldots,m,$ to be trivial for the given system $\{F=0\}$. Note that the substitution

$$\Psi\Big|_{w=\varphi} = 0 \text{ for any } \Psi\in\mathcal{I}_F \text{ and } \varphi\in \mathrm{Sol}\{F=0\},$$

because in this case the integral submanifold $S(\varphi)\subset F$ (see Section 10).

Example 13.5. Consider the algebra $\mathcal{L}=\mathcal{C}^\infty_{ext}(F)$ with the quotient differentiations ∂_μ, $\mu=1,\ldots,m$. Here

$$\Lambda^q\mathcal{C}^\infty_{ext}(F) = \frac{\Lambda^q\mathcal{C}^\infty(M)}{\Lambda^q\mathcal{I}_F}, \quad q = 0,1,\ldots.$$

If the system $\{F=0\}$ has a standard realization as a reduction of $\mathcal{M}$ then

$$\Lambda^q\mathcal{C}^\infty_{ext}(F) = \frac{E^{0,q}}{RE^{0,q}} = E_0^{0,q}(F), \quad q = 0,1,\ldots,$$

(cf. Proposition 9.14 and Lemma 9.12). Further,

$$\mathcal{H}^q(\mathcal{C}^\infty_{ext}(F)) = \frac{\operatorname{Ker} \partial : \Lambda^q\mathcal{C}^\infty_{ext}(F) \longrightarrow \Lambda^{q+1}\mathcal{C}^\infty_{ext}(F)}{\operatorname{Im} \partial : \Lambda^{q-1}\mathcal{C}^\infty_{ext}(F) \longrightarrow \Lambda^q\mathcal{C}^\infty_{ext}(F)}$$

$$= \frac{\left\{\Omega \in \Lambda^q\mathcal{C}^\infty(M) \; : \; D\Omega \in \Lambda^{q+1}\mathcal{I}_F\right\}}{\left\{\Omega=\Phi+D\Psi \; ; \; \Phi \in \Lambda^q\mathcal{I}_F, \; \Psi \in \Lambda^{q-1}\mathcal{C}^\infty(M)\right\}},$$

$q=0,1,\ldots$. If the system $\{F=0\}$ has a standard realization as a reduction of $\mathcal{M}$ then

$$\mathcal{H}^q(\mathcal{C}^\infty_{ext}(F)) = \frac{\operatorname{Ker} \partial : E_0^{0,q}(F) \longrightarrow E_0^{0,q+1}(F)}{\operatorname{Im} \partial : E_0^{0,q-1}(F) \longrightarrow E_0^{0,q}(F)}$$

$$= E_1^{0,q}(F) = \operatorname{Cons}^q\mathcal{F}, \quad q = 0,1,\ldots,m-1,$$

(see Section 5, cf. Corollary 9.18). In the general case *the spaces* $\mathcal{H}^q(\mathcal{C}^\infty_{ext}(F))$ *are the natural substitutes for the spaces* $\operatorname{Cons}^q\mathcal{F}$, and its elements can be considered as conservation laws of the system $\{F=0\}$, because in any case

$$d\left(\Omega\Big|_{w=\varphi}\right) = \left(D\Omega\right)\Big|_{w=\varphi} = 0, \quad [\Omega]\in\mathcal{H}^q(\mathcal{C}^\infty_{ext}(F)),$$

and

$$\left(\Phi + D\Psi\right)\Big|_{w=\varphi} = d\left(\Psi\Big|_{w=\varphi}\right), \quad \Phi \in \Lambda^q\mathcal{I}_F, \; \Psi \in \Lambda^{q-1}\mathcal{C}^\infty(M),$$

for every solution $\varphi\in\operatorname{Sol}\{F=0\}$. Taking these arguments into account we denote

$$\operatorname{Cons}^q_{ext}\mathcal{F} = \mathcal{H}^q(\mathcal{C}^\infty_{ext}(F)), \quad q = 0,1,\ldots,m-1,$$

and call elements of $\operatorname{Cons}^q_{ext}\mathcal{F}$ *conservation laws* of a system $\{F=0\}$.

Example 13.6. Consider the linear space $\mathscr{L}$=DO of all linear differential operators on $\mathscr{C}^{\infty}(M)$ (see Section 10), and let us define the linear mappings $D_{\mu}\circ$:DO⟶DO by the composition rule:

$$\mathrm{DO} \ni P \mapsto D_{\mu}\circ P \in \mathrm{DO}, \quad \mu=1,\ldots,m.$$

By the definition,

$$P(\Phi) \in \mathscr{C}^{\infty}(M) \quad \text{and} \quad \left(D_{\mu}\circ P\right)(\Phi) = D_{\mu}\left(P(\Phi)\right)$$

for all $P\in$DO, $\Phi\in\mathscr{C}^{\infty}(M)$ and $\mu=1,\ldots,m$. Hence,

$$Q(\Phi) \in \Lambda^{q}C^{\infty}(M) \quad \text{and} \quad \left(D\circ Q\right)(\Phi) = D\left(Q(\Phi)\right)$$

for all $Q\in\Lambda^{q}$DO and $\Phi\in\mathscr{C}^{\infty}(M)$. The *characteristic* $\chi(P)\in\mathscr{C}^{\infty}(M)$ of a differential operator

$$P = \sum_{k\in K} a^{k}(x,\mathbf{w})D_{k} \in \mathrm{DO}$$

is defined by the formula

$$\chi(P) = \sum_{k\in K} (-1)^{|k|}D_{k}a^{k}$$

(see Appendix). This rule determines the linear mapping

$$\chi : \mathrm{DO} \longrightarrow \mathscr{C}^{\infty}(M),$$

and further the linear mappings

$$\chi : \Lambda^{q}\mathrm{DO} \longrightarrow \Lambda^{q}\mathscr{C}^{\infty}(M), \quad q = 0,1,\ldots,$$

with the componentwise action. Consider the complex of the linear spaces

$$0 \longrightarrow \Lambda^{0}\mathrm{DO} \xrightarrow{D\circ} \Lambda^{1}\mathrm{DO} \xrightarrow{D\circ} \ldots \xrightarrow{D\circ} \Lambda^{m}\mathrm{DO} \xrightarrow{\chi} \Lambda^{m}\mathscr{C}^{\infty}(M) \longrightarrow 0,$$

and denote by $\mathcal{H}^q(\mathrm{DO})$, $q=0,\dots,m,m+1$, its cohomolgy spaces, where

$$\mathcal{H}^{m+1}(\mathrm{DO}) = \frac{\Lambda^m\mathcal{C}^\infty(M)}{\mathrm{Im}\ \chi\ :\ \Lambda^m\mathrm{DO}\longrightarrow\Lambda^m\mathcal{C}^\infty(M)}.$$

By Theorem A.5 (see also Remark 13.1),

$$\mathcal{H}^q(\mathrm{DO}) = 0, \quad q=0,\dots,m+1,$$

i.e., the cohomology spaces of the above complex are trivial. In particular, the constant subspace $\mathbb{K}=0$ in this case.

Example 13.7. For the linear space $\mathcal{L}=\mathrm{DO}_s$ the operations $D_\mu\circ$, $D\circ$ and χ are extended componentwise, and we get the complex

$$0 \longrightarrow \Lambda^0\mathrm{DO}_s \xrightarrow{D\circ} \Lambda^1\mathrm{DO}_s \xrightarrow{D\circ} \dots \xrightarrow{D\circ} \Lambda^m\mathrm{DO}_s \xrightarrow{\chi} \Lambda^m\mathcal{C}^\infty(M)_s \longrightarrow 0.$$

By the same reasons,

$$\mathcal{H}^q(\mathrm{DO}_s) = 0, \quad q=0,\dots,m+1.$$

Clear $D_\mu\circ P\in\mathrm{DI}_F$ for any $\mu=1,\dots,m$ and $P\in\mathrm{DI}_F$. Thus the subcomlex

$$0 \longrightarrow \Lambda^0\mathrm{DI}_F \xrightarrow{D\circ} \Lambda^1\mathrm{DI}_F \xrightarrow{D\circ} \dots \xrightarrow{D\circ} \Lambda^m\mathrm{DI}_F \xrightarrow{\chi} \Lambda^m\mathcal{C}^\infty(M)_s \longrightarrow 0.$$

is defined. The cohomology spaces $\mathcal{H}^q(\mathrm{DI}_F)$, $q=0,\dots,m+1$ don't need to be trivial in the general case. Here, for example,

$$\mathcal{H}^q(\mathrm{DI}_F) = \frac{\left\{Q \in \Lambda^q\mathrm{DI}_F\ :\ D\circ Q=0\right\}}{\left\{Q=D\circ R\ ;\ R \in \Lambda^{q-1}\mathrm{DI}_F\right\}}, \quad q = 1,\dots,m-1.$$

We call a differential identity $Q\in\Lambda^q\mathrm{DI}_F$ *closed* if $D\circ Q=0$, and *exact* if $Q=D\circ R$, $R\in\Lambda^{q-1}\mathrm{DI}_F$.

Theorem 13.8 (Zharinov [7]). *Let $\{F=0\}$ be a generating system of PDE. Then there are isomorphisms:*

$$Z=Z^q : \mathrm{Cons}^q_{ext}\mathcal{F} \simeq \mathcal{H}^{q+2}(\mathrm{DI}_F), \quad q = 0,\dots,m-2.$$

Proof. First, by Lemma A.8, Example 13.3 and Remark 13.1,

$$\mathrm{Cons}^q_{ext}\mathcal{F} \simeq \mathcal{H}^{q+1}(\mathcal{I}_F), \quad q = 0,\dots,m-2,$$

and then, by Lemma A.11, Example 13.7 and again Remark 13.1,

$$\mathcal{H}^{q+1}(\mathcal{I}_F) \simeq \mathcal{H}^{q+2}(\mathrm{DI}_F), \quad q = 0,\dots,m-2.$$

A simple analysis of Lemmas A.8 and A.11 shows that the isomorphisms Z act as follows. Let an equivalence class $[\omega]\in\mathrm{Cons}^q_{ext}\mathcal{F}$. then $\omega=[\Omega]\in\Lambda^q\mathcal{C}^\infty_{ext}(F)$, and $\partial\omega=0$, i.e., $\Omega\in\Lambda^q\mathcal{C}^\infty(M)$, and $D\Omega\in\Lambda^{q+1}\mathcal{I}_F$. Now, $D\Omega=Q(F)$, $Q\in\Lambda^{q+1}\mathrm{DO}_s$, because the system $\{F=0\}$ supposed to be generating. Further, $D\circ Q(F)=DD\Omega=0$, i.e., $D\circ Q\in\Lambda^{q+2}\mathrm{DI}_F$, and we define $Z[\omega]=[D\circ Q]\in\mathcal{H}^{q+2}(\mathrm{DI}_F)$. One can easily check that this definition is correct.

Remark.13.9. (a) The theorem gives no restrictions on conservation laws of the maximal dimension $q=m-1$ in accordance with the fact that the space $\mathcal{H}^m(\mathcal{C}^\infty(M))\neq 0$ (see Theorem 6.4 and Lemma A.8). (b) The shift $q\mapsto q+2$ has a homological origin, Lemmas A.8 and A.11 are proved by pure diagram methods, and I don't know any other explanation of this phenomenon. (c) If the system $\{F=0\}$ has a standard realization then $\mathcal{C}^\infty_{ext}(F)=\mathcal{C}^\infty(F)$ and $\mathrm{Cons}^q_{ext}\mathcal{F}=\mathrm{Cons}^q\mathcal{F}$ are the internal objects while the spaces $\mathcal{H}^q(\mathrm{DI}_F)$ are external. Thus, Theorem 13.8 implies that *equivalent systems of* PDE *with the same* *solution* manifold F *have the same cohomology spaces of differential identities* (note that we assume here that involve systems of PDE have standard realizations).

The following model example demonstrates how low-dimensional conservation laws can arise.

Example 13.10. Let $x=(x^{\mu})$, $\mu=1,\ldots m$, be independent variables and w be a dependent variable ($a=1$). Let us fix a number s, $1<s<m$, and consider the following system of linear PDE

$$\frac{\partial w}{\partial x^{\sigma}} = 0, \quad \sigma = 1,\ldots,s.$$

In this case the system is determined by the functions $F_{\sigma}=w_{(\sigma)}$, $\sigma=1,\ldots,s$, the order of the system is equal to 1, and the system has a standard realization. In more detail, let $K=\mathbb{N}^m$, $\mathbf{w}=(w_k;k\in K)$, $M=\{x,\mathbf{w}\}$. Further, let

$$I = \{k \in \mathbb{N}^m : k_1+\ldots+k_s = 0\}, \quad J = \{k \in \mathbb{N}^m : k_1+\ldots+k_s > 0\},$$

then $K=I\cup J$, $I\cap J=\varnothing$, $\mathbf{w}=(w_k;k\in K)$, and

$$F = \Big\{x,\mathbf{w} : w_k = 0,\ k \in J\Big\} = \Big\{x,\mathbf{u},\mathbf{v} : \mathbf{v}=0\Big\},$$

where $\mathbf{w}=(\mathbf{u},\mathbf{v})$, $\mathbf{u}=(u_i=w_i;i\in I)$, $\mathbf{v}=(v_j=w_j;j\in J)$ (here $D_kF_{\sigma}=w_{k+(\sigma)}$). Now, the variables $x,\mathbf{u}$ are global coordinates on F, and

$$\begin{aligned}\partial_{\mu} = D_{\mu}\Big|_F &= \frac{\partial}{\partial x^{\mu}} && \text{if } \mu = 1,\ldots,s,\\ &= \frac{\partial}{\partial x^{\mu}} + \sum_{i\in I} u_{i+(\mu)}\frac{\partial}{\partial u_i} && \text{if } \mu = s+1,\ldots,m.\end{aligned}$$

Consider a $(m-s)$-form

$$\omega = \varphi(x,\mathbf{u})dx^{s+1}\wedge\ldots\wedge dx^m \in \Lambda^{m-s}\mathcal{C}^{\infty}(F), \quad \varphi \in \mathcal{C}^{\infty}(F)$$

(here $\mathcal{C}^\infty(F)=\mathcal{C}^\infty_{ext}(F)$). In this case

$$\partial\omega = \sum_{\sigma=1}^{s} \frac{\partial\varphi}{\partial x^\sigma}\, dx^\sigma \wedge dx^{s+1}\wedge\ldots\wedge dx^m,$$

and $\partial\omega=0$ iff the function $\varphi(x,\mathbf{u})=\varphi(x^{s+1},\ldots,x^m,\mathbf{u})$ does not depend on $x^1,\ldots,x^s$. Suppose that this is the case, and hence ω is a $(m-s)$-current. When is this current trivial? Consider an auxiliary trivial differential manifold $\mathcal{N}$ with coordinates $y=(x^{s+1},\ldots,x^m)$, $\mathbf{u}$ and basis Cartan fields $\partial_{s+1},\ldots,\partial_m$ from defined above (cf. Section 6). Then the form $\omega\in\Lambda^{m-s}(N)$, and we are interested when $\omega=\partial\theta$, $\theta\in\Lambda^{m-s-1}(N)$. Proposition 6.8 gives the answer:

$$\omega=\partial\theta \quad \text{iff} \quad \frac{\delta\varphi}{\delta u} = \sum_{i\in I} (-1)^{|i|}\partial_i \frac{\partial\varphi}{\partial u_i} = 0.$$

In particular, *the linear space* $\mathrm{Cons}^{m-s}\mathcal{F}$ *is not trivial.*

Let $[\omega]\in\mathrm{Cons}^{m-s}\mathcal{F}$, where

$$\omega = \varphi(y,\mathbf{u})d^{m-s}y, \quad d^{m-s}y = dx^{s+1}\wedge\ldots\wedge dx^m,$$

is a nontrivial $(m-s)$-current. The image $Z[\omega]=[R]\in\mathcal{H}^{m-s}(\mathrm{DI}_F)$, where the differential identity $(m-s+2)$-form R is calculated as follows. Let $\Omega=\omega\in\Lambda^{m-s}\mathcal{C}^\infty(M)$, then

$$D\Omega = \sum_{\sigma=1}^{s} \left(D_\sigma\varphi\right) dx^\sigma\wedge d^{m-s}y = \sum_{\sigma=1}^{s}\sum_{i\in I} \frac{\partial\varphi}{\partial u_i}\, w_{i+(\sigma)}\, dx^\sigma\wedge d^{m-s}y$$

$$= \sum_{\sigma=1}^{s} P(F)_\sigma\, dx^\sigma\wedge d^{m-s}y = Q(F),$$

where

$$P(\Phi)_\sigma = \sum_{i\in I} \frac{\partial\varphi}{\partial u_i} D_i\Phi_\sigma, \quad \sigma = 1,\dots,s, \quad \Phi \in \mathcal{C}^\infty(M)_s.$$

Hence,

$$\begin{aligned} R(\Phi) &= (D\circ Q)(\Phi) = D(Q(\Phi)) \\ &= \sum_{\rho=1}^{s} D_\rho\left(\sum_{\sigma=1}^{s} \sum_{i\in I} \frac{\partial\varphi}{\partial u_i} D_i\Phi_\sigma\right) dx^\rho\wedge dx^\sigma\wedge d^{m-s}y \\ &= \sum_{\rho,\sigma} R(\Phi)_{\rho\sigma}\, dx^\rho\wedge dx^\sigma\wedge d^{m-s}y, \quad \Phi \in \mathcal{C}^\infty(M)_s, \end{aligned}$$

where

$$R(\Phi)_{\rho\sigma} = \left(\sum_{i,k\in I} \frac{\partial^2\varphi}{\partial u_i\partial u_k}\left(D_iF_{[\rho}\right)\left(D_k\Phi_{\sigma]}\right) + \sum_{i\in I}\frac{\partial\varphi}{\partial u_i}\left(D_{i+[(\rho)}\Phi_{\sigma]}\right)\right),$$

and the brackets [...] denote the skew-symmetrization in indices $\rho,\sigma=1,\dots,s$.

Example 13.10(a). Let us pass from variables x,w to new variables t,r with transition formulas

$$\begin{cases} t^\mu = T^\mu(x,w), \\ r = R(x,w), \end{cases} \quad \begin{cases} x^\mu = X^\mu(t,r), \\ w = W(t,r), \end{cases} \quad \mu = 1,\dots,m.$$

Then the above system

$$\frac{\partial w}{\partial x^\sigma} = 0, \quad \sigma = 1,\dots,s,$$

transforms into the equivalent quasi-linear system

$$\sum_{\mu=1}^{m} A^\mu_\sigma \frac{\partial r}{\partial t^\mu} = B_\sigma, \quad \sigma = 1,\dots,s,$$

where

$$A^{\mu}_{\sigma}(t,r) = \left.\frac{\partial T^{\mu}}{\partial x^{\sigma}}\right|_{\substack{x=X(t,r),\\ w=W(t,r)}} \quad B_{\sigma}(t,r) = \left.\frac{\partial R}{\partial x^{\sigma}}\right|_{\substack{x=X(t,r),\\ w=W(t,r)}}$$

$\mu=1,\dots,m$, $\sigma=1,\dots,s$. Clear that the new system has the same spaces of conservation laws as the source system.

Example 13.11. Let us present a complex counterpart of Example 13.10. Namely, consider the Cauchy-Riemann system of PDE

$$\frac{\partial w}{\partial \bar{z}^{\nu}} = 0, \quad \nu = 1,\dots,n,$$

where $z^1,\dots,z^n$ are complex independent variables, w is a complex dependent variable. In our geometrical approach we could split up complex variables into their real and imaginary parts, and then study the system of $2n$ equations for 2 real functions of $2n$ real arguments, but we prefer to work directly with complex variables. In this case the ambient trivial differential manifold

$$M = \mathbb{R}^{2n}\times\mathbb{R}^{2}_{I\times I} = \mathbb{C}^{n}\times\mathbb{C}_{I\times I}, \quad I = \mathbb{N}^{n},$$

has global coordinates

$$z=(z^{\nu}),\ \bar{z}=(\bar{z}^{\nu}),\ \nu=1,\dots,n,\ \mathbf{w}=(w_{ij}),\ \bar{\mathbf{w}}=(\bar{w}_{ij}),\ i,j\in I,$$

where

$$w_{ij} \longleftrightarrow \frac{\partial^{|i+j|}w}{(\partial z)^{i}(\partial \bar{z})^{j}}, \quad \bar{w}_{ij} \longleftrightarrow \frac{\partial^{|i+j|}\bar{w}}{(\partial \bar{z})^{i}(\partial z)^{j}},$$

and basis Cartan fields

$$D_{\nu} = D_{z^{\nu}} = \frac{\partial}{\partial z^{\nu}} + \sum_{i,j\in I}\left(w_{i+(\nu),j}\frac{\partial}{\partial w_{ij}} + \bar{w}_{i,j+(\nu)}\frac{\partial}{\partial \bar{w}_{ij}}\right),$$

$$\bar{D}_{\nu} = D_{\bar{z}^{\nu}} = \frac{\partial}{\partial \bar{z}^{\nu}} + \sum_{i,j\in I}\left(w_{i,j+(\nu)}\frac{\partial}{\partial w_{ij}} + \bar{w}_{i+(\nu),j}\frac{\partial}{\partial \bar{w}_{ij}}\right),$$

where $\nu=1,\ldots,n$. The Cauchy-Riemann system is determined by the functions $F_\nu = w_{0,(\nu)} = \bar{D}_\nu w$, $\bar{F}_\nu = \bar{w}_{0,(\nu)} = D_\nu \bar{w}$, $\nu=1,\ldots,n$ (note that here $m=2n$, $a=2$, $s=2n$). Further, the solution manifold

$$F = \left\{ z, \bar{z}, \mathbf{w}, \bar{\mathbf{w}} \; : \; w_{ij} = \bar{w}_{ij} = 0, \; i,j \in I, \; |j|>0 \right\}.$$

$$= \left\{ z, \bar{z}, \mathbf{u}, \bar{\mathbf{u}}, \mathbf{v}, \bar{\mathbf{v}} \; : \; \mathbf{v} = \bar{\mathbf{v}} = 0 \right\},$$

where $\mathbf{u}=(u_i=w_{i,0})$, $\bar{\mathbf{u}}=(\bar{u}_i=\bar{w}_{i,0})$, $\mathbf{v}=(v_{ij}=w_{ij})$, $\bar{\mathbf{v}}=(\bar{v}_{ij}=\bar{w}_{ij})$, $i,j\in I$, $|j|>0$. The variables $z,\bar{z},\mathbf{u},\bar{\mathbf{u}}$ are global coordinates on F, and

$$\partial_\nu = \frac{\partial}{\partial z^\nu} + \sum_{i\in I} u_{i+(\nu)} \frac{\partial}{\partial u_i}, \quad \bar{\partial}_\nu = \frac{\partial}{\partial \bar{z}^\nu} + \sum_{i\in I} \bar{u}_{i+(\nu)} \frac{\partial}{\partial \bar{u}_i}, \quad \nu = 1,\ldots,n,$$

are basis Cartan fields on F.

Consider a n-form

$$\omega = \varphi(z,\bar{z},\mathbf{u},\bar{\mathbf{u}}) d^n z \in \Lambda^n \mathscr{C}^\infty(F), \quad \varphi \in \mathscr{C}^\infty(F),$$

where

$$d^n z = dz^1 \wedge \ldots \wedge dz^n.$$

Then

$$\partial\omega = \sum_{\nu=1}^{n} \bar{\partial}_\nu \varphi \; d\bar{z}^\nu \wedge d^n z = 0 \quad \text{iff} \quad \bar{\partial}_\nu \varphi = 0, \quad \nu = 1,\ldots,n,$$

Simplifying the situation we shall denote solutions of the last system by $\varphi=\varphi(z,\mathbf{u})$ (it is correct in the real-analytic category). Now, let

$$\omega = \varphi(z,\mathbf{u}) d^n z$$

be an n-current. When is this current trivial? Consider the auxiliary trivial differential manifold $\mathcal{N}$ with the global coordinates

z, $\mathbf{u}$ and the basis Cartan fields ∂_ν, $\nu=1,\dots,n$, defined above (cf. Section 6). Then the form $\omega\in\Lambda^n(N)$, and we are interested when $\omega=\partial\theta$, $\theta\in\Lambda^{n-1}(N)$. Proposition 6.8 gives the answer:

$$\omega=\partial\theta \quad \text{iff} \quad \frac{\delta\varphi}{\delta u} = \sum_{i\in I} (-1)^{|i|}\partial_i \frac{\partial\varphi}{\partial u_i} = 0.$$

In particular, *the linear space* $\mathrm{Cons}^n\mathcal{F}$ *is not trivial.* Note that there are nontrivial n-currents of the form

$$\kappa = \psi(\bar{z},\bar{\mathbf{u}})d^n\bar{z}, \quad \text{where} \quad \frac{\delta\psi}{\delta\bar{u}} = \sum_{i\in I} (-1)^{|i|}\bar{\partial}_i \frac{\partial\varphi}{\partial\bar{u}_i} \neq 0.$$

Let $[\omega]\in\mathrm{Cons}^n\mathcal{F}$, where

$$\omega = \varphi(z,\mathbf{u})d^n z,$$

is a nontrivial n-current. The image $Z[\omega]=[R]\in\mathcal{H}^{n+2}(\mathrm{DI}_F)$, where the differential identity $(n+2)$-form R is calculated as follows. Let $\Omega=\omega\in\Lambda^n\mathcal{C}^\infty(M)$, then

$$D\Omega = \sum_{\nu=1}^{n} \bar{D}_\nu\varphi \; d\bar{z}^\nu\wedge d^n z = \sum_{\nu=1}^{n} \sum_{i\in I} \frac{\partial\varphi}{\partial u_i} w_{i,(\nu)} \; d\bar{z}^\nu\wedge d^n z$$

$$= \sum_{\nu=1}^{n} P(F,\bar{F})_\nu \; d\bar{z}^\nu\wedge d^n z = Q(F,\bar{F}),$$

where

$$P(\Phi,\bar{\Phi})_\nu = \sum_{i\in I} \frac{\partial\varphi}{\partial u_i} D_i\Phi_\nu, \quad \Phi,\bar{\Phi} \in \mathcal{C}^\infty(M)_n,$$

(cf. Section 6, remind that $F_\nu=\bar{D}_\nu w=w_{0,(\nu)}$). Hence,

$$R(\Phi,\Phi) = (D\circ Q)(\Phi,\Phi) = D(Q(\Phi,\Phi))$$

$$= \sum_{\mu,\nu=1}^{n} \bar{D}_{\mu}\left(P(\Phi,\bar{\Phi})_{\nu}\right)\ d\bar{z}^{\mu}\wedge d\bar{z}^{\nu}\wedge d^{n}z$$

$$= \sum_{\mu,\nu=1}^{n} R(\Phi,\bar{\Phi})_{\mu\nu}\ d\bar{z}^{\mu}\wedge d\bar{z}^{\nu}\wedge d^{n}z,$$

where

$$R(\Phi,\bar{\Phi})_{\mu\nu} = \sum_{i,j\in I} \frac{\partial^{2}\varphi}{\partial u_{i}\partial u_{j}}\left(D_{i}F_{[\mu}\right)\left(D_{j}\Phi_{\nu]}\right) + \sum_{i\in I}\frac{\partial\varphi}{\partial u_{i}}D_{i}\left(\bar{D}_{[\mu}\Phi_{\nu]}\right),$$

and the brackets [...] denote the skew-symmetrization in indices $\mu,\nu=1,\ldots,n$.

Remind, that according to Theorem 7.8 evolution systems don't have low-dimensional conservation laws. Also any system of PDE, which can be reduced to an evolution system by a suitable change of independent and dependent variables, have no such laws. In particular, this is true for pseudo Cauchy-Kowalevski systems, normal systems and regular systems (see Section 7). A great number of PDE of mathematical physics are of this type, hence they don't have low-dimensional conservation laws, and this fact can explain the lack of interest in the latter. Nevertheless, there are important in applications systems of PDE which are not normal. For instance, gauge invariant systems (Yang-Mills, Einstein-Hilbert and so on) are of this class.

Example 13.12. The simple example of a gauge invariant system of PDE having low-dimensional conservation laws is the Yang-Mills equation in the 2-dimensional space-time. Let $\mathfrak{A}$ be a Lie algebra with a bracket $[\cdot,\cdot]$ and a Killing metric $(\cdot,\cdot)$, so

$$([f,g],h) + (g,[f,h]) = 0, \quad f,g,h \in \mathfrak{A}.$$

In the standard situation $\mathfrak{A}=su(n)$ is the Lie algebra of all skew-Hermitian $n\times n$-matrices with the zero trace,

$$[f,g] = fg - gf, \quad (f,g) = \mathrm{Tr}\, fg^*, \quad f,g \in \mathfrak{A},$$

where the asterisk denotes the Hermitian conjugation (note that $g^*=-g$ for all $g\in\mathfrak{A}$, by the definition). Further, let $x=(x^0,x^1)\in\mathbb{R}^2$ be independent variables and $w=(w_0,w_1)\in\mathfrak{A}_2$ be dependent variables. Consider the Lagrangian system of PDE

$$F^\mu = \frac{\delta\mathcal{L}}{\delta w_\mu} = \sum_{\nu=0}^{1} \nabla_\nu f^{\nu\mu}, \quad \mu = 0,1,$$

where

$$f_{\mu\nu} = \frac{\partial w_\nu}{\partial x^\mu} - \frac{\partial w_\mu}{\partial x^\nu} + \left[w_\mu, w_\nu\right] = -f^{\mu\nu}, \quad \mu,\nu = 0,1,$$

(the last equality is valid in the 2-dimensional space-time)

$$\mathcal{L} = -\frac{1}{4} \sum_{\mu,\nu=0}^{1} \left(f_{\mu\nu}, f^{\mu\nu}\right) = \frac{1}{2}(f,f), \quad f = f_{01},$$

$$\nabla_\mu g = \frac{\partial g}{\partial x^\mu} + \left[w_\mu, g\right], \quad \mu = 0,1, \quad g(x) \in \mathfrak{A}.$$

This system is called the *Yang-Mills system*, and satisfies to the differential identity

$$\sum_{\mu=0}^{1} \nabla_\mu F^\mu \equiv 0,$$

because,

$$\sum_{\mu=0}^{1}\nabla_\mu F^\mu = \sum_{\mu,\nu=0}^{1}\nabla_\mu\nabla_\nu f^{\mu\nu} = \sum_{\mu,\nu=0}^{1}\left[\nabla_\mu,\nabla_\nu\right]f^{\mu\nu} = \sum_{\mu,\nu=0}^{1}\left[f_{\mu\nu},f^{\mu\nu}\right] \equiv 0.$$

In this case the ambient trivial differential manifold $\mathcal{M}=(M,D)$, $M=\mathbb{R}^2\times\mathfrak{A}_{K\times 2}$, $K=\mathbb{N}^2$, has global coordinates $x=(x_\mu)\in\mathbb{R}^2$, $\mathbf{w}=(w_{k;\mu})\in\mathfrak{A}_{K\times 2}$, $\mu=0,1$, $k\in K$, and basis Cartan fields

$$D_\mu = \frac{\partial}{\partial x^\mu} + \sum_{\nu=0}^{1}\sum_{k\in K} w_{k+(\mu);\nu}\frac{\partial}{\partial w_{k;\nu}}, \quad \mu = 0,1,$$

acting in the algebras $\mathcal{C}^\infty(M)$ and $\mathcal{C}^\infty(M)^{\mathfrak{A}}$ (the latter is the algebra of all smooth $\mathfrak{A}$-valued functions on M). Let us prolong the operators ∇_μ, $\mu=0,1$, on the algebra $\mathcal{C}^\infty(M)^{\mathfrak{A}}$ by the formula

$$\nabla_\mu\Phi = D_\mu\Phi + \left[w_{0;\mu},\Phi\right], \quad \mu = 0,1, \quad \Phi \in \mathcal{C}^\infty(M)^{\mathfrak{A}}.$$

The Yang-Mills equations are determined by the functions

$$F^\mu = \sum_{\nu=0}^{1}\nabla_\nu f^{\nu\mu}, \quad \mu=0,1,$$

where

$$f_{\mu\nu} = -f^{\mu\nu} = w_{(\mu);\nu}-w_{(\nu);\mu}+\left[w_{0;\mu},w_{0;\nu}\right] = e_{\mu\nu}f, \quad f = f_{01},$$

$e_{\mu\nu}=-e^{\mu\nu}$ is the sign of the permutation $\{0,1\}\mapsto\{\mu,\nu\}$, $\mu,\nu=0,1$. The system $\{F=0\}$ has a standard realization, but we shall do our study without using this realization explicitly.

Consider a 0-form

$$\Omega = \vartheta(\mathcal{L}), \quad \vartheta \in \mathcal{C}^\infty(\mathbb{R}), \quad \mathcal{L} = -\frac{1}{4}\sum_{\mu,\nu=0}\left(f_{\mu\nu},f^{\mu\nu}\right).$$

Here

$$D\Omega = \sum_{\mu=0}^{1} D_\mu \vartheta \; dx^\mu = \vartheta' \sum_{\mu=0}^{1} D_\mu \mathcal{L} \; dx^\mu$$

$$= \vartheta' \sum_{\mu,\nu=0}^{1} \left(f_{\mu\nu}, F^\mu\right) \; dx^\nu = \sum_{\nu=0}^{1} P(F)_\nu \; dx^\nu = Q(F),$$

where

$$P(\Phi)_\nu = \vartheta' \sum_{\mu=0}^{1} \left(f_{\mu\nu}, \Phi^\mu\right), \quad \nu = 0,1, \quad \Phi \in \mathcal{C}^\infty(M)^{2\times\mathfrak{A}},$$

and

$$D_\mu \mathcal{L} = \left(f, D_\mu f\right) = \left(f, \nabla_\mu f\right) - \left(f, \left[w_{0;\mu}, f\right]\right)$$

$$= -\sum_{\nu=0}^{1} e_{\mu\nu}\left(f, F^\nu\right) = \sum_{\nu=0}^{1} \left(f_{\nu\mu}, F^\nu\right),$$

because

$$\left(f, \left[w_{0;\mu}, f\right]\right) = \left(w_{0;\mu}, \left[f, f\right]\right) = 0, \quad \text{and} \quad \nabla_\mu f = -\sum_{\nu=0} e_{\mu\nu} F^\nu.$$

Hence, Ω is a 0-current. This current is nontrivial if ϑ is not constant (i.e., $\vartheta' \neq 0$). In particular, *the space* Cons$^0\mathcal{F}$ *is not trivial.*

Let $\Omega = \vartheta(\mathcal{L})$ be a nontrivial 0-current. The corresponding differential identity 2-form R is calculated as follows.

$$R(\Phi) = (D\circ Q)(\Phi) = D(Q(\Phi))$$

$$= D\left(\vartheta' \sum_{\mu,\nu=0}^{1} \left(f_{\mu\nu}, \Phi^\mu\right) dx^\nu\right) = D\left(\sum_{\mu,\nu=0}^{1} e_{\mu\nu}\left(\vartheta' f, \Phi^\mu\right) dx^\nu\right)$$

$$= \sum_{\lambda,\mu,\nu=0}^{1} e_{\mu\nu} D_\lambda \left(\vartheta' f, \Phi^\mu\right) \; dx^\lambda \wedge dx^\nu = S(\Phi) \; d^2x,$$

where

$$S(\Phi) = \sum_{\mu=0}^{1} D_\mu\left(\vartheta' f, \Phi^\mu\right), \quad \Phi \in \mathscr{C}^\infty(M)^{2\times\mathfrak{A}}.$$

In particular,

$$S(F) = \sum_{\mu=0}^{1} \left(\vartheta'' \left(D_\mu \mathscr{L}\right)\left(f, F^\mu\right) + \vartheta' \left(\left(\nabla_\mu f, F^\mu\right) + \left(f, \nabla_\mu F^\mu\right)\right)\right)$$

$$= \sum_{\mu=0}^{1} \left(\vartheta'' \sum_{\nu=0}^{1} \left(f_{\nu\mu}, F^\nu\right)\left(f, F^\mu\right) + \vartheta' \left(- \sum_{\nu=0}^{1} e_{\mu\nu} F^\nu, F^\mu\right)\right)$$

$$= \sum_{\mu,\nu=0}^{1} e_{\mu\nu}\left(\vartheta'' \left(f, F^\mu\right)\left(f, F^\nu\right) + \vartheta' \left(F^\mu, F^\nu\right)\right) = 0,$$

where we used the previous calculations and the simple equality

$$D_\mu\left(\varphi, \psi\right) = \left(\nabla_\mu \varphi, \psi\right) + \left(\varphi, \nabla_\mu \psi\right), \quad \mu = 0, 1, \quad \varphi, \psi \in \mathscr{C}^\infty(M)^{\mathfrak{A}}.$$

This means that indeed $R \in \Lambda^2 \mathrm{DI}_F$, and $[R] \in \mathscr{H}^2(\mathrm{DI}_F)$.

Let us return to the general situation and give the external description of conservation laws of the maximal dimension (cf. Olver [1, Sect.4.3]). So, let $\{F=0\}$ be a generating system of PDE and $\mathcal{M}=(M,D)$ the corresponding ambient trivial differential manifold. We want to study the space

$$\mathrm{Cons}_{ext}^{m-1}\mathscr{F} = \frac{\left\{\Omega \in \Lambda^{m-1}\mathscr{C}^\infty(M) \ : \ D\Omega \in \Lambda^m \mathscr{I}_F\right\}}{\left\{\Omega = \Phi + D\Psi \ ; \ \Phi \in \Lambda^{m-1}\mathscr{I}_F, \ \Psi \in \Lambda^{m-2}\mathscr{C}^\infty(M)\right\}}.$$

Remind that every $(m-1)$-form $\Omega \in \Lambda^{m-1}\mathscr{C}^\infty(M)$ has the representation

$$\Omega = \sum_{\mu=1}^{m} J^\mu(x, \mathbf{w})\, d^{m-1}x_\mu, \quad J \in \mathscr{C}^\infty(M)^m,$$

(see Section 6), and

$$D\Omega = \left(\sum_{\mu=1}^{m} D_\mu J^\mu \right) d^m x \in \Lambda^m \mathcal{C}^\infty(M).$$

Consider the linear space

$$\text{PreChar } \mathcal{F} = \left\{ \Gamma \in \mathcal{C}^\infty(M)_s \; : \; (\Gamma, F) = \Sigma D_\mu R^\mu, \; R \in \mathcal{C}^\infty(M)^m \right\},$$

where

$$(\Gamma, F) = \sum_{\sigma=1}^{s} \Gamma_\sigma F^\sigma, \quad \Sigma D_\mu R^\mu = \sum_{\mu=1}^{m} D_\mu R^\mu.$$

Elements of this space we call *current precharacteristics* (or *characteristics*, for brevity) of the system $\{F=0\}$. Further we call a precharacteristic Γ *trivial* if $(\Gamma, F) = \Sigma D_\mu R^\mu$ for some trivial current $R \in \mathcal{I}_F{}^m$, and denote by TriPreChar$\mathcal{F}$ the space of all trivial precharacteristics. At last, we denote by Char$\mathcal{F}$ the corresponding quotient space, and call its elements (i.e., equivalence classes $[\Gamma]$) *current characteristics* (or simply *characteristics*) of the system $\{F=0\}$.

Theorem 13.13 (Zharinov [7]). *Let $\{F=0\}$ be a generating system of PDE. Then there is an isomorphism*

$$X \; : \; \text{Cons}_{ext}^{m-1} \mathcal{F} \simeq \text{Char } \mathcal{F}.$$

Proof. Let an equivalence class $[\Omega] \in \text{Cons}_{ext}^{m-1}\mathcal{F}$, so

$$\sum_{\mu=1}^{m} D_\mu J^\mu = P(F) \quad \text{for some } P \in \text{DO}_s.$$

Let us put $\Gamma = \chi(P)$, then by Proposition A.2,

$$(\Gamma,F) = (\chi(P),F) = P(F) - \sum_{\mu=1}^{m} D_\mu\left(h^\mu(P)(F)\right)$$

$$= \sum_{\mu=1}^{m} D_\mu\left(J^\mu - h^\mu(P)(F)\right) = \sum_{\mu=1}^{m} D_\mu R^\mu, \quad R \in \mathcal{C}^\infty(M)^m.$$

The operator P is determined up to a differential identity $Q \in \mathrm{DI}_F$, and

$$(\chi(Q),F) = Q(F) - \sum_{\mu=1}^{m} D_\mu\left(h^\mu(Q)(F)\right)$$

$$= \sum_{\mu=1}^{m} D_\mu\left(-h^\mu(P)(F)\right) = \sum_{\mu=1}^{m} D_\mu R^\mu, \quad R \in \mathcal{I}_F{}^m,$$

because $Q(F)=0$, by the definition. Now, let $\Omega=\Phi+D\Psi$ be a trivial $(m-1)$-current, then $D\Omega=D\Phi$, where

$$\Phi = \sum_{\mu=1}^{m} Q^\mu(F)d^{m-1}x_\mu, \quad Q \in \mathrm{DO}_S^m, \quad D\Phi = P(F)d^m x, \quad P = \sum_{\mu=1}^{m} D_\mu \circ Q^\mu,$$

and, as one can easily check, $\chi(P)=0$. Further, let $\Gamma \in \mathcal{C}^\infty(M)_S$, where $(\Gamma,F)=\Sigma D_\mu R^\mu$, $R \in \mathcal{C}^\infty(M)^m$. Then $[\Gamma]=X[\Omega]$, where $\Omega=\Sigma R^\mu d^{m-1}x_\mu$. At last, suppose that $X[\Omega]=0$, i.e.,

$$\Omega = \sum_{\mu=1}^{m} J^\mu d^{m-1}x_\mu, \quad \sum_{\mu=1}^{m} D_\mu J^\mu = P(F),$$

and

$$(\Gamma,F) = (\chi(P),F) = \sum_{\mu=1}^{m} D_\mu R^\mu, \quad R = Q(F) \in \mathcal{I}_F{}^m.$$

Then, by Proposition A.2,

$$\sum_{\mu=1}^{m} D_\mu J^\mu = (\chi(P),F) + \sum_{\mu=1}^{m} D_\mu \circ h^\mu(P)(F)$$

$$= \sum_{\mu=1}^{m} D_\mu \circ \left(Q^\mu + h^\mu(P)\right)(F) = \sum_{\mu=1}^{m} D_\mu \circ H^\mu(F),$$

or

$$\sum_{\mu=1}^{m} D_\mu \left(J^\mu - H^\mu(F)\right) = 0.$$

In other words, $D(\Omega-\Phi)=0$, where $\Phi=\Sigma H^\mu(F)d^{m-1}x_\mu \in \Lambda^{m-1}\mathcal{J}_F$. By Theorem 6.4, $\Omega-\Phi=D\Psi$, $\Psi\in\Lambda^{m-2}\mathcal{C}^\infty(M)$, i.e., the current Ω is trivial. This completes the proof.

Note that Olver ([1, Sect.4.3]) calls a function $\Gamma=\chi(P)$ a *characteristic* of a conservation law $[\Omega]$.

According to Proposition 6.8,

$$(\Gamma,F) = \sum_{\mu=1}^{m} D_\mu R^\mu \quad \text{for some} \quad R \in \mathcal{C}^\infty(M)^m \quad \text{iff} \quad \frac{\delta(\Gamma,F)}{\delta w} = 0,$$

i.e., iff

$$\Gamma^*(F) + F^*(\Gamma) = 0,$$

where

$$\Gamma^*(F)_\alpha = \sum_{\sigma=1}^{s} \sum_{k\in K} (-1)^{|k|} D_k\left(\frac{\partial\Gamma_\sigma}{\partial w_k^\alpha} F_\sigma\right), \quad F^*(\Gamma)_\alpha = \sum_{\sigma=1}^{s} \sum_{k\in K} (-1)^{|k|} D_k\left(\frac{\partial F^\sigma}{\partial w_k^\alpha} \Gamma_\sigma\right),$$

$\alpha = 1,\dots,a$. Denote by $\overline{F^*}:\mathcal{C}^\infty_{ext}(F)_s \longrightarrow \mathcal{C}^\infty_{ext}(F)_a$ the "restriction" of the differential operator $F^*:\mathcal{C}^\infty(M)_s \longrightarrow \mathcal{C}^\infty(M)_a$ defined by the rule

$$\mathcal{C}^\infty_{ext}(F)_s \ni \phi=[\Phi] \longmapsto \overline{F^*}(\phi)=[F^*\Phi] \in C^\infty_{ext}(F)_a$$

(see the beginning of this Section). Let $\mathrm{Ker}\overline{F^*}$ be the kernel of this restricted operator, i.e., the linear space of all solutions of the equation $\overline{F^*}(\psi)=0$, $\psi\in\mathcal{C}^\infty_{ext}(F)_s$.

Now, let $\gamma=\Gamma\big|_F$ be the restriction of a precharacteristic Γ of a conservation law $[\Omega]$, then the above criterion yields

$$\overline{F^*}(\gamma) = 0$$

(clear that $\Gamma^*(F)\in\mathcal{I}_F$). Note that the quotient function γ is not uniquely determined by $[\Omega]$.

Proposition 13.14. *Let $\{F=0\}$ be a generating well-determined system of PDE. Then every characteristic $[\Gamma]$ is uniquely determined by the restriction $\gamma=\Gamma\big|_F$, and thus the linear mapping*

$$\chi : \mathrm{Cons}^{m-1}_{ext}\mathcal{F} \longrightarrow \mathrm{Ker}\ \overline{F^*}$$

is defined.

Proof. Indeed, let $\Gamma\in\mathcal{C}^\infty(M)_s$, $(\Gamma,F)=\Sigma D_\mu R^\mu$, $R\in\mathcal{I}_F{}^m$. In view of Theorem 13.13, we should prove that $\Gamma\big|_F=0$. In our case $R=Q(F)$ for some differential operator

$$Q = \left(Q^\mu_\sigma = \sum_{k\in K} Q^{\mu;k}_\sigma D_k\right) \in \mathrm{DO}^m_s,$$

because the system $\{F=0\}$ is assumed to be generating. Hence, we have the differential identity

$$\left(\sum_{\mu=1}^m D_\mu\circ Q^\mu_\sigma - \Gamma_\sigma\right) \in \mathrm{DI}_F,$$

whose coefficients should belong to $\mathcal{I}_F$, since the system $\{F=0\}$ is well-determined by the hypothesis. Let us use the multi-index notation: $k=(\mu_1)+(\mu_2)+\ldots+(\mu_r)$ if $|k|=r$ (see the proof of Lemma 6.6), then (p is the order of the differential operator Q):

$$\sum_{\mu=1}^{m} D_\mu \circ Q_\sigma^\mu - \Gamma_\sigma = \sum_\mu D_\mu \circ \sum_{r=0}^{p} \sum_{|k|=r} Q_\sigma^{\mu;\mu_1\ldots\mu_r} D_{\mu_1}\ldots D_{\mu_r} - \Gamma_\sigma$$

$$= \sum_{|k|=p+1} Q_\sigma^{\{\mu_1;\mu_2\ldots\mu_{p+1}\}} D_{\mu_1}\ldots D_{\mu_{p+1}}$$

$$+ \sum_{r=1}^{p} \sum_{|k|=r} \left(\sum_\mu D_\mu Q_\sigma^{\mu;\mu_1\ldots\mu_r} + Q_\sigma^{\{\mu_1;\mu_2\ldots\mu_r\}} \right) D_{\mu_1}\ldots D_{\mu_r}$$

$$+ \sum_\mu D_\mu Q_\sigma^{\mu;0} - \Gamma_\sigma,$$

where $\sigma=1,\ldots,s$. Thus we have:

(a) $Q_\sigma^{\{\mu_1;\mu_2\ldots\mu_{p+1}\}} = 0 \mod \mathcal{I}_F,$

(b) $\sum_\mu D_\mu Q_\sigma^{\mu;\mu_1\ldots\mu_r} + Q_\sigma^{\{\mu_1;\mu_2\ldots\mu_r\}} = 0 \mod \mathcal{I}_F,$

(c) $\sum_\mu D_\mu Q_\sigma^{\mu;0} - \Gamma_\sigma = 0 \mod \mathcal{I}_F,$

where $\sigma=1,\ldots,s$, the curly braces $\{\ldots\}$ denote symmetrization in indices $\mu_1,\ldots,\mu_r$. Taking into account that the ideal $\mathcal{I}_F$ is differential (i.e., $D_\mu f \in \mathcal{I}_F$ for all $f \in \mathcal{I}_F$ and $\mu=1,\ldots,m$), we get

$$\Gamma_\sigma = \sum_\mu D_\mu Q_\sigma^{\mu;0} \mod \mathcal{I}_F$$

$$= - \sum_{\mu_1,\mu_2} D_{\mu_1} D_{\mu_2} Q_\sigma^{\{\mu_1;\mu_2\}} \mod \mathcal{I}_F$$

$$\cdots\cdots\cdots\cdots\cdots\cdots\cdots\cdots\cdots\cdots$$

$$= (-1)^p \sum_{\mu_1\ldots\mu_{p+1}} D_{\mu_1}\ldots D_{\mu_{p+1}} Q_\sigma^{\{\mu_1;\mu_2\ldots\mu_{p+1}\}} \mod \mathcal{I}_F$$

$$= 0 \mod \mathcal{I}_F,$$

$\sigma=1,\ldots,s$. Thus $\Gamma\Big|_F=0$, and the proof is completed.

Suppose that a well-determined system $\{F=0\}$ has a standard realization. Then $\mathrm{Cons}_{ext}^{m-1}\mathcal{F}=\mathrm{Cons}^{m-1}\mathcal{F}$, and we have two characteristic mappings from $\mathrm{Cons}^{m-1}\mathcal{F}$ into $\mathrm{Ker}\overline{F^*}$, one defined in Section 11 (we denote it by χ_g here) and other defined above (we denote it by χ_a here).

Proposition 13.15. *If a well-determined system $\{F=0\}$ has s standard realization then the mappings χ_g and χ_a coincide.*

Proof. Let $\Omega=\Sigma J^\mu d^{m-1}x_\mu$ be a $(m-1)$-current, then

$$\sum_{\mu=1}^m D_\mu J^\mu = P(F) = \sum_{\sigma=1}^s \sum_{k\in K} A_\sigma^k(x,\mathbf{w})D_k F^\sigma = \sum_{\sigma=1}^s \sum_{k\in K} A_\sigma^k(x,\mathbf{u},\mathbf{v})v_k^\sigma$$

(see Propositions 10.8 and 10.10). Now,

$$\Gamma_\sigma = \chi(P)_\sigma = \sum_{k\in K} (-1)^{|k|} D_k A_\sigma^k, \quad \sigma = 1,\ldots,s.$$

On the other side (see Proposition 11.3),

$$\frac{\partial}{\partial v_k^\sigma}\left(\Sigma D_\mu J^\mu\right) = A_\sigma^k(x,\mathbf{u},\mathbf{v}) \mod \mathcal{I}_F.$$

Hence,

$$\chi_g[\Omega] = \Gamma\Big|_F = \left(\sum_{k\in K}(-1)^{|k|}D_kA_\sigma^k\right)\Big|_F = \chi_a[\Omega].$$

Remark 13.16. The realization of a system $\{F=0\}$ under study as a reduction of an ambient trivial differential manifold $\mathcal{M}$ and further work in F-coordinates $x,\mathbf{u},\mathbf{v}$ usually involves considerable computational difficulties. From this point of view the technic developed in this section has certain advantages, since it uses only ambient coordinates $x,\mathbf{w}$. Moreover, one can apply this technic even when the solution variety F has no structure of a smooth submanifold of M, and usual methods don't work.

14. BACKLUND CORRESPONDENCE

Isomorphisms in the category of differential manifolds are Lie-Backlund isomorphisms. Two differential manifolds connected by a Lie-Backlund isomorphism can be identified in practice. A transition from a given differential manifold to an isomorphic one can be considered as a transition to a different equivalent atlas. Lie-Backlund isomorphisms are encountered very rarely in practice. A more frequent case is a differential relation between two differential manifold, which has come to be called a *Backlund transformation*. Since such a relation is essentially ambiguous in both directions, we regard the term transformation as lame in this case and adhere to the vaguer term *correspondence*.

Namely, we define *a Backlund correspondence* as a quintuple $(\mathcal{R},\mathcal{M},\mathcal{N},\pi,\rho)$, where $\pi:\mathcal{R}\longrightarrow\mathcal{M}$ and $\rho:\mathcal{R}\longrightarrow\mathcal{N}$ are Lie-Backlund submersions (see Section 6). It is convenient to display such a quintuple as a diagram

$$
\begin{array}{ccc}
 & \mathcal{R} & \\
{}^{\pi}\swarrow & & \searrow^{\rho} \\
\mathcal{M} & \sim & \mathcal{N}
\end{array}
$$

The differential relation arising here between the differential manifolds $\mathcal{M}$ and $\mathcal{N}$, which we indicate by a tilde, is a geometrical expression of a traditional Backlund correspondence (Zharinov [4]). Let us reveal this relation.

Proposition 14.1. *Let $\pi:\mathcal{R}\longrightarrow\mathcal{M}$ be a Lie-Backlund submersion. Then for each point $b\in M$ the inverse image $\pi^{-1}(b)$ gives rise to a differential submanifold of $\mathcal{R}$ whose Cartan dimension $l=r-m$, where r and m are Cartan dimensions of $\mathcal{R}$ and $\mathcal{M}$, respectively* (cf. Vinogradov, Krasil'shchik, Lychagin [1, Sect.7.5.9]).

Proof. Indeed, let a point $a\in F=\pi^{-1}(b)$. Let $(V,\psi,\mathbb{R}^{m+l+I+J},\nabla)$, $a\in V$, be a standard defining chart on $\mathcal{R}$ with coordinates x^μ, u^i, y^λ, v^j, and basis Cartan fields ∇_{x^μ}, ∇_{y^λ}, and $(U,\varphi,\mathbb{R}^{m+I},\partial)$ be a chart on $\mathcal{M}$ with coordinates x^μ, u^i, and fields ∂_{x^μ}, $\mu=1,\dots,m$, $i\in I$, $\lambda=1,\dots,l$, $j\in J$ (see Proposition 2.11). It can be assumed without los of generality that the point b has coordinates $x=0$, $u=0$. Then

$$F\cap V = \left\{x,u,y,v\ :\ x=0,u=0\right\},$$

and hence F is a submanifold of R. Further, the restrictions of the basis fields $\nabla_{y^1},\dots,\nabla_{y^l}$ on $F\cap V$, which are equal to

$$\partial_{y^\lambda} = \nabla_{y^\lambda}\Big|_{F\cap V} = \frac{\partial}{\partial y^\lambda} + \sum_{j\in J} C^j_\lambda(0,0,y,v)\,\frac{\partial}{\partial v^j},\quad \lambda = 1,\dots,l,$$

are tangent vector fields on $F\cap V$ that specify an l-dimensional involutive bundle, which we take as a Cartan distribution CTF on $F\cap V$. We show that this definition does not depend on a choice of charts. Let $(V',\psi',\mathbb{R}^{m+l+I'+J'},\nabla')$, $a\in V$, and $(U',\varphi',\mathbb{R}^{m+I'},\partial')$ be another standard defining charts on $\mathcal{R}$ and $\mathcal{M}$, respectively, where the new coordinates of the point b are again $x'=0,\ u'=0$. In this new coordinates

$$F\cap V' = \left\{x',u',y',v' \ : \ x'=0, u'=0\right\},$$

and the old basis fields are

$$\nabla_{y^\lambda} = \sum_{\mu=1}^{m} \nabla_{y^\lambda}\left(x'^{\mu}\right)\nabla'_{x',\mu} + \sum_{\kappa=1}^{l} \nabla_{y^\lambda}\left(y'^{\kappa}\right)\nabla_{y',\kappa'}, \quad \lambda = 1,\dots,l,$$

where $x'^{\mu}, y'^{\lambda}(x,u,y,v)$ are the components of the transition functions in the new pair of charts, and $x'^{\mu}(0,0,y,v)=0$ by construction. The restrictions satisfy

$$\nabla_{y^\lambda}\Big|_H = \sum_{\mu=1}^{m} \nabla_{y^\lambda}\left(x'^{\mu}\right)\Big|_H \nabla'_{x',\mu}\Big|_H + \sum_{\kappa=1}^{l} \nabla_{y^\lambda}\left(y'^{\kappa}\right)\Big|_H \nabla_{y',\kappa'}\Big|_H$$

$$= \sum_{\kappa=1}^{l} \nabla_{y^\lambda}\left(y'^{\kappa}\right)\Big|_H \nabla_{y',\kappa'}\Big|_{H'} \quad \lambda = 1,\dots,l,$$

since the restrictions $\nabla_{y^\lambda}\left(x'^{\mu}\right)\Big|_H=0$ on $H=F\cap V\cap V'$. Thus, the vector fields ∂_{y^λ} and $\partial'_{y',\lambda}$ determine the same distribution on H, as it was required to prove.

Proposition 14.2. *Let $\pi:\mathcal{R}\longrightarrow\mathcal{M}$ be a Lie-Backlund submersion. Then every integral submanifold P of $\mathcal{M}$ gives rise to a reduction $\mathcal{S}=(S,CTS)$ of $\mathcal{R}$, where $S=\pi^{-1}(P)$ and CTS is the restriction of the Cartan distribution CTR on S.*

Proof. Indeed, let a point $a\in S$ and let $(V,\psi,\mathbb{R}^{m+l+I+J},\nabla)$ and $(U,\varphi,\mathbb{R}^{m+I},\partial)$, $a\in V$, be standard defining charts, the same as in the proof of Proposition 14.1. According to Proposition 1.8,

$$P\cap U = \left\{x,u \ : \ u=s(x)\right\} \quad \text{hence} \quad S\cap V = \left\{x,u,y,v \ : \ u=s(x)\right\}.$$

Let us pass in V from the coordinates x^{μ}, u^{i}, y^{λ}, v^{j} to the coordinates x^{μ}, w^{i}, y^{λ}, v^{j}, where $w^{i}=u^{i}-s^{i}(x)$, $i\in I$. In the new coordinates we have

$$S\cap V = \left\{x,w,y,v \ : \ w=0\right\},$$

and hence S is a submanifold of R. The basis Cartan fields on V in the new coordinates take the form

$$\nabla_{x^{\mu}} = \frac{\partial}{\partial x^{\mu}} + \sum_{i\in I} \left(A^{i}_{\mu}(x,w+s)-\frac{\partial s^{i}}{\partial x^{\mu}}\right) \frac{\partial}{\partial w^{i}} + \sum_{j\in J} B^{j}_{\mu}(x,w+s,y,v) \frac{\partial}{\partial v^{j}},$$

$$\nabla_{y^{\lambda}} = \frac{\partial}{\partial y^{\lambda}} + \sum_{j\in J} C^{j}_{\lambda}(x,w+s,y,v) \frac{\partial}{\partial v^{j}},$$

$\mu=1,\ldots,m$, $\lambda=1,\ldots,l$. By the contact equations in Proposition 1.8, their restrictions on $S\cap V$B, which are equal to

$$\nabla_{x^{\mu}}\Big|_{F\cap V} = \frac{\partial}{\partial x^{\mu}} + \sum_{j\in J} B^{j}_{\mu}(x,s,y,v) \frac{\partial}{\partial v^{j}}, \quad \mu = 1,\ldots,m,$$

$$\nabla_{y^{\lambda}}\Big|_{F\cap V} = \frac{\partial}{\partial y^{\lambda}} + \sum_{j\in J} C^{j}_{\lambda}(x,s,y,v) \frac{\partial}{\partial v^{j}}, \quad \lambda = 1,\ldots,l,$$

are tangent vector fields on $S\cap V$ specifying a $(m+l)$-dimensional Cartan distribution CTS. Thus, the pair $(\mathcal{S},CTS)$ is a reduction of the differential manifold $\mathcal{R}$.

Proposition 14.3. *Let $\rho:\mathcal{R}\to\mathcal{N}$ be a Lie-Backlund submersion. Then locally the image of every integral submanifold of $\mathcal{R}$ is an integral submanifold of $\mathcal{N}$* (cf. Proposition 2.8).

Proof. Indeed, let S be an integral submanifold of the differential manifold $\mathcal{R}$, and let $a\in S$. Again, let $(V,\psi,\mathbb{R}^{m+l+I+J},\nabla)$ and $(U,\varphi,\mathbb{R}^{m+I},\partial)$, $a\in V$, be standard defining charts, the same as in the proof of Proposition 14.1. Then, in view of Proposition 1.8,

$$S\cap V = \left\{x,u,y,v \ : \ u=u(x,y),\ v=t(x,y)\right\},$$

where the smooth functions $s^i(x,y)$, $t^j(x,y)$, $i\in I$, $j\in J$, satisfy the contact equations

$$\frac{\partial s^i}{\partial x^\mu} = A^i_\mu(x,s), \qquad \frac{\partial s^i}{\partial y^\lambda} = 0,$$

$$\frac{\partial t^j}{\partial x^\mu} = B^j_\mu(x,s,y,t), \quad \frac{\partial t^j}{\partial y^\lambda} = C^j_\lambda(x,s,y,t),$$

$\mu=1,\dots,m$, $i\in I$, $\lambda=1,\dots,l$, $j\in J$. In particular, the functions s^i do not depend on y (i.e., $s^i=s^i(x)$, $i\in I$). Further, by Proposition 2.11,

$$\rho(S\cap V) = \left\{x,u \ : \ u=s(x)\right\},$$

and it follows from the first of the four subsystems written out above and Proposition 1.8 that $\rho(S\cap V)$ is an integral submanifold of the differential manifold $\mathcal{N}$.

Now, let $(\mathcal{R},\mathcal{M},\mathcal{N},\pi,\rho)$ be a Backlund correspondence. Let P be an integral submanifold of the differential manifold $\mathcal{M}$. In view of Proposition 14.2, P determines the reduction $\mathcal{S}=(S,CTS)$ of $\mathcal{R}$, where $S=\pi^{-1}(P)$ and CTS is the restriction of the Cartan distribution CTR on S. In turn, integral submanifolds of the differential manifold $\mathcal{S}$, being simultaneously integral submanifolds of $\mathcal{R}$, locally give rise to integral submanifolds of $\mathcal{N}$. Thus, a Backlund correspondence relates each integral submanifold of $\mathcal{M}$ with a set of local integral submanifolds of $\mathcal{N}$, parameterized by auxiliary coordinates in R. In applications the manifold S is usually finite dimensional, hence in order to construct integral submanifolds of $\mathcal{S}$ one should solve a maximally overdetermined system of PDE, and generally this is much more easy to do than to solve a system of PDE, corresponding to the differential manifold $\mathcal{N}$. Analogously, a differential relation acts also in the inverse direction, i.e., from the differential manifold $\mathcal{N}$ to the differential manifold $\mathcal{M}$. We remark that the domain of application of Backlund correspondences is not limited to the construction of integral submanifolds of differential manifolds, but includes also the description of their nonlocal symmetries and conservation laws (cf. Vinogradov, Krasil'shchik, Lychagin [1, Sect.7.5.17] and Tsujishita [1]).

Let $\mathcal{M}$ and $\mathcal{N}$ be two differential manifolds, and $\mathcal{M}\times\mathcal{N}$ be their direct product. Denote by $\pi:\mathcal{M}\times\mathcal{N}\longrightarrow\mathcal{M}$ and $\rho:\mathcal{M}\times\mathcal{N}\longrightarrow\mathcal{N}$ the natural projections. Then the quintuple $(\mathcal{M}\times\mathcal{N},\mathcal{M},\mathcal{N},\pi,\rho)$ will be a Backlund correspondence. However, the differential relation arising here is purely formal, it in no way restricts the differential manifolds $\mathcal{M}$ and $\mathcal{N}$, and does not allow us to carry information over from one differential manifold to the other. The opposite limiting case appears when we have a Lie-Backlund isomorphism $f:\mathcal{M}\simeq\mathcal{N}$. Indeed, consider a differential submanifold $\mathcal{G}_f$ of the direct product $\mathcal{M}\times\mathcal{N}$

determined by the rule: a point $c = (a,b) \in M\times N$ belongs to G iff $b=f(a)$ (i.e., G_f is the graph of the isomorphism f), and let $\pi_f:\mathcal{G}_f\longrightarrow\mathcal{M}$ and $\rho_f:\mathcal{G}_f\longrightarrow\mathcal{N}$ be the restrictions of the projections π and ρ, respectively. Then the quintuple $(\mathcal{G}_f,\mathcal{M},\mathcal{N},\pi_f,\rho_f)$ will be a Backlund correspondence, and the differential relation between $\mathcal{M}$ and $\mathcal{N}$ reduces in this case to the original Lie-Backlund isomorphism $f:\mathcal{M}\simeq\mathcal{N}$. The remaining cases lie between these two extreme cases. More precisely, let $(\mathcal{R},\mathcal{M},\mathcal{N},\pi,\rho)$ be a Backlund correspondence. We define a Lie-Backlund mapping $\sigma:\mathcal{R}\longrightarrow\mathcal{M}\times\mathcal{N}$ by the formula

$$R \ni c \longmapsto \sigma(c) = (\pi(c),\rho(c)) \in M\times N.$$

Then in the first case the image $\sigma(R)$ is the whole set $M\times N$, and in the second it is the graph of the isomorphism f, while in the remaining cases it is proper subsets of $M\times N$ characterizing the rigidity of the corresponding differential relation.

These arguments lead to the idea of a realization of a Backlund correspondence as a differential submanifold of the direct product of the given differential manifolds. Namely, let $\mathcal{M}$ and $\mathcal{N}$ be two differential manifolds, and suppose that we have constructed a differential submanifold $\mathcal{R}\subset\mathcal{M}\times\mathcal{N}$ such that the restrictions $\pi_R:R\longrightarrow M$ and $\rho_R:R\longrightarrow N$ of the projections $\pi:M\times N\longrightarrow M$ and $\rho:M\times N\longrightarrow N$ on R are Lie-Backlund submersions. Then the quintuple $(\mathcal{R},\mathcal{M},\mathcal{N},\pi_R,\rho_R)$ will be a Backlund correspondence relating $\mathcal{M}$ and $\mathcal{N}$. The practical implementation of this program leads to very strong restrictions on the form of the submanifold $\mathcal{R}$, so that the existence of a nontrivial differential relation is a fairly rare occurrence that is by no means inherent to every pair of differential manifolds.

Example 14.4. Historically, Backlund correspondences are connected with the partial differential equation now called the sine-Gordon equation (see, for example, Ibragimov [1, Sect.14.4], Whitham [1, Sect.17.12], and the original papers cited there). We show how this equation and closely related equations arise in our approach. More precisely, we determine what equations of the form $u_{xy}=\varphi(u)$ and $v_{xy}=\psi(v)$ have nontrivial differential relations, and establish their form.

Thus, we consider the partial differential equation $u_{xy}=\varphi(u)$. The corresponding differential manifold $\mathcal{M}=(M,\partial_x,\partial_y)$, $M=\mathbb{R}^2\times\mathbb{R}_I$, with coordinates $x,y,u,u_x,u_y,\dots,u_{x^i},u_{y^i},\dots,$ $i\in\{1,2,..\}$ (in particular, $I=\{0\}+2\{1,2,\dots\}$, $u=u_0$), and basis Cartan fields

$$\partial_x = \frac{\partial}{\partial x} + u_x\frac{\partial}{\partial u} + \sum_{i=1}^{\infty}\left(u_{x^{i+1}}\frac{\partial}{\partial u_{x^i}} + \varphi_{y^{i-1}}\frac{\partial}{\partial u_{y^i}}\right),$$

$$\partial_y = \frac{\partial}{\partial y} + u_y\frac{\partial}{\partial u} + \sum_{i=1}^{\infty}\left(\varphi_{x^{i-1}}\frac{\partial}{\partial u_{x^i}} + u_{y^{i+1}}\frac{\partial}{\partial u_{y^i}}\right),$$

where $\varphi_z=\varphi' u_z$, $\varphi_{zz}=\varphi'' u_z^2+\varphi' u_{zz}$, etc., is the total derivative of corresponding order of the composite function $\varphi(u(z))$ with respect to the argument $z=x,y$. We remark that

$$u_{z^i} \longleftrightarrow \frac{\partial^i u}{(\partial z)^i}, \quad z = x,y, \; i = 1,2,\dots.$$

Similarly, the differential manifold $\mathcal{N}=(N,\partial_r,\partial_s)$, $N=\mathbb{R}^2\times\mathbb{R}_I$, corresponding to the partial differential equation $v_{rs}=\psi(v)$, has coordinates $r,s,v,v_r,v_s,\ldots,v_{r^i},v_{s^i},\ldots,$ $i\in\{1,2,..\}$, and basis Cartan fields

$$\partial_r = \frac{\partial}{\partial r} + v_r \frac{\partial}{\partial v} + \sum_{i=1}^{\infty}\left(v_{r^{i+1}} \frac{\partial}{\partial v_{r^i}} + \psi_{s^{i-1}} \frac{\partial}{\partial v_{s^i}}\right),$$

$$\partial_s = \frac{\partial}{\partial s} + v_s \frac{\partial}{\partial v} + \sum_{i=1}^{\infty}\left(\psi_{r^{i-1}} \frac{\partial}{\partial v_{r^i}} + v_{s^{i+1}} \frac{\partial}{\partial v_{s^i}}\right).$$

The coordinates $x,y,r,s,u,v,u_x,u_y,v_r,v_s,,u_{xx},u_{yy},v_{rr},v_{ss},\ldots,$ are defined on the direct product $M\times N$, and the Cartan distribution is given by the fields $\partial_x,\partial_y,\partial_r,\partial_s$. Following the above scheme, we should define a system of equations on $M\times N$, determining a submanifold R of $M\times N$. According to Section 8, this system should have initial "seed" equations and their differential prolongations of all orders. For simplicity we confine ourselves to the case $r=x$, $s=y$ (these are the first two seed equations), and we pass to the new coordinates $x,y,\alpha,\beta,u,v,u_x,u_y,v_x,v_y,u_{xx},u_{yy},v_{xx},v_{yy},\ldots,$ where $\alpha=x-r$, $\beta=y-s$, $v_x=v_r$, $v_y=v_s,\ldots,$ and to the new basis Cartan fields $\nabla_x=\partial_x+\partial_r$, $\nabla_y=\partial_y+\partial_s$, $\nabla_\alpha=-\partial_r$, $\nabla_\beta=-\partial_s$. In the new coordinates the fields ∇_x and ∇_y have the form

$$\nabla_x = \frac{\partial}{\partial x} + u_x \frac{\partial}{\partial u} + v_x \frac{\partial}{\partial v} + u_{xx} \frac{\partial}{\partial u_x} + \varphi \frac{\partial}{\partial u_y} + v_{xx} \frac{\partial}{\partial v_x} + \psi \frac{\partial}{\partial v_y} +\ldots,$$

$$\nabla_y = \frac{\partial}{\partial y} + u_y \frac{\partial}{\partial u} + v_y \frac{\partial}{\partial v} + \varphi \frac{\partial}{\partial u_x} + u_{yy} \frac{\partial}{\partial u_y} + \psi \frac{\partial}{\partial v_x} + v_{yy} \frac{\partial}{\partial v_y} +\ldots.$$

Clearly they admit the restriction to the submanifold $\{\alpha=0,\beta=0\}$

and can be used for determining a Cartan distribution on the submanifold we want to construct.

To the equations

$$\alpha = 0, \quad \beta = 0, \tag{1}$$

we add two other seed equations

$$v_x = f(u, v, u_x, u_y), \tag{2}$$

$$v_y = g(u, v, u_x, u_y), \tag{3}$$

where f and g are unknown smooth functions to be found along with the functions φ and ψ (again for simplicity we confine ourselves to a first-order differential relation that does not depend explicitly on x and y). In the traditional notation the problem appears as follows: to find a Backlund correspondence of the form (2)-(3) that relates the equations

$$u_{xy} = \varphi(u) \quad \text{and} \quad v_{xy} = \psi(v). \tag{4}$$

According to Section 8, we now add to the seed equations (1)-(3) the differential prolongations of them obtained by the successive actions of the fields ∇_x and ∇_y. Since $\nabla_x(\alpha)=\nabla_y(\alpha)=\nabla_x(\beta)=\nabla_y(\beta)\equiv 0$, we should in essence take into account only the prolongations of the equations (2) and (3),

$$\nabla_x^i \nabla_y^j \left(v_x - f \right) = 0, \tag{5}$$

$$\nabla_x^i \nabla_y^j \left(v_y - g \right) = 0, \tag{6}$$

where $i, j \in \mathbb{N}$. In other words,

$$R = \left\{\alpha = 0,\ \beta = 0,\ \nabla_x^i\nabla_y^j(v_x - f) = 0,\ \nabla_x^i\nabla_y^j(v_y - g) = 0,\ i,j \in \mathbb{N}\right\}.$$

Now, $\nabla_x^k v_x = v_{x^{k+1}}$ and $\nabla_y^k v_y = v_{y^{k+1}}$, hence we are able to express $v_{x^{k+1}}$ and $v_{y^{k+1}}$ in terms of $u, v, u_x, u_y, \ldots, u_{x^k}, u_{y^k}$. Therefore, we can take the variables $x, y, u, v, u_x, u_y, u_{xx}, u_{yy}, \ldots$ as global coordinates on R (cf. Section 10), then the remaining equations of the system (5)-(6) will be equations in the functions φ, ψ, f and g. For the Backlund correspondence to be invertible it must be possible for us to take the variables $x, y, u, v, v_x, v_y, v_{xx}, v_{yy}, \ldots$ also as coordinates on R. And for this it is necessary to require solvability of the system (2)-(3) with respect to the variables u_x and u_y. With this goal we impose the condition

$$\det \begin{Vmatrix} f_{u_x} & f_{u_y} \\ g_{u_x} & g_{u_y} \end{Vmatrix} \neq 0. \tag{7}$$

In the prolongations (5)-(6) the decisive roles for the functions φ, ψ, f and g are played by the first prolongations $\nabla_y(v_x - f) = 0$ and $\nabla_x(v_y - g) = 0$. With the seed equations (2)-(3) taken into account they yield

$$\psi - \left(u_y f_u + g f_v + \varphi f_{u_x} + u_{yy} f_{u_y}\right) = 0, \tag{8}$$

$$\psi - \left(u_x g_u + f g_v + u_{xx} g_{u_x} + \varphi g_{u_y}\right) = 0. \tag{9}$$

If now the functions φ, ψ, f and g make up a solution of the system (8)-(9) (i.e., when they are substituted in this system, it becomes an identity in the variables $x, y, u, v, u_x, u_y, \ldots$), then all the prolongations of the system (8)-(9) by the fields ∇_x, ∇_y are also identically satisfied. Further, it is possible to show that the subset R will be a submanifold of $M\times N$ in this case, and the fields ∇_x and ∇_y will have the restrictions

$$\bar{\nabla}_x = \frac{\partial}{\partial x} + u_x \frac{\partial}{\partial u} + f \frac{\partial}{\partial v} + u_{xx} \frac{\partial}{\partial u_x} + \varphi \frac{\partial}{\partial u_y} + \dots,$$

$$\bar{\nabla}_y = \frac{\partial}{\partial y} + u_y \frac{\partial}{\partial u} + g \frac{\partial}{\partial v} + \varphi \frac{\partial}{\partial u_x} + u_{yy} \frac{\partial}{\partial u_y} + \dots,$$

to R that determine a Cartan distribution CTR on R. This completely determines a differential manifold $\mathcal{R}=(R, CTR)$. The projections π and ρ onto $\mathcal{M}$ and $\mathcal{N}$ are now defined in an obvious way.

Returning to the defining equations (8)-(9), we observe first that they must be satisfied identically for all u_{xx} and u_{yy}, which implies that $f_{u_y}=0$ and $g_{u_x}=0$, i.e., $f=f(u,v,u_x)$ and $g=g(u,v,u_y)$, and equations (8)-(9) take the form

$$\psi - \left(u_y f_u + g f_v + \varphi f_{u_x}\right) = 0, \tag{10}$$

$$\psi - \left(u_x g_u + f g_v + \varphi g_{u_y}\right) = 0. \tag{11}$$

Analysis of system (10)-(11) is elementary, though fairly tedious. We write out an answer for a case when the functions $\varphi(u)$ and $\psi(v)$ are nonlinear:

$$\varphi = \frac{m}{n}\left(\beta\delta e^{-2nmu} - \alpha\gamma e^{2nmu}\right), \quad f = nu_x + \alpha e^{m(nu+v)} + \beta e^{-m(nu+v)},$$

$$\psi = m\left(\alpha\delta e^{2mv} - \beta\gamma e^{-2mv}\right), \quad g = -nu_y + \gamma e^{m(nu-v)} + \delta e^{-m(nu-v)},$$

where $m,n,\alpha,\beta,\gamma,\delta$ are complex numbers restricted by the condition that the functions φ,ψ,f,g be real, $mn\neq 0$. Selecting these numbers appropriately, we can get, for example, well-known Backlund correspondences for the sine-Gordon and Liouville equations.

Example 14.14(a) (sine-Gordon equation). Let us choose $\mu=\frac{i}{2}$, $n=-1$, $\alpha=ia$, $\beta=-ia$, $\gamma=\frac{1}{ia}$, $\delta=-\frac{1}{ia}$, then the above formulas yield equations

$$u_{xy} = \sin u, \quad v_{xy} = \sin v,$$

connected by the differential relation

$$v_x = -u_x + 2a \sin \frac{u-v}{2}, \quad v_y = u_y - \frac{2}{a} \sin \frac{u+v}{2}, \tag{12}$$

where $a\neq 0$ is an arbitrary real number. This is the original Backlund correspondence for sine-Gordon equation. It can be used to generate exact solutions of sine-Gordon equation as follows. Let $u=0$ be the trivial solution of $u_{xy}=\sin u$, then formulas (12) reduce to

$$v_x = -2a \sin \frac{v}{2}, \quad v_y = -\frac{2}{a} \sin \frac{v}{2}.$$

Taking into account the equation $v_{xy}=\sin v$ (which is the compatibility condition in this context), one can easily integrate this maximally overdetermined system to get

$$v = 4 \arctan \left\{c \exp \left[-\left(ax + \frac{y}{a}\right)\right]\right\}$$

the new solution of sine-Gordon equation. In turn, this solution can be used to generate another one, and so on (see, for example, Whitham [1, Sect.17.12]).

Example 14.15. We show next how to construct a Backlund correspondences for evolution equations. For simplicity we take the equation $u_t=\varphi(u,u_x)$. The corresponding differential manifold $\mathcal{M}$ has global coordinates $t,x,u,u_x,u_{xx},\ldots$, and basis Cartan fields

$$\partial_x = \frac{\partial}{\partial x} + u_x \frac{\partial}{\partial u} + u_{xx} \frac{\partial}{\partial u_x} + \ldots,$$

$$\partial_t = \frac{\partial}{\partial t} + \varphi \frac{\partial}{\partial u} + \partial_x\varphi \frac{\partial}{\partial u_x} + \ldots,$$

(see Section 7).

As the second equation we take the simplest equation of the same type as the first, namely, $v_s=v_y$ (s is the time variable, and y is the space variable). The corresponding differential manifold $\mathcal{N}$ has global coordinates $s,y,v,v_y,v_{yy},\ldots$, and basis Cartan fields

$$\partial_y = \frac{\partial}{\partial y} + v_y \frac{\partial}{\partial v} + v_{yy} \frac{\partial}{\partial v_y} + \ldots,$$

$$\partial_s = \frac{\partial}{\partial s} + v_y \frac{\partial}{\partial v} + v_{yy} \frac{\partial}{\partial v_y} + \ldots.$$

The global coordinates $x,t,y,s,u,v,u_x,v_y,u_{xx},v_{yy},\ldots$ are defined on the direct product $M\times N$, and the Cartan distribution is given by the fields $\partial_x,\partial_t,\partial_y,\partial_s$. In constructing a differential submanifold $\mathcal{R}$ we again confine ourselves to the case $y=x$, $s=t$. On $M\times N$ we pass to the new coordinates $x,t,\alpha,\beta,u,v,u_x,v_x,u_{xx},v_{xx},\ldots$, where $\alpha=x-y$, $\beta=t-s$, $v_x=v_y$, $v_{xx}=v_{yy},\ldots$, and to the new basis fields $\nabla_x=\partial_x+\partial_y$, $\nabla_t=\partial_t+\partial_s$, $\nabla_\alpha=-\partial_y$, $\nabla_\beta=-\partial_s$. The fields

$$\nabla_x = \frac{\partial}{\partial x} + u_x \frac{\partial}{\partial u} + v_x \frac{\partial}{\partial v} + u_{xx} \frac{\partial}{\partial u_x} + v_{xx} \frac{\partial}{\partial v_x} + \ldots,$$

$$\nabla_t = \frac{\partial}{\partial t} + \varphi \frac{\partial}{\partial u} + v_x \frac{\partial}{\partial v} + \partial_x\varphi \frac{\partial}{\partial u_x} + v_{xx} \frac{\partial}{\partial v_x} + \dots$$

(note that $\partial_x\varphi=\nabla_x\varphi$ for $\varphi=\varphi(u,u_x)$), admit restriction to the submanifold $\{\alpha=0,\ \beta=0\}$, and we use them for constructing a differential submanifold $\mathcal{R}$.

To the seed equations

$$\alpha = 0, \quad \beta = 0, \tag{13}$$

we add the equation

$$v_x = f(u,v,u_x), \tag{14}$$

where the condition $f_{u_x} \neq 0$ is imposed on the function f to ensure that the Backlund correspondence to be constructed is invertible. In traditional notation the problem is written as follows: for the equations $u_t=\varphi(u,u_x)$ and $v_t=v_x$ it is required to find a Backlund correspondence $v_x=f(u,v,u_x)$. This is a problem on constructing the functions φ and f simultaneously.

To the seed equations (13)-(14) we should add prolongation of them by the fields ∇_x and ∇_t. As in Example 14.14, it is enough to use prolongations of the form

$$\nabla_x^i\nabla_t^j\left(v_x - f\right) = 0, \quad i,j\in\mathbb{N}. \tag{15}$$

The equations $\nabla_x^k(v_x-f)=0$ allow us to also here to express $v_{x^{k+1}}$ in terms of $u,v,u_x,u_{xx},\dots,u_{x^{k+1}}$. Starting from this, we take the variables $x,t,u,v,u_x,u_{xx},\dots$ as coordinates on the manifold R, then the remaining equations (15) will be equations in the unknown functions φ and f. The decisive role is again played by the first prolongation, which takes the form $\nabla_x f - \nabla_t f = 0$, or, in greater detail,

$$u_x f_u + u_{xx} f_{u_x} - \varphi f_u - (\partial_x \varphi) f_{u_x} = 0.$$

Using the fact that $\partial_x \varphi = u_x \varphi_u + u_{xx} \varphi_{u_x}$, we get

$$u_{xx}\left(1-\varphi_{u_x}\right) f_{u_x} + \left(u_x - \varphi\right) f_u - u_x \varphi_u f_{u_x} = 0. \qquad (16)$$

Suppose that the functions φ and f satisfy this partial differential equation, then the remaining prolongations are satisfied identically, and the same arguments as in Example 14.14 are valid. As a result we get a differential manifold $\mathcal{R}$ with global coordinates $x, t, u, v, u_x, u_{xx}, \ldots,$ and basis Cartan fields

$$\bar{\nabla}_x = \frac{\partial}{\partial x} + f \frac{\partial}{\partial v} + u_x \frac{\partial}{\partial u} + u_{xx} \frac{\partial}{\partial u_x} + \ldots,$$

$$\bar{\nabla}_t + \frac{\partial}{\partial t} + f \frac{\partial}{\partial v} + \varphi \frac{\partial}{\partial u} + \partial_x \varphi \frac{\partial}{\partial u_x} + \ldots .$$

Equation (16) is easily solved and leads to two series of solutions. The first has the form

$$\varphi = u_x, \quad f(u, v, u_x) \text{ is an arbitrary smooth function.}$$

In the second case

$$\varphi = u_x + g(u), \quad f = h\left(v, g(u) u_x^{-1}\right),$$

where $g(u)$ and $h(v,w)$ are arbitrary smooth functions.

In conclusion we note that in both examples we took the simplest possible seed equations, because the computational labor increases sharply with any increasing complexity of the problem, and the determination of Backlund correspondences for actual equations in mathematical physics is a problem in itself. We hope that our proposed scheme gives a systematic base for its solution.

CHAPTER III. FURTHER STUDIES

15. Lagrangian Formalism

In this section we show how Lagrangian formalism is treated within the frameworks of the geometrical approach to PDE (see also Vinogradov [4]).

Let $\mathcal{M}=(M,D)$, $M=\mathbb{R}^m\times\mathbb{R}^a_K$, $K=\mathbb{N}^m$, be a trivial differential manifold, and $E_1^{0m}(\mathcal{M})$, $E_1^{1m}(\mathcal{M})$ and $\delta=d_1^{0m}:E_1^{0m}(\mathcal{M})\longrightarrow E_1^{1m}(\mathcal{M})$ be elements of the C-spectral sequence (see Section 6). Remind that

$$E_1^{0m}(\mathcal{M}) = \frac{\left\{\omega = \mathcal{L}d^m x \;; \mathcal{L} \in \mathcal{C}^\infty(\mathcal{M})\right\}}{\left\{\omega = \sum\limits_\mu D_\mu J^\mu d^m x \;;\; J \in C^\infty(\mathcal{M})^m\right\}},$$

$$E_1^{1m}(\mathcal{M}) = \left\{\omega = \sum_\alpha G_\alpha \delta w^\alpha \wedge d^m x \;;\; G \in \mathcal{C}^\infty(M)_a\right\},$$

$$\delta\left[\mathcal{L}d^m x\right] = \sum_\alpha \frac{\delta\mathcal{L}}{\delta w^\alpha}\delta w^\alpha \wedge d^m x, \quad \text{where} \quad \frac{\delta\mathcal{L}}{\delta w^\alpha} = \sum_i (-1)^{|i|} D_i \frac{\partial\mathcal{L}}{\partial w_i^\alpha}.$$

Here an equivalence class $[\mathcal{L}d^m x]\in E_1^{0m}(\mathcal{M})$ is called *a functional*, a m-form $\mathcal{L}d^m x$ is called *a Lagrangian*, a function $\mathcal{L}$ is called *a Lagrangian density*, the mapping δ is called *the Euler-Lagrange operator*, and $\delta\mathcal{L}/\delta w$ is called *a variational derivative*. We shall write $[\mathcal{L}]$ instead of $[\mathcal{L}d^m x]$ for brevity.

According to Proposition 6.8,

$$\frac{\delta \mathcal{L}}{\delta w} \equiv 0 \quad \text{iff} \quad \mathcal{L} = \sum_{\mu} D_{\mu} J^{\mu} \text{ for some } J \in \mathcal{C}^{\infty}(M)^{m},$$

and according to Lemma 6.15,

$$F = \frac{\delta \mathcal{L}}{\delta w} \text{ for some } \mathcal{L} \quad \text{iff} \quad F_{*} = F^{*}.$$

Now, let $\mathcal{L} \in \mathcal{C}^{\infty}(M)$ be a Lagrangian density, and

$$F_{\alpha}(x, \mathbf{w}) = \frac{\delta \mathcal{L}}{\delta w^{a}}, \quad \alpha = 1, \ldots, a,$$

be its variational derivative. The corresponding system of PDE

$$F_{\alpha}\left(x, w, \frac{\partial w}{\partial x}, \ldots, \frac{\partial^{r} w}{\partial x^{r}}\right) = 0, \quad \alpha = 1, \ldots, a,$$

is called *the Euler-Lagrange* or *variational system*. Its solution variety

$$F = \left\{x, \mathbf{w} \, : \, D_{k} F_{\alpha}(x, \mathbf{w}) = 0, \ k \in K, \ \alpha = 1, \ldots, a\right\}.$$

Variational systems possess nice special properties. We present some of them (see Olver [1,Sect 5.3] and Ibragimov [1,Sect 22] for more detail and the original references).

Remind that the Lie derivative $L_{[\zeta]}[\omega] \in E_{1}^{pq}(\mathcal{M})$ is defined for every symmetry $[\zeta] \in \mathrm{Sym}\mathcal{M}$ and every element $[\omega] \in E_{1}^{pq}(\mathcal{M})$ (see Proposition 5.5). In our case, symmetries are in fact evolution differentiations, i.e., $[\zeta] = \mathrm{E}_{\Phi}$, $\Phi \in \mathcal{C}^{\infty}(M)^{a}$, for every $[\zeta] \in \mathrm{Sym}\mathcal{M}$ (see Proposition 6.2).

One can easily check that $L_{\mathrm{E}_{\Phi}}[\mathcal{L}] = [\mathrm{E}_{\Phi}(\mathcal{L})]$ for every symmetry E and every functional $[\mathcal{L}]$, where

$$\mathrm{E}_\Phi(\mathcal{L}) = \mathcal{L}_*(\Phi) = \sum_{k,\alpha} \frac{\partial\mathcal{L}}{\partial w^\alpha_k} D_k\Phi^\alpha.$$

The natural imbedding $i:\mathcal{F}\longrightarrow\mathcal{M}$ induces the restriction mappings $\mathrm{Sym}\mathcal{M}\longrightarrow\mathrm{Sym}\mathcal{M}_F$ and $E^{pq}_r(\mathcal{M})\longrightarrow E^{pq}_r(\mathcal{F})$ for any differential submanifold $\mathcal{F}$ of $\mathcal{M}$ (see Section 9, in our case $\mathcal{F}$ is the solution manifold of a variational system). We shall denote by $\mathrm{E}_\Phi\big|_F\in\mathrm{Sym}\mathcal{M}_F$ the restriction of a symmetry $\mathrm{E}_\Phi\in\mathrm{Sym}\mathcal{M}$, and by $[\omega]\big|_F\in E^{pq}_r(\mathcal{F})$ the restriction of an element $[\omega]\in E^{pq}_r(\mathcal{M})$. In our case,

$$\mathrm{E}_\Phi\Big|_F = \sum_{k,\alpha}\left(\partial_k\phi^\alpha\right)\frac{\partial}{\partial w^\alpha_k} \quad\text{for}\quad \mathrm{E}_\Phi = \sum_{k,\alpha}\left(D_k\Phi^\alpha\right)\frac{\partial}{\partial w^\alpha_k},$$

where $\phi=\Phi\big|_F\in\mathcal{C}^\infty(F)^a$ for $\Phi\in\mathcal{C}^\infty(M)^a$ (see Section 10).

Let $[\mathcal{L}]$ be a functional, then

$$\left(L_{\mathrm{E}_\Phi}[\mathcal{L}]\right)\Big|_F = \left[\left(\mathrm{E}_\Phi(\mathcal{L})\right)\Big|_F\right],$$

where the equivalence class at the right-hand side is taken in the sense of the space $E^{0m}_1(\mathcal{F})$, and

$$\left(\mathrm{E}_\Phi(\mathcal{L})\right)\Big|_F = \left(\mathcal{L}_*(\Phi)\right)\Big|_F = \overline{\mathcal{L}_*}(\phi) = \sum_{k,\alpha}\frac{\partial\mathcal{L}}{\partial w^\alpha_k}\Big|_F\,\partial_k\phi^\alpha, \quad \phi=\Phi\Big|_F.$$

Proposition 15.1. *Let* $[\mathcal{L}]$ *be a functional and* F *the solution manifold of the corresponding variational system. Then*

$$\left(L_{\mathrm{E}_\Phi}[\mathcal{L}]\right)\Big|_F = 0$$

for any symmetry $\mathrm{E}_\Phi\in\mathrm{Sym}\mathcal{M}$.

Proof. Indeed, according to Noether's identity (see Proposition 6.21),

$$\mathcal{L}_*(\Phi) = \sum_\alpha F_\alpha \Phi^\alpha + \sum_\mu D_\mu H^\mu, \quad H \in \mathcal{C}^\infty(M)^m.$$

Hence,

$$\overline{\mathcal{L}_*}(\phi) = \sum_\mu \partial_\mu h^\mu, \quad h = H\Big|_F \in \mathcal{C}^\infty(F)^m,$$

and $[\overline{\mathcal{L}_*}(\phi)]=0$, by the definition of $E_1^{0m}(\mathcal{F})$ (see Section 5).

We denote by

$$\text{Sym}\ [\mathcal{L}] = \left\{\mathrm{E}_\Phi \in \text{Sym}\ \mathcal{M} : L_{\mathrm{E}_\Phi}[\mathcal{L}] = 0\right\}$$

$$= \left\{\Phi \in \mathcal{C}^\infty(M)^a : \mathcal{L}_*(\Phi) = \sum_\mu D_\mu H^\mu,\ H \in \mathcal{C}^\infty(M)^m\right\}$$

the Lie algebra of *symmetries of a functional* $[\mathcal{L}]$.

Remind that elements of the linear space $E_1^{1,m}(\mathcal{M})$ are in fact $(m+1)$-forms $\Omega=\sum_\alpha G_\alpha \delta w^\alpha \wedge d^m x$, $G\in\mathcal{C}^\infty(M)_a$ (see Proposition 6.7).

Lemma 15.2. *Let* $\mathrm{E}_\Phi\in\text{Sym}\mathcal{M}$ *and* $\Omega\in E_1^{1,m}(M)$, *where* $\Omega=\sum_\alpha G_\alpha\delta w^\alpha\wedge d^m x$. *Then*

$$L_{\mathrm{E}_\Phi}\Omega = \sum_\alpha \left(G_*(\Phi)_\alpha + \Phi^*(G)_\alpha\right)\delta w^\alpha \wedge d^m x,$$

where

$$G_*(\Phi)_\alpha = \sum_{k,\beta} \frac{\partial G_\alpha}{\partial w_k^\beta} D_k \Phi^\beta \quad \text{and} \quad \Phi^*(G)_\alpha = \sum_{k,\beta} (-1)^{|k|} D_k\left(\frac{\partial \Phi^\beta}{\partial w_k^\alpha} G_\beta\right).$$

Proof. Indeed,

$$L_{\mathrm{E}_\Phi}\Omega = \sum_{\alpha=1}^a \left(\mathrm{E}_\Phi(G_\alpha)\delta w^\alpha + G_\alpha d\mathrm{E}_\Phi(w^\alpha)\right)\wedge d^m x$$

$$= \sum_{\alpha=1}^a \left(G_*(\Phi)_\alpha \delta w^\alpha + G_\alpha d\Phi^\alpha\right)\wedge d^m x$$

$$= \sum_{\alpha=1}^{a} \left(G_*(\Phi)_\alpha \delta w^\alpha + G_\alpha \sum_{k\in K} \sum_{\beta=1}^{a} \frac{\partial \Phi^\alpha}{\partial w_k^\beta} dw_k^\beta \right) \wedge d^m x$$

$$= \sum_{\alpha=1}^{a} \left(G_*(\Phi)_\alpha \delta w^\alpha + \sum_{k\in K} \sum_{\beta=1}^{a} G_\beta \frac{\partial \Phi^\beta}{\partial w_k^\alpha} dw_k^\alpha \right) \wedge d^m x$$

(note that $\delta w_k^\alpha \wedge d^m x = dw_k^\alpha \wedge d^m x$). Further, according to Lemma 6.5,

$$\sum_{\alpha=1}^{a} \sum_{k\in K} \sum_{\beta=1}^{a} G_\beta \frac{\partial \Phi^\beta}{\partial w_k^\alpha} dw_k^\alpha \wedge d^m x$$

$$= \sum_{\alpha=1}^{a} \sum_{k\in K} \sum_{\beta=1}^{a} (-1)^{|k|} D_k \left(G_\beta \frac{\partial \Phi^\beta}{\partial w_k^\alpha} \right) \delta w^\alpha \wedge d^m x \mod \partial E^{1,m-1}$$

$$= \sum_{\alpha=1}^{a} \Phi^*(G)_\alpha \, \delta w^\alpha \wedge d^m x,$$

and the proof is finished.

Remind that the Lie algebra of external symmetries

$$\mathrm{Sym}_{ext}\mathcal{F} = \left\{ \phi \in \mathcal{C}^\infty(F)^a \; : \; \overline{F_*}(\phi) = 0 \right\} \subset \mathrm{Sym}\, \mathcal{M}_{\mathrm{F}}$$

for any reduction $\mathcal{F}$ of the trivial differential manifold $\mathcal{M}$ (see Section 10).

Proposition 15.3. *Suppose that the variational system $\{F=0\}$, determined by a functional $[\mathcal{L}]$, has a standard realization. Let $\Phi\in\mathrm{Sym}[\mathcal{L}]$, then the restriction $\phi=\Phi\big|_F\in\mathrm{Sym}_{ext}\mathcal{F}$.*

Proof. Indeed, according to Proposition 5.6(b),

$$L_{\mathrm{E}_\Phi}\left(d_1[\mathcal{L}]\right) = d_1\left(L_{\mathrm{E}_\Phi}[\mathcal{L}]\right) = 0,$$

and by Lemma 15.2,

$$F_*(\Phi) + \Phi^*(F) = 0, \quad F = \frac{\delta\mathcal{L}}{\delta w}.$$

Hence, $\overline{F_*(\phi)}=0$, because clearly $\Phi^*(F)\Big|_F=0$.

A symmetry $\phi\in\mathrm{Sym}_{ext}\mathcal{F}$ is called *variational* if $\phi=\Phi\Big|_F$ for some $\Phi\in\mathrm{Sym}[\mathcal{L}]$ (cf. Olver [1, Sect.5.3]), and we denote by $\mathrm{Sym}_{var}\mathcal{F}$ the Lie subalgebra of all variational symmetries of a variational system $\{F=0\}$.

Theorem 15.4 (Noether's theorem). *Suppose that a variational system* $\left\{F=\delta\mathcal{L}\big/\delta w = 0\right\}$ *is generating. Then*

$$\mathrm{Sym}\ [\mathcal{L}] = \mathrm{PreChar}\ \mathcal{F}$$

(see Theorem 13.13). In other words, *a function* $\Phi\in\mathcal{C}^\infty(M)^a$ *is a symmetry of the Lagrangian* $[\mathcal{L}]$ *iff it is a precharacteristic of some current of the corresponding variational system.*

Proof. Let $\Phi\in\mathcal{C}^\infty(M)^a\in\mathrm{Sym}[\mathcal{L}]$, i.e.,

$$\mathcal{L}_*(\Phi) = \sum_\mu D_\mu R^\mu \quad \text{for some} \quad R \in \mathcal{C}^\infty(M)^m.$$

Then by Noether's identity (see Proposition 6.21),

$$\left(\frac{\delta\mathcal{L}}{\delta w}, \Phi\right) = \mathcal{L}_*(\Phi) - \sum_\mu D_\mu J^\mu = \sum_\mu D_\mu\left(R^\mu - J^\mu\right),$$

hence $\Phi\in\mathrm{PreChar}\mathcal{F}$. Conversely, let $\Phi\in\mathrm{PreChar}\mathcal{F}$, i.e.,

$$\left(\frac{\delta\mathcal{L}}{\delta w}, \Phi\right) = \sum_\mu D_\mu H^\mu \quad \text{for some} \quad H \in \mathcal{C}^\infty(M)^m.$$

Again by Noether's identity,

$$\mathcal{L}_*(\Phi) = \left(\frac{\delta\mathcal{L}}{\delta w}, \Phi\right) + \sum_\mu D_\mu J^\mu = \sum_\mu D_\mu\left(H^\mu + J^\mu\right),$$

and $\Phi\in\mathcal{C}^\infty(M)^a\in\mathrm{Sym}[\mathcal{L}]$.

This theorem establishes the correspondence between conservation laws and variational symmetries of a given variational system. If some additional assumptions are fulfilled, this correspondence is one-to-one. For instance, according to Corollary 11.7, this is the case if the variational system under study can be transformed into an evolution system by a suitable change of independent and dependent variables (see also Theorem 5.42 in Olver [1], where another sufficient conditions are given).

Example 15.5. In the space $M=\mathbb{R}^m\times\mathbb{R}_K$, $K=\mathbb{N}^m$, let us consider a Lagrangian density

$$\mathcal{L} = -\frac{1}{2}\sum_{\mu=1}^{m} w_{(\mu)}^2 + V(w),$$

where a smooth function $V(w)\in\mathcal{C}^\infty(\mathbb{R})$ is a potential. In this case the variational equation is

$$F = \sum_{\mu} w_{2(\mu)} + V'(w) = \Delta w + V'(w) = o.$$

Further, the differential operator $\mathcal{L}_*:\mathcal{C}^\infty(M)\longrightarrow\mathcal{C}^\infty(M)$ is acting by the formula

$$\mathcal{L}_*(\Phi) = V'(w)\Phi - \sum_{\mu} w_{(\mu)}D_\mu\Phi, \quad \Phi\in\mathcal{C}^\infty(M).$$

Now, a function $\Phi\in\mathcal{C}^\infty(M)$ is a symmetry of the Lagrangian $[\mathcal{L}]$ if

$$\mathcal{L}_*(\Phi) = \sum_{\mu} D_\mu H^\mu \quad \text{for some } H\in\mathcal{C}^\infty(M)^m,$$

i.e., if

$$V'(w)\Phi - \sum_{\mu} w_{(\mu)}D_\mu\Phi = \sum_{\mu} D_\mu H^\mu,$$

i.e., if

$$\Big(V'(w)+\sum_\mu w_{2(\mu)}\Big)\Phi-\sum_\mu D_\mu\Big(w_{(\mu)}\Phi\Big)=\sum_\mu D_\mu H^\mu,$$

i.e., if

$$\sum_\mu D_\mu J^\mu=\Phi F,\quad\text{where}\quad J^\mu=H^\mu+\sum_\mu w_{(\mu)}\Phi$$

(note that here $a=1$). Thus, a function $\Phi\in\mathcal{C}^\infty(M)$ is a symmetry of the functional $[\mathcal{L}]$ iff Φ is a precharacteristic of some current of the variational equation $\Delta w+V'(w)=0$.

Let us discuss now the variational theory with constraints in geometrical setting.

Let $\mathcal{M}=(M,D)$, $M=\mathbb{R}^m\times\mathbb{R}^a_K$, $K=\mathbb{N}^m$, be a trivial differential manifold with global coordinates $x=(x^\mu)$, $\mathbf{w}=(w^\alpha_k)$, and let a system $\{C=0\}$ of nonlinear partial differential equations (*constraints*)

$$C^\sigma\Big(x,w,\frac{\partial w}{\partial x},\ldots,\frac{\partial^r w}{\partial w^r}\Big)=0,\quad\sigma=1,\ldots,s,$$

be given (see Section 10). Suppose that this system has a standard realization as a reduction $\mathcal{C}=(C,\nabla)$ of $\mathcal{M}$, where

$$M=\Big\{x,\mathbf{w}\ :\ D_kC^\sigma(x,\mathbf{w})=0,\ k\in K,\ \sigma=1,\ldots,s\Big\},$$

and $\nabla_\mu=D_\mu\big|_C$ are restrictions of the basis Cartan fields D_μ to C, $\mu=1,\ldots,m$. In this case

$$\mathrm{Sym}_{ext}\mathcal{C}=\Big\{\phi=\Phi\big|_C,\ \Phi\in\mathcal{C}^\infty(M)^a\ :\ C_*(\Phi)\big|_C=0\Big\}.$$

Let $[\mathcal{L}]\in E_1^{0m}(\mathcal{M})$ be a functional. We denote

$$\mathrm{Sym}_C[\mathcal{L}]=\Big\{\Phi\in\mathcal{C}^\infty(M)^a\ :\ \Big(L_{\mathrm{E}_\Phi}[\mathcal{L}]\Big)\Big|_C=0\Big\}$$

the Lie algebra of *constrained symmetries of* $[\mathcal{L}]$. Note that

$$\mathrm{Sym}_C[\mathcal{L}] \supset \mathrm{Sym}\,[\mathcal{L}].$$

Proposition 15.6. *A function* $\Phi\in\mathcal{C}^\infty(M)^a$ *is a constrained symmetry of* $[\mathcal{L}]$ *iff*

$$\mathcal{L}_*(\Phi) = \sum_{\mu=1}^{m} D_\mu R^\mu + \sum_{\sigma=1}^{s} \Psi_\sigma C^\sigma,$$

for some $R\in\mathcal{C}^\infty(M)^m$ *and* $\Psi\in\mathcal{C}^\infty(M)_s$.

Proof. Indeed, a function $\Phi\in\mathcal{C}^\infty(M)^a$ is a constrained symmetry of $[\mathcal{L}]$ iff

$$\mathcal{L}_*(\Phi)\Big|_C = \sum_\mu \nabla_\mu h^\mu \quad \text{for some } h \in \mathcal{C}^\infty(C),$$

i.e., iff

$$\mathcal{L}_*(\Phi) = \sum_\mu D_\mu H^\mu + V \quad \text{for some } H \in \mathcal{C}^\infty(M)^m \text{ and } V \in \mathcal{I}_C,$$

where $\mathcal{I}_C$ is the defining ideal of the manifold C. On the other hand, every function $V\in\mathcal{I}_C$ can be written in the form

$$V = \sum_\sigma \Psi_\sigma C^\sigma + \sum_\mu D_\mu J^\mu \quad \text{with some } \Psi \in \mathcal{C}^\infty(M)_s \text{ and } J \in \mathcal{C}^\infty(M)^m$$

(see Lemma A.9, remind that every system, having a standard realization is generating, see Proposition 10.10). Thus, the representation we claimed is valid with $R=H+J$.

For a functional $[\mathcal{L}]$ and a constraints $\{C=0\}$ let us call the joint system $\{C=0,\ F=0\}$, where $F=\delta\mathcal{L}/\delta w$, *a constrained variational system.*

Example 15.7 (Maxwell's equations). Here $m=a=4$, independent variables $x=(x^0,x^1,x^2,x^3)$ and dependent variables (electromagnetic potential) $A=(A^0,A^1,A^2,A^3)$ belong to Minkowski space $\mathbb{M}^4$ with the metric

$$g_{\mu\nu} = \begin{cases} 1, & \mu=\nu=0, \\ -1, & \mu=\nu=1,2,3, \\ 0, & \mu\neq\nu, \end{cases}$$

the differential variables $\mathbf{A}=(A^\mu_k)\in\mathbb{M}^4_K$, $K=\mathbb{N}^4$, the space $N=\mathbb{M}^4\times\mathbb{M}^4_K$, and the basis Cartan fields are total derivatives D_0,D_1,D_2,D_3. The constraint equation is

$$H\left(x,\frac{\partial A}{\partial x}\right) = \sum_{\mu=0}^{3}\frac{\partial A^\mu}{\partial x^\mu} = 0$$

(Lorentz gauge frame, $s=1$). The Lagrangian density

$$\mathcal{L}(\mathbf{A}) = -\frac{1}{4}\sum_{\mu,\nu=0}^{3} f_{\mu\nu}f^{\mu\nu},$$

where the Maxwell tensor of electromagnetic field

$$f_{\mu\nu} = D_\mu A_\nu - D_\nu A_\mu,$$

we are using the standard tensor notation, for example,

$$A_\mu = \sum_{\lambda=0}^{3} g_{\mu\lambda}A^\lambda, \quad f^{\mu\nu} = \sum_{\lambda,\rho=0}^{3} g^{\mu\lambda}g^{\nu\rho}f_{\lambda\rho}, \quad g^{\mu\nu} = g_{\mu\nu}.$$

Further, the variational derivatives

$$\frac{\delta\mathcal{L}}{\delta A^\mu} = -\sum_{\lambda=0}^{3} D_\lambda\frac{\partial\mathcal{L}}{\partial D_\lambda A^\mu} = \sum_{\lambda=0}^{3} D_\lambda f^\lambda{}_\mu = \sum_{\lambda=0}^{3} D_\lambda\left(D^\lambda A_\mu - D_\mu A^\lambda\right)$$

$$= \left(\sum_{\lambda=0}^{3} D_\lambda D^\lambda\right) A_\mu - D_\mu\left(\sum_{\lambda=0}^{3} D_\lambda A^\lambda\right) = \Box A_\mu - D_\mu H, \quad \mu = 0,1,2,3,$$

where

$$\Box = D_0^2 - D_1^2 - D_2^2 - D_3^2$$

is the D'Alembert operator, hence,

$$F_\mu = \left(\frac{\delta \mathcal{L}}{\delta A^\mu}\right)\bigg|_M = \Box A_\mu, \quad \mu = 0,1,2,3.$$

Thus the solution manifold for the Maxwell equations in the Lorentz frame is

$$F = \left\{(x, \mathbf{A}) \in N : D_k H=0,\ D_k F_\mu=0,\ k \in K,\ \mu = 0,1,2,3\right\}.$$

Suppose that a constrained variational system $\{C=0,\ F=0\}$, where $F=\delta\mathcal{L}\big/\delta w$, has a standard realization as a reduction $\mathcal{F}=(F,\partial)$ of $\mathcal{M}$ with the solution manifold

$$F = \left\{x, \mathbf{w} : D_k C^\sigma=0,\ D_k F_\alpha=0,\ k \in K,\ \sigma = 1,\ldots,s\ \alpha = 1,\ldots,a\right\}$$

and the basis Cartan fields $\partial_\mu = D_\mu\big|_F$, $\mu=1,\ldots,m$.

Proposition 15.8. *Suppose that* $\Phi \in \mathrm{Sym}_C[\mathcal{L}]$ *and* $\Phi\big|_C \in \mathrm{Sym}_{ext}\mathcal{C}$. *Then* $\Phi\big|_C \in \mathrm{Sym}_{ext}\mathcal{F}$.

Proof. Indeed,

$$\mathrm{Sym}_{ext}\mathcal{F} = \left\{\phi=\Phi\big|_F,\ \Phi \in \mathcal{C}^\infty(M)^a : C_*(\Phi)\big|_F=0,\ F_*(\Phi)\big|_F=0\right\}.$$

In our case, $C_*(\Phi)\big|_F=0$, because $C_*(\Phi)\big|_C=0$ by the assumption and $F\subset C$. Further (see the proof of Proposition 15.3),

$$\left(L_{\mathrm{E}_\Phi} d_1^{0m}[\mathcal{L}]\right)\Big|_C = d_1^{0m}\left(L_{\mathrm{E}_\Phi}[\mathcal{L}]\Big|_C\right) = 0,$$

i.e., by Lemma 15.2,

$$\left(F_*(\Phi) + \Phi^*(F)\right)\Big|_C = 0,$$

and hence, $F_*(\Phi)\Big|_F=0$, because clearly $\Phi^*(F)\Big|_F=0$. This completes the proof.

Theorem 15.9. (Noether's theorem with constraints). *Suppose that a constrained variational system $\{C=0,\ F=0\}$ has a standard realization. Let a function $\Phi\in\mathscr{C}^\infty(M)^a$, then $\Phi\in\mathrm{Sym}_C[\mathscr{L}]$ iff the pair $(\Psi,\Phi)\in\mathrm{PreChar}\mathscr{F}$ for some function $\Psi\in\mathscr{C}^\infty(M)_s$.*

Proof. Indeed, let $\Phi\in\mathrm{Sym}_C[\mathscr{L}]$. Then by Proposition 15.6,

$$\mathscr{L}_*(\Phi) = \sum_\mu D_\mu R^\mu - \sum_\sigma \Psi_\sigma C^\sigma,$$

for some $R\in\mathscr{C}^\infty(M)^m$ and $\Psi\in\mathscr{C}^\infty(M)_s$, and by Noether's identity,

$$\sum_\mu D_\mu R^\mu - \sum_\sigma \Psi_\sigma C^\sigma - \sum_\alpha \Phi^\alpha F_\alpha = \sum_\mu D_\mu H^\mu.$$

Hence, the pair (Ψ,Φ) is a precharacteristic of the current $J=R-H$. Conversely, suppose that there is a function $\Psi\in\mathscr{C}^\infty(M)_s$, such that the pair (Ψ,Φ) is a precharacteristic of some current J, i.e.,

$$\sum_\mu D_\mu J^\mu = \sum_\sigma \Psi_\sigma C^\sigma + \sum_\alpha \Phi^\alpha F_\alpha.$$

Then, again by Noether's identity,

$$\mathscr{L}_*(\Phi) - \sum_\mu D_\mu J^\mu + \sum_\sigma \Psi_\sigma C^\sigma = \sum_\mu D_\mu H^\mu,$$

i.e.,

$$\mathscr{L}_*(\Phi) = \sum_\mu D_\mu\left(H^\mu + J^\mu\right) - \sum_\sigma \Psi_\sigma C^\sigma,$$

and by Proposition 15.6, $\Phi\in\mathrm{Sym}_C[\mathscr{L}]$.

16. Hamiltonian Equations

In this section we shall discuss some elements of the geometrical description of Hamiltonian systems of PDE (a detailed exposition of the latter can be found in Olver [1; Ch.7]), which are the special case of the evolution systems (see Section 7). To get a better understanding of the situation we show first how evolution systems arise in the geometrical approach.

We shall use the notion of a *deformation* of a given differential manifold. Let us explain what it is. Let $\mathcal{M}^0=(M,D^0)$ be an initial "seed" differential manifold with a smooth manifold M and the Cartan distribution defined by basis fields D^0_μ, $\mu=1,..,m$, where $[D^0_\mu,D^0_\nu]=0$, $\mu,\nu=1,...,m$. Then the *deformed* differential manifold $\mathcal{M}=(M,D)$ has the same smooth manifold M, but the Cartan distribution is defined by deformed basis fields $D_\mu=D^0_\mu+\nabla_\mu$, $\mu=1,...,m$, where deformations ∇_μ, $\mu=1,..,m$, are vector fields on M, satisfying the *commutation relations*:

$$\left[D_\mu,D_\nu\right] = \left[D^0_\mu,\nabla_\nu\right] + \left[\nabla_\mu,D^0_\nu\right] + \left[\nabla_\mu,\nabla_\nu\right] = 0, \quad \mu,\nu = 1,..,m.$$

In the simplest case, when only one of the basis fields is deformed the last Lie bracket is absent. For example, if only $\nabla \neq 0$, we have

$$\left[D_1,D^0_\mu\right] = \left[\nabla_1,D^0_\mu\right] = 0, \quad \mu = 2,\dots,m.$$

Let us turn now to evolution systems. The basic object of our construction (the "seed" manifold) here is a trivial evolution differential manifold $\mathcal{M}^0=(M,D^0_t,D_x)$, where the space $M=\mathbb{R}\times\mathbb{R}^m\times\mathbb{R}^a_I$, $I=\mathbb{N}^m$, has global coordinates $t\in\mathbb{R}$, $x=(x^\mu)\in\mathbb{R}^m$, $\mathbf{u}=(u^\alpha_i)\in\mathbb{R}^a_I$, and the Cartan distribution D^0_t,D_x is defined by the basis fields

$$D^0_t = \frac{\partial}{\partial t},$$

$$D_\mu = D_{x^\mu} = \frac{\partial}{\partial x^\mu} + \sum_{\alpha,\,i} u^\alpha_{i+(\mu)} \frac{\partial}{\partial u^\alpha_i}, \quad \mu = 1,\dots,m,$$

(cf. Section 7). For further convenience let us introduce also an auxiliary trivial differential manifold $\mathcal{M}_x=(M_x,D_x)$, the x-space $M_x=\mathbb{R}^m\times\mathbb{R}^a_I$ (so $M=\mathbb{R}\times M_x$), and the Cartan distribution D_x is defined by the basis fields D_μ, $\mu=1,\dots,m$ (cf. Section 6).

Integral submanifolds of the manifold $\mathcal{M}^0$ are easily computed, using Proposition 1.8, and have the form

$$S = \left\{t,x,\mathbf{u} \,:\, u^\alpha_i=D_i\varphi^\alpha(x),\ \alpha = 1,\dots,a,\ i \in I\right\}$$

$$= \mathbb{R}\times S_x,$$

where $\varphi(x)\in\mathcal{C}^\infty(\mathbb{R}^m)^a$ is an arbitrary smooth function, S_x is an integral submanifold of $\mathcal{M}_x$. Thus, in this case the evolution is indeed trivial - all integral submanifolds are *stationary*, i.e., constant along t-axis.

Proposition 7.3 yields the description of infinitesimal symmetries of the manifold $\mathcal{M}^0$,

$$\operatorname{Sym} \mathcal{M}^0 = \left\{ \zeta = \mathrm{E}_\varphi \ ; \ \varphi(x,\mathbf{u}) \in \mathcal{C}^\infty(M_x)^a \right\}$$

$$= \operatorname{Sym} \mathcal{M}_x,$$

remind that evolution differentiation

$$\mathrm{E}_\varphi = \sum_{i,\alpha} \left(D_i \varphi^\alpha \right) \frac{\partial}{\partial u_i^\alpha}$$

(see Section 6).

Using the description of the spectral spaces, given in Section 7, one can easily check that

$$E_1^{0m}(\mathcal{M}^0) = \frac{\left\{ \omega = T(x,\mathbf{u}) d^m x \ ; \ T \in \mathcal{C}^\infty(M_x) \right\}}{\left\{ \omega = \Sigma D_\mu J^\mu(x,\mathbf{u}) d^m x \ ; \ J \in \mathcal{C}^\infty(M_x)^m \right\}} = E_1^{0m}(\mathcal{M}_x),$$

$$E_1^{1m}(\mathcal{M}^0) = \left\{ \Omega = \Sigma \varphi_\alpha du^\alpha \wedge d^m x \ ; \ \varphi(x,\mathbf{u}) \in \mathcal{C}^\infty(M_x)_a \right\}$$

$$= \left\{ \varphi(x,\mathbf{u}) \in \mathcal{C}^\infty(M_x)_a \right\} = E_1^{1m}(\mathcal{M}_x),$$

and the mapping $\delta = d_1^{0m} : E_1^{0m}(\mathcal{M}^0) \longrightarrow E_1^{1m}(\mathcal{M}^0)$,

$$E_1^{0m}(\mathcal{M}^0) \ni [T] \ \mapsto \ \delta[T] = \frac{\delta T}{\delta u} \in E_1^{1m}(\mathcal{M}^0),$$

coincides with the mapping $d_1^{0m} : E_1^{0m}(\mathcal{M}_x) \longrightarrow E_1^{1m}(\mathcal{M}_x)$, where variational derivatives

$$\frac{\delta T}{\delta u^\alpha} = \sum_i (-1)^{|i|} D_i \frac{\partial T}{\partial u_i^\alpha}, \quad T(x,\mathbf{u}) \in \mathcal{C}^\infty(M_x), \ \alpha = 1,\dots,a.$$

In particular,

$$\operatorname{Im} \delta = \left\{ \varphi(x,\mathbf{u}) \in \mathcal{C}^\infty(M_x)_a \ : \ \varphi_* = \varphi^* \right\}$$

(see Proposition 6.17). Note, that the equality $E_1^{0m}(\mathcal{M}^0)=E_1^{0m}(\mathcal{M}_x)$ allows us to call elements of $E_1^{0m}(\mathcal{M}^0)$ *functionals* (cf. Section 6).

To get a nontrivial dynamics on the manifold M, we shall deform the trivial evolution field D_t^0 by a suitable vector field

$$\nabla = \sum_{\alpha, i} f_i^\alpha(t,x,\mathbf{u}) \frac{\partial}{\partial u_i^\alpha}, \quad f_i^\alpha \in \mathcal{C}^\infty(M).$$

To be a basis field for the new Cartan distribution the deformed evolution field $D_t=D_t^0+\nabla$ should satisfy the conditions

$$\left[D_t, D_\mu\right] = \left[\nabla, D_\mu\right] = 0, \quad \mu = 1,\ldots,m.$$

According to Proposition 6.2 (t is a parameter here), these conditions are satisfied iff

$$\nabla = \mathrm{E}_f = \sum_{i,\alpha} \left(D_i f^\alpha(t,x,\mathbf{u})\right) \frac{\partial}{\partial u_i^\alpha}, \quad f \in \mathcal{C}^\infty(M)^a.$$

In other words, *the linear space of all evolution deformations of the trivial evolution differential manifold $\mathcal{M}^0$ is*

$$\mathrm{EvDef}\ \mathcal{M}^0 = \left\{\nabla = \mathrm{E}_f\ ;\ f(t,x,\mathbf{u}) \in \mathcal{C}^\infty(M)^a\right\} = \mathcal{C}^\infty(M)^a.$$

Every function $f\in\mathrm{EvDef}\mathcal{M}^0$ determines the the evolution differential manifold $\mathcal{M}^f=(M,D_t^f,D_x)$, where $D_t^f=D_t^0+\mathrm{E}_f$. If the function f is fixed we can omit the superscript ^{f} and write $\mathcal{M}=(M,D_t,D_x)$, as it was done in Section 7.

If a defining function $f=f(x,\mathbf{u})$ does not depend on t then the evolution system $\{u_t=f\}$ is called *autonomous*. In this case the Lie bracket $[\nabla,D_t^0]=0$, i.e., $f\in\mathrm{Sym}\mathcal{M}^0$, and this allows the deformed evolution system to inherit some objects from the trivial one.

Proposition 16.1. *Let* $f \in \mathrm{Sym}\mathcal{M}^0$. *Then* $\varphi \in \mathrm{Sym}\mathcal{M}^0$ *is a symmetry of* $\mathcal{M}^f$ *iff it is f-invariant, i.e.,*

$$L_{\mathrm{E}_f}\left(\mathrm{E}_\varphi\right) = \left[\mathrm{E}_f, \mathrm{E}_\varphi\right] = 0.$$

In other words,

$$\left\{\varphi \in \mathrm{Sym}\ \mathcal{M}^0 \ : \ \{f,\varphi\}=0\right\} \subset \mathrm{Sym}\ \mathcal{M}^f$$

(remind that the Jacoby bracket $\{f,\varphi\}=\mathrm{E}_f(\varphi)-\mathrm{E}_\varphi(f)$, see Remark 6.3).

For instance, $f \in \mathrm{Sym}\mathcal{M}^f$.

Proof. According to Proposition 7.3, an evolution differentiation $\mathrm{E}_\varphi \in \mathrm{Sym}\ \mathcal{M}^f$ iff the function $\varphi \in \mathcal{C}^\infty(M)^a$ satisfies the equation $D_t^f\varphi - f_*(\varphi)=0$. The latter can be written as

$$\frac{\partial\varphi}{\partial t} + \mathrm{E}_f(\varphi) - \mathrm{E}_\varphi(f) = 0,$$

or

$$\frac{\partial\varphi}{\partial t} + \left\{f,\varphi\right\} = 0$$

Thus a function $\varphi(x,\mathbf{u}) \in \mathrm{Sym}\mathcal{M}^0$ will determine a symmetry of $\mathcal{M}^f$ iff $\{f,\varphi\}=0$, or

$$\left[\mathrm{E}_f, \mathrm{E}_\varphi\right] = 0.$$

Proposition 16.2. *Let* $f \in \mathrm{Sym}\mathcal{M}^0$. *Then a functional* $[T] \in E_1^{0m}(\mathcal{M}^0)$ *is a conservation law of the evolution system* $\{u_t=f\}$ *iff it is f-invariant, i.e.,*

$$L_{\mathrm{E}_f}[T] = 0.$$

In other words,

$$\left\{[T] \in E_1^{0m}(\mathcal{M}^0) \ : \ L_{\mathrm{E}_f}[T]=0\right\} \subset E_1^{0m}(\mathcal{M}^f).$$

Proof. According to Section 7, a function $T(t,x,\mathbf{u})$ determines a conservation law for $\mathcal{M}^f$ iff $D_t^f T=\Sigma D_\mu X^\mu$ with some $X\in\mathcal{C}^\infty(M)^m$. In our case $T=T(x,\mathbf{u})$, and the divergency equation reads

$$\mathrm{E}_f(T) = \sum D_\mu X^\mu \quad \text{for some } X \in \mathcal{C}^\infty(M_x)^m,$$

or $L_{\mathrm{E}_f}[T]=0$ (cf. Section 15).

Thus, evolution systems are nothing but deformations of trivial systems. From this point of view Hamiltonian evolution systems are a special class of such deformations. To describe them we need some preliminaries.

According to Proposition 5.4, a *pairing*

$$\langle\cdot,\cdot\rangle \ : \ \mathrm{Sym}\mathcal{M}^0\times E_1^{0m}(\mathcal{M}^0) \longrightarrow E_1^{0m}(\mathcal{M}^0)$$

is determined by the rule

$$\langle\varphi,[T]\rangle = \mathrm{E}_\varphi \rfloor_1 \delta[T], \quad \varphi \in \mathrm{Sym}\ \mathcal{M}^0, \ [T] \in E_1^{0m}(\mathcal{M}^0).$$

Remark 16.3. Here

$$\langle\varphi,[T]\rangle = L_{\mathrm{E}_\varphi}[T] = [\mathrm{E}_\varphi(T)] = [T_*(\varphi)],$$

according to Proposition 5.6(a) (see also Section 15).

Proposition 16.4 (Olver [1; Ex.5.42]). *The pairing* $\langle\cdot,\cdot\rangle$ *is nondegenerated, i.e.,* $\langle\varphi,[T]\rangle=0$ *for all* $[T]\in E_1^{0m}(\mathcal{M}^0)$ *only if* $\varphi=0$, *and* $\langle\varphi,[T]\rangle=0$ *for all* $\varphi\in\mathrm{Sym}\mathcal{M}^0$ *only if* $[T]=0$.

Proof. Indeed, suppose first that $\varphi \in \mathrm{Sym}\mathcal{M}^0$ and $\langle\varphi,[T]\rangle=0$ for all $[T]\in E_1^{0m}(\mathcal{M}^0)$. Remind that $\langle\varphi,[T]\rangle=[T_*(\varphi)]=0$ iff

$$\frac{\delta T_*(\varphi)}{\delta u^\alpha} = 0, \quad \alpha = 1,\dots,a,$$

(see Proposition 6.8). Let us fix an index $\beta=1,\dots,a$, and let p be an order of the differential function $\varphi^\beta(x,\mathbf{u})$. For $T(x,\mathbf{u})=h(x)u_0^\beta$, where

$$h(x) = \frac{(x-z)^k}{k!}, \quad z \in \mathbb{R}^m, \quad k \in I = \mathbb{N}^m, \quad |k| = p,$$

(we use the multi-index notation, see Section 6), we shall have $T_*(\varphi)=h(x)\varphi^\beta$, and

$$\frac{\delta}{\delta u^\alpha}\left(h(x)\varphi^\beta(x,\mathbf{u})\right)\Big|_{x=z} = \frac{\partial\varphi^\beta(z,\mathbf{u})}{\partial u_k^\alpha},$$

for all $z\in\mathbb{R}^m$, $\alpha=1,\dots,a$, $|k|=p$, because

$$\left(D_i h(x)\right)\Big|_{x=z} = \begin{cases} 1, & i=k, \\ 0, & i\neq k. \end{cases}$$

Hence, by the above condition, the order of φ^β is $\leq(p-1)$, and by induction $\varphi^\beta=\varphi^\beta(x)$. Now, let $T=\frac{1}{2}\left(u_0^\beta\right)^2$, then $T_*(\varphi)=u_0^\beta\varphi^\beta(x)$,

$$\frac{\delta}{\delta u^\beta}\left(u_0^\beta\varphi^\beta(x)\right) = \varphi^\beta(x),$$

and $\varphi^\beta(x)=0$, by the same condition. Thus $\varphi=0$. On the other hand, suppose that $[T]\in E_1^{0m}(\mathcal{M}^0)$ and $\langle\varphi,[T]\rangle=0$ for all $\varphi\in\mathrm{Sym}\mathcal{M}^0$. By the definition,

$$\langle\varphi,[T]\rangle = \left[\mathrm{E}_\varphi \lrcorner_1 \delta[T]\right] = \left[\sum_\alpha \varphi^\alpha \frac{\delta T}{\delta u^\alpha}\right].$$

Let us fix an index $\beta=1,\ldots,a$, and let $p-1$ be an order of the variational derivative $\delta T/\delta u$. Choose $\varphi^{\beta}=\frac{1}{2}\left(u^{\beta}_{p(1)}\right)^{2}$ and $\varphi^{\alpha}=0$ for $\alpha\neq\beta$, then the assumption $<\varphi,[T]>=0$ implies

$$\frac{\delta}{\delta u^{\beta}}\left(\frac{1}{2}\left(u^{\beta}_{p(1)}\right)^{2}\frac{\delta T}{\delta u^{\beta}}\right) = (-1)^{p}u^{\beta}_{2p(1)}\frac{\delta T}{\delta u^{\beta}}+\ldots = 0,$$

where dotes denote low-order terms. Hence, $\delta T/\delta u=0$, and $[T]=0$, by Proposition 6.8.

Every linear operator $\mathcal{D}:E^{1m}_{1}(\mathcal{M}^{0})\longrightarrow \mathrm{Sym}\mathcal{M}^{0}$ determines two operations on the trivial manifold $\mathcal{M}^{0}$. The first one is the composition

$$\mathcal{D}\circ\delta\ :\ E^{0m}_{1}(\mathcal{M}^{0})\longrightarrow \mathrm{Sym}\ \mathcal{M}^{0},$$

where

$$E^{0m}_{1}(\mathcal{M}^{0})\ni[T]\ \mapsto\ \mathcal{D}\circ\delta[T]=\mathcal{D}\left(\delta[T]\right)\in \mathrm{Sym}\ \mathcal{M}^{0}.$$

The second one is the bilinear mapping on $E^{0m}_{1}(\mathcal{M}^{0})$,

$$\{\cdot,\cdot\}\ :\ E^{0m}_{1}(\mathcal{M}^{0})\times E^{0m}_{1}(\mathcal{M}^{0})\longrightarrow E^{0m}_{1}(\mathcal{M}^{0}),$$

where

$$E^{0m}_{1}(\mathcal{M}^{0})\ni[R],[S]\ \longmapsto\left\{[R],[S]\right\}=<\mathcal{D}\circ\delta[R],[S]>\in E^{0m}_{1}(\mathcal{M}^{0})$$

(see Proposition 5.4). There is a simple but important relation between these operations.

Lemma 16.5. *Let $f=\mathcal{D}\circ\delta[H]$ with some $[H]\in E^{0m}_{1}(\mathcal{M}^{0})$. Then*

$$L_{E_f}[T]=\left\{[H],[T]\right\}\ \textit{for all}\ [T]\in E^{0m}_{1}(\mathcal{M}^{0}).$$

Proof. Follows directly from the definitions and Remark 16.3.

Corollary 16.6. *Let $f=\mathcal{D}\circ\delta[H]$, $[H]\in E^{0m}_{1}(\mathcal{M}^{0})$. Then a functional*

$[T]\in E_1^{0m}(\mathcal{M}^0)$ *is a conservation law of the evolution system* $\{u_t=f\}$ *iff* $\{[H],[T]\}=0$. *In other words,*

$$\left\{[T]\in E_1^{0m}(\mathcal{M}^0)\ :\ \left\{[H],[T]\right\}=0\right\}\subset E_1^{0m}(\mathcal{M}^f).$$

Proof. It is enough to use Proposition 16.2 and Lemma 16.5.

A linear operator $\mathcal{D}:E_1^{1m}(\mathcal{M}^0)\longrightarrow \mathrm{Sym}\mathcal{M}^0$ is called *nondegenerated* if $\mathrm{Ker}\mathcal{D}=0$; in this case $\mathrm{Ker}\mathcal{D}\circ\delta=0$, also. A functional $[T]\in E_1^{0m}(\mathcal{M}^0)$ is called *distinguished* for $\mathcal{D}$ if $[T]\in\mathrm{Ker}\mathcal{D}\circ\delta$, i.e., if $\mathcal{D}\circ\delta[T]=0$.

Proposition 16.7. *Let a functional* $[T]\in E_1^{0m}(\mathcal{M}^0)$ *be distinguished for* $\mathcal{D}$, *then*

$$\left\{[T],[R]\right\}=0\quad \textit{for all}\ [R]\in E_1^{0m}(\mathcal{M}^0).$$

Proof. Follows directly from the definitions.

A linear operator $\mathcal{D}:E_1^{1m}(\mathcal{M}^0)\longrightarrow \mathrm{Sym}\mathcal{M}^0$ is called *Hamiltonian* if the bilinear operation $\{\cdot,\cdot\}$ is in fact *Poisson bracket*, i.e., if it has the following properties:

(a) $\left\{[R],[S]\right\}+\left\{[S],[R]\right\}=0$

(*skew-symmetry*), and

(b) $\left\{[R],\left\{[S],[T]\right\}\right\}+\left\{[S],\left\{[T],[R]\right\}\right\}+\left\{[T],\left\{[R],[S]\right\}\right\}=0$

(the *Jacoby identity*), for all $[R],[S],[T]\in E_1^{0m}(\mathcal{M}^0)$.

We shall not touch here upon a problem of description of Hamiltonian operators, because it is rather technical and geometrical point of view adds nothing new to it. An efficient, simple computational algorithm for determining when an operator $\mathcal{D}$ is Hamiltonian can be find in Olver [1; Theorem 7.8].

Every Hamiltonian operator $\mathcal{D}:E_1^{1m}(\mathcal{M}^0)\longrightarrow\mathrm{Sym}\mathcal{M}^0$ turns the linear space $E_1^{1m}(\mathcal{M}^0)$ into a Lie algebra with the Poisson bracket as a multiplication.

Theorem 16.8 (Olver [1; Pr.7.14]). *Let $\mathcal{D}:E_1^{1m}(\mathcal{M}^0)\longrightarrow\mathrm{Sym}\mathcal{M}^0$ be a Hamiltonian operator. Then the composition mapping $\mathcal{D}\circ\delta$ is in fact a homomorphism of the Lie algebra $E_1^{0m}(\mathcal{M}^0)$ into the Lie algebra $\mathrm{Sym}\mathcal{M}^0$.*

Proof. We need only to prove that

$$\mathcal{D}\circ\delta\Big\{[R],[S]\Big\} = \Big\{\varphi,\psi\Big\}$$

for all $[R],[S]\in E_1^{0m}(\mathcal{M}^0)$, where $\varphi=\mathcal{D}\circ\delta[R],\psi=\mathcal{D}\circ\delta[S]$. So, let us put $\chi=\mathcal{D}\circ\delta\{[R],[S]\}$, then according to Proposition 16.4, it is enough to prove that $<\chi-\{\varphi,\psi\},[T]>=0$ for all $[R],[S],[T]\in E_1^{0m}(\mathcal{M}^0)$. But, by the definition,

$$<\chi,[T]> = \Big\{\Big\{[R],[S]\Big\},[T]\Big\},$$

and

$$<\{\varphi,\psi\},[T]> = L_{\mathrm{E}_{\{\varphi,\psi\}}}[T],$$

by Remark 16.3. Further,

$$\mathrm{E}_{\{\varphi,\psi\}} = \Big[\mathrm{E}_\varphi,\mathrm{E}_\psi\Big]$$

(see Remark 6.3), hence by Lemma 16.5,

$$\begin{aligned}<\{\varphi,\psi\},[T]> &= L_{\mathrm{E}_\varphi}L_{\mathrm{E}_\psi}[T] - L_{\mathrm{E}_\psi}L_{\mathrm{E}_\varphi}[T]\\ &= L_{\mathrm{E}_\varphi}\Big\{[S],[T]\Big\} - L_{\mathrm{E}_\psi}\Big\{[R],[T]\Big\}\end{aligned}$$

$$= \Big\{[R],\Big\{[S],[T]\Big\}\Big\} - \Big\{[S],\Big\{[R],[T]\Big\}\Big\}.$$

Thus,

$$<\chi-\{\varphi,\psi\},[T]> =$$

$$= \Big\{\Big\{[R],[S]\Big\},[T]\Big\} - \Big\{[R],\Big\{[S],[T]\Big\}\Big\} + \Big\{[S],\Big\{[R],[T]\Big\}\Big\} = 0,$$

due to the properties (a) and (b) of the Poisson brackets.

Let $\mathcal{D}:E_1^{1m}(\mathcal{M}^0)\longrightarrow\mathrm{Sym}\mathcal{M}^0$ be a Hamiltonian operator. A symmetry $\varphi\in\mathrm{Sym}\mathcal{M}^0$ is called *Hamiltonian* if $\varphi\in\mathrm{Im}\mathcal{D}\circ\delta$, i.e., if $\varphi=\mathcal{D}\circ\delta[T]$ for some $[T]\in E_1^{0m}(\mathcal{M}^0)$. An evolution system $\{u_t=f\}$ is called *Hamiltonian* if f is a Hamiltonian symmetry.

Proposition 16.9. *Let $\mathcal{D}:E_1^{1m}(\mathcal{M}^0)\longrightarrow\mathrm{Sym}\mathcal{M}^0$ be a Hamiltonian operator. Then*

$$\mathrm{Ker}\mathcal{D}\circ\delta \subset E_1^{0m}(\mathcal{M}^f), \quad f = \mathcal{D}\circ\delta[H],$$

for any functional $[H]\in E_1^{0m}(\mathcal{M}^0)$. In other words, every distinguished functional is a conservation law for any Hamiltonian system.

Proof. Indeed, let $[T]\in\mathrm{Ker}\mathcal{D}\circ\delta$, then

$$\Big\{[H],[T]\Big\} = -\Big\{[T],[H]\Big\} = -<\mathcal{D}\circ\delta[T],[H]> = 0,$$

where we used the skew-symmetry of the Poisson brackets. Hence, $[T]\in E_1^{0m}(\mathcal{M}^f)$, by Corollary 16.6.

There is a fundamental relation between conservation laws and Hamiltonian symmetries of Hamiltonian systems.

Theorem 16.10 (Olver [1; Th.7.15]). *Let $\mathcal{D}$ be a Hamiltonian operator, and $f=\mathcal{D}\circ\delta[H]$, $[H]\in E_1^{0m}(\mathcal{M}^0)$. If $[T]\in E_1^{0m}(\mathcal{M}^0)$ is a conserva-*

tion law of the system $\{u_t=f\}$, *then* $\varphi=\mathcal{D}\circ\delta[T]$ *is a symmetry of this system. If the operator* $\mathcal{D}$ *is nondegenerated, and* $\varphi=\mathcal{D}\circ\delta[T]\in Sym\mathcal{M}^0$ *is a symmetry of the system* $\{u_t=f\}$, *then* $[T]$ *is a conservation law of this system.*

Proof. Let $[T]$ be a conservation law, then by Theorem 16.8 and Corollary 16.6,

$$\Big\{f,\varphi\Big\} = \mathcal{D}\circ\delta\Big\{[H],[T]\Big\} = 0,$$

and it is enough to use Proposition 16.1. Conversely, let $\mathcal{D}$ be nondegenerated and $\varphi=\mathcal{D}\circ\delta[T]$ be a symmetry, then

$$\mathcal{D}\circ\delta\Big\{[H],[T]\Big\} = \Big\{f,\varphi\Big\} = 0,$$

and $\{[H],[T]\}=0$, because $\mathrm{Ker}\mathcal{D}\circ\delta=0$, by the hypothesis.

Remark 16.11. If $\mathrm{Ker}\mathcal{D}\neq 0$ (i.e., when the operator $\mathcal{D}$ is degenerated), we can only conclude that $\{[H],[T]\}=[R]\in\mathrm{Ker}\mathcal{D}$ is a distinguished functional. In this case $[T]$ is no longer a conservation law of the system $\{u_t=f\}$. Nevertheless, the linear combination $[S]=[T-tR]$ will be a solution of the equation

$$\frac{\partial}{\partial t}[S] + \Big\{[H],[S]\Big\} = 0,$$

and hence will be a conservation law of $\{u_t=f\}$.

Note that the last conservation law explicitly depends on the time variable t. Thus, even for autonomous Hamiltonian systems we naturally come to objects with an explicit time dependence. Such objects have no counterparts in the seed trivial evolution system. To overcome this difficulty we propose another interpretation of evolution systems.

This time we shall use the notion of an *extension* of a given differential manifold. Let $\mathcal{M}^{in}=(M,D^{in})$ be an initial "seed" differential manifold with the smooth manifold M and the Cartan distribution D^{in}, determined by basis fields $D_1,\dots,D_m$, $[D_\mu,D_\nu]=0$ for $\mu,\nu=1,\dots,m$. Let D_0 be a Lie-Backlund field on $\mathcal{M}^0$ such that

(a) $\left[D_0,D_\mu\right] = 0, \quad \mu = 1,..,m,$

(b) vector fields $D_0,D_1,\dots,D_m$ form a $(m+1)$-dimensional distribution on M,

(the second condition means that vectors $D_0(z),D_1(z),\dots,D_m(z)$ are linearly independent in every point $z\in M$). In this case *the extended* differential manifold $\mathcal{M}^{ext}=(M,D^{ext})$ is determined, with the same smooth manifold M and the Cartan distribution D^{ext}, defined by the basis fields $D_0,D_1,\dots,D_m$.

We apply this procedure to evolution systems. Here the initial differential manifold $\mathcal{M}^{in}=(M,D_x)$, where $M=\mathbb{R}\times M_x$, has global coordinates $t\in\mathbb{R}$, $x\in\mathbb{R}^m$, $\mathbf{u}\in\mathbb{R}^a_I$ (so, $M_x=\mathbb{R}^m\times\mathbb{R}^a_I$, $I=\mathbb{N}^m$), and the Cartan distribution D_x, defined by basis fields

$$D_\mu = D_{x^\mu} = \frac{\partial}{\partial x^\mu} + \sum_{\alpha,i} u^\alpha_{i+(\mu)} \frac{\partial}{\partial u^\alpha_i}, \quad \mu = 1,\dots,m.$$

In this case integral submanifolds of $\mathcal{M}^{in}$ have the form

$$S = \left\{t,x,\mathbf{u} : t = \phi(x),\ u^\alpha_i = D_i\varphi^\alpha(x),\ i\in I,\ \alpha = 1,\dots,a\right\},$$

where $\phi,\varphi^\alpha\in\mathcal{C}^\infty(\mathbb{R}^m)$, $\alpha=1,\dots,a$, are arbitrary smooth functions (see Proposition 1.8). In other words, the evolution here is completely uncontrolled - every point $x\in\mathbb{R}^m$ lives by its own time! From this point of view, an additional Cartan field D_0 serves to establish a time order.

Let us list some properties of the seed manifold $\mathcal{M}^{in}$ we shall use below.

Proposition 3.1 yields the description of infinitesimal symmetries of $\mathcal{M}^{in}$,

$$\mathrm{Sym}\ \mathcal{M}^{in} = \left\{\zeta = \tau(t)\ \frac{\partial}{\partial t} + \sum_{\alpha,i} \left(D_i\Phi^{\alpha}(t,x,\mathbf{u})\right)\ \frac{\partial}{\partial u_i^{\alpha}}\right\}$$
$$= \left\{\zeta = \tau\ \frac{\partial}{\partial t} + \mathrm{E}_{\Phi}\right\},$$

where $\tau\in\mathcal{C}^{\infty}(\mathbb{R})$ and $\Phi\in\mathcal{C}^{\infty}(M)^{a}$ are arbitrary smooth functions.

Lemma 16.12. *Let*

$$\xi = R\frac{\partial}{\partial t} + \mathrm{E}_{\Phi},\quad \vartheta = S\frac{\partial}{\partial t} + \mathrm{E}_{\Psi} \in \mathrm{Sym}\ \mathcal{M}^{in},$$

then the Lie bracket

$$\left[\xi,\vartheta\right] = \tau\frac{\partial}{\partial t} + \mathrm{E}_{\chi},$$

where $\tau=AS_t-SR_t$ *and* $\chi=R\Psi_t-S\Phi_t+\{\Phi,\Psi\}$ (here and below the subscript $_t$ denotes the partial derivative in t, i.e., $\frac{\partial}{dt}$).

Remind that *the* Jacoby bracket $\{\Phi,\Psi\}=\mathrm{E}_{\Phi}(\Psi)-\mathrm{E}_{\Psi}(\Phi)$.

Proof. The proof is done by direct computations.

Consider a subset

$$\mathrm{Sym}_0\mathcal{M}^{in} = \left\{\zeta = \mathrm{E}_{\Phi}\ ;\ \Phi \in \mathcal{C}^{\infty}(M)^{a}\right\} \qquad (\tau=0)$$
$$= \mathcal{C}^{\infty}(\mathbb{R})\otimes\mathrm{Sym}\ \mathcal{M}_{x}$$

of the Lie algebra $\mathrm{Sym}\mathcal{M}^{in}$. Clear that $\mathrm{Sym}_0\mathcal{M}^{in}$ is a Lie algebra, isomorphic to the Lie algebra $\mathcal{C}^\infty(M)^a$ with the Jacoby bracket $\{\cdot,\cdot\}$ as a multiplication. We shall identify them.

Corollary 16.13. $\mathrm{Sym}_0\mathcal{M}^{in}$ *is an ideal of* $\mathrm{Sym}\mathcal{M}^{in}$.

Proof. Indeed, by Lemma 16.12,

$$\left[\tau\frac{\partial}{\partial t} + E_\Phi, E_\Psi\right] = E_\chi, \quad \chi=\tau\Psi_t+\{\Phi,\Psi\}.$$

Consider also a subset

$$\begin{aligned}\mathrm{Sym}_1\mathcal{M}^{in} &= \left\{\zeta = \frac{\partial}{\partial t} + E_\Phi \;;\; \Phi \in \mathcal{C}^\infty(M)^a\right\} \qquad (\tau=1)\\ &= \frac{\partial}{\partial t} + \mathrm{Sym}_0\mathcal{M}^{in}.\end{aligned}$$

Corollary 16.14. $\left[\mathrm{Sym}_1\mathcal{M}^{in},\mathrm{Sym}_1\mathcal{M}^{in}\right] \subset \mathrm{Sym}_0\mathcal{M}^{in}$.

Proof. Indeed, by Lemma 16.12,

$$\left[\frac{\partial}{\partial t} + E_\Phi, \frac{\partial}{\partial t} + E_\Psi\right] = E_\chi, \quad \chi=\Psi_t-\Phi_t+\{\Phi,\Psi\}.$$

Let us turn now to elements of the spectral sequence. Simple calculations (cf. Section 6) show that

$$E_1^{0m}(\mathcal{M}^{in}) = \frac{\left\{\omega = T(t,x,\mathbf{u})d^m x\right\}}{\left\{\omega = \Sigma D_\mu J^\mu(t,x,\mathbf{u})d^m x\right\}} = \mathcal{C}^\infty(\mathbb{R})\otimes E_1^{0m}(\mathcal{M}_x),$$

$$\begin{aligned}E_1^{1m}(\mathcal{M}^{in}) &= \frac{\left\{\Omega = T(t,x,\mathbf{u})dt\wedge d^m x + \Sigma\Psi_\alpha(t,x,\mathbf{u})\delta u^\alpha\wedge d^m x\right\}}{\left\{\Omega = \Sigma D_\mu J^\mu(t,x,\mathbf{u})dt\wedge d^m x\right\}}\\ &= \mathcal{C}^\infty(\mathbb{R})\otimes\left(dt\wedge E_1^{0m}(\mathcal{M}_x) + E_1^{1m}(\mathcal{M}_x)\right),\end{aligned}$$

where $T, J^\mu, \Psi_\alpha \in \mathcal{C}^\infty(M)$ are arbitrary smooth functions, and $\mathcal{M}_x=(M_x,D_x)$

(see the first part of this section).

The linear mapping $d_1=d_1^{0m}:E_1^{0m}(\mathcal{M}^{in})\to E_1^{1m}(\mathcal{M}^{in})$ is acting by the rule

$$E_1^{0m}(\mathcal{M}^{in}) \ni [T] \longmapsto d_1[T] = dt\wedge[T_t]+\delta[T] \in E_1^{1m}(\mathcal{M}^{in}),$$

where

$$T_t = \frac{\partial T}{\partial t}, \quad \delta[T] = \sum_\alpha \frac{\delta T}{\delta u^\alpha}\delta u^\alpha\wedge d^m x, \quad \frac{\delta T}{\delta u^\alpha} = \sum_i (-1)^{|i|}D_i\frac{\partial T}{\partial u_i^\alpha}.$$

By Proposition 6.8, the equality $\delta[T]=0$ for some $[T]\in E_1^{0m}(\mathcal{M}^{in})$ implies $[T]=0$ (and hence $[T_t]=0$). In particular, $\mathrm{Ker}\, d_1=0$.

According to Proposition 5.4, a *pairing*

$$\langle\cdot,\cdot\rangle : \mathrm{Sym}\mathcal{M}^{in}\times E_1^{0m}(\mathcal{M}^{in}) \longrightarrow E_1^{0m}(\mathcal{M}^{in})$$

is determined by the rule

$$\langle\zeta,[T]\rangle = \zeta\rfloor_1 d_1[T], \quad \zeta\in\mathrm{Sym}\mathcal{M}^{in}, \quad [T]\in E_1^{0m}(\mathcal{M}^{in}).$$

Remark 16.15. Let

$$\zeta = \tau\frac{\partial}{\partial t} + \mathrm{E}_\Phi \in \mathrm{Sym}\ \mathcal{M}^{in} \quad \text{and} \quad [T]\in E_1^{0m}(\mathcal{M}^{in}),$$

then

$$\begin{aligned}\langle\zeta,[T]\rangle &= [\tau T_t] + \mathrm{E}_\Phi\rfloor_1\delta[T] = [\tau T_t] + L_{\mathrm{E}_\Phi}[T] \\ &= [\tau T_t + \mathrm{E}_\Phi(T)] = \left[\tau T_t + \sum_\alpha \Phi^\alpha \frac{\delta T}{\delta u^\alpha}\right]\end{aligned}$$

(cf. Remark 16.3).

Proposition 16.16. *The pairing $\langle\cdot,\cdot\rangle$ is nondegenerated, i.e., $\langle\zeta,[T]\rangle=0$ for all $[T]\in E_1^{0m}(\mathcal{M}^{in})$ only if $\zeta=0$, and $\langle\zeta,[T]\rangle=0$ for all*

$\zeta\in\mathrm{Sym}\mathcal{M}^{in}$ *only if* $[T]=0$.

Proof. Indeed, suppose that $\langle\zeta,[T]\rangle=0$ for all $[T]\in E_1^{0m}(\mathcal{M}^{in})$, where $\zeta=\tau\frac{\partial}{dt}+\mathrm{E}_\Phi$. In particular, $\mathrm{E}_\Phi\rfloor_1\delta[T]=0$ for all $T=T(x,\mathbf{u})$, i.e., for all $[T]\in E_1^{0m}(\mathcal{M}^0)$, hence $\Phi=0$, by Proposition 16.4. Thus, we are left with $\langle\zeta,[T]\rangle=[\tau T_t]=\tau[T_t]=0$ for all $[T]\in E_1^{0m}(\mathcal{M}^{in})$, and this clearly implies $\tau=0$. On the other hand, suppose that $\langle\zeta,[T]\rangle=0$ for all $\zeta\in\mathrm{Sym}\mathcal{M}^{in}$, where $[T(t,x,\mathbf{u})]\in E_1^{0m}(\mathcal{M}^{in})$. Choose $\zeta=\mathrm{E}_\Phi$, then we get the equality $\langle\zeta,[T]\rangle=\mathrm{E}_\Phi\rfloor_1\delta[T]=0$ for all $\Phi\in\mathrm{Sym}_0\mathcal{M}^{in}$, and $[T]=0$, again by Proposition 16.4.

Let us now discuss extensions of the initial differential manifold $\mathcal{M}^{in}$. A vector field $\tau\frac{\partial}{\partial t}+\mathrm{E}_\Phi\in\mathrm{Sym}\mathcal{M}^{in}$ will be linearly independent from the original basis fields $D_1,\ldots,D_m$ if the smooth function $\tau(t)\neq0$ (i.e., $\tau(t)>0$ or $\tau(t)<0$) for all $t\in\mathbb{R}$. In this case the resulting extended differential manifold $\mathcal{M}^{ext}$ will not depend on the concrete choice of τ, and without loss of generality we may assume $\tau=1$. Thus, *evolution extensions* (i.e., extensions with $\tau\neq0$) are described by the subset $\mathrm{Sym}_1\mathcal{M}^{in}$ of $\mathrm{Sym}\mathcal{M}^{in}$. Taking into account the decomposition $\mathrm{Sym}_1\mathcal{M}^{in}=\frac{\partial}{\partial t}+\mathrm{Sym}_0\mathcal{M}^{in}$, we shall identify evolution extensions with evolution differentiations $\mathrm{E}_\Phi\in\mathrm{Sym}_0\mathcal{M}^{in}$, and put

$$\mathrm{EvExt}\ \mathcal{M}^{in}=\mathrm{Sym}_0\mathcal{M}^{in}.$$

Remark 16.17. Let $\mathcal{M}^f=(M,D_t^f,D_x)$ be the extended differential manifold corresponding to an evolution extension field $D_t^f=\frac{\partial}{\partial t}+\mathrm{E}_f$, $f\in\mathrm{Sym}\ \mathcal{M}^{in}$. Then according to Section 7,

$$\mathrm{Sym}\,\mathcal{M}^f = \left\{\Phi \in \mathrm{Sym}_0\mathcal{M}^{in} : \Phi_t+\{f,\Phi\}=0\right\}$$

(cf. Proposition 16.1), and

$$E_1^{om}(\mathcal{M}^f) = \left\{[T] \in E_1^{0m}(\mathcal{M}^{in}) : [D_t^f(T)]=0\right\}$$

(cf. Proposition 16.2). Note that $[\zeta(T)]=L_\zeta[T]$ for any $\zeta\in\mathrm{Sym}\mathcal{M}^{in}$.

Let $\mathcal{D}:E_1^{1m}(\mathcal{M}^{in})\longrightarrow\mathrm{Sym}_0\mathcal{M}^{in}$ be a linear operator. Then two operations on $\mathcal{M}^{in}$ are defined. The first one is the composition

$$\mathcal{D}\circ d_1 : E_1^{0m}(\mathcal{M}^{in}) \longrightarrow \mathrm{Sym}_0\mathcal{M}^{in}.$$

The second one is is the bilinear mapping on $E_1^{0m}(\mathcal{M}^{in})$,

$$\{\cdot,\cdot\} : E_1^{0m}(\mathcal{M}^{in})\times E_1^{0m}(\mathcal{M}^{in}) \longrightarrow E_1^{0m}(\mathcal{M}^{in}),$$

where

$$E_1^{0m}(\mathcal{M}^{in}) \ni [R],[S] \longmapsto \left\{[R],[S]\right\} = \langle\mathcal{D}\circ d_1[R],[S]\rangle$$

$$= L_{\mathrm{E}_\Phi}[S] \in E_1^{0m}(\mathcal{M}^{in}),$$

with $\Phi=\mathcal{D}\circ d_1[R]\in Sym_0\mathcal{M}^{in}$ (see Remark 16.13).

Proposition 16.18. *Let* $f=\mathcal{D}\circ d_1[H]$, *where* $\mathcal{D}:E_1^{1m}(\mathcal{M}^{in})\longrightarrow\mathrm{Sym}_0\mathcal{M}^{in}$ *is a linear operator,* $[H]\in E_1^{0m}(\mathcal{M}^{in})$. *Then*

$$E_1^{0m}(\mathcal{M}^f) = \left\{[T] \in E_1^{0m}(\mathcal{M}^{in}) : [T_t]+\{[H],[T]\}=0\right\}.$$

Proof. Follows from the definition and Remark 16.17.

We use below the definitions given in the first part of this section, if there is no danger of a misunderstanding. Thus, we call a linear operator $\mathcal{D}:E_1^{1m}(\mathcal{M}^{in})\longrightarrow\mathrm{Sym}_0\mathcal{M}^{in}$ Hamiltonian if the bilinear operation $\{\cdot,\cdot\}$ is the Poisson bracket. Theorem 16.8 reads now as follows.

Theorem 16.19. *Let* $\mathcal{D}: E_1^{1m}(\mathcal{M}^{in}) \longrightarrow \mathrm{Sym}_0 \mathcal{M}^{in}$ *be a Hamiltonian operator. Then the composition mapping* $\mathcal{D} \circ d_1$ *is a homomorphism of the Lie algebra* $E_1^{0m}(\mathcal{M}^{in})$ *into the Lie algebra* $\mathrm{Sym}_0 \mathcal{M}^{in}$.

Proof. The proof of Theorem 16.8 is valid here, because the pairing $\langle \cdot , \cdot \rangle$ is again nondegenerated (see proposition 16.16).

Clear that the partial derivative $\frac{\partial}{\partial t}$ and the partial integration $\int_0^t$ are correctly defined on the spaces $\mathrm{Sym}_0 \mathcal{M}^{in}$, $E_1^{0m}(\mathcal{M}^{in})$ and $E_1^{1m}(\mathcal{M}^{in})$ by their action on $\mathcal{C}^{\infty}(\mathbb{R})$-multiplier. Moreover, the commutation relations $\frac{\partial}{\partial t} \circ d_1 = d_1 \circ \frac{\partial}{\partial t}$, and $\int_0^t \circ d_1 = d_1 \circ \int_0^t$ are valid.

In all known examples the Hamiltonian operator $\mathcal{D} = \mathbb{I} \otimes \mathcal{D}_x$, where $\mathbb{I}$ is the identity operator on the space $\mathcal{C}^{\infty}(\mathbb{R})$,

$$\mathcal{D}_x : \left(dt \wedge E_1^{0m}(\mathcal{M}_x) + E_1^{1m}(\mathcal{M}_x) \right) \longrightarrow \mathrm{Sym}\, \mathcal{M}_x .$$

In this case the skew-symmetry property (a) of the Poisson bracket implies the vanishing of the linear operator $\mathcal{D}_x$ on the subspace $dt \wedge E_1^{0m}(\mathcal{M}_x)$, so in fact

$$\mathcal{D}_x : E_1^{1m}(\mathcal{M}_x) \longrightarrow \mathrm{Sym}\, \mathcal{M}_x .$$

Comparing this situation with the deformation approach, proposed in the first part of this section, one can see that in both cases the classes of Hamiltonian operators are essentially the same (remind that $E_1^{1m}(\mathcal{M}^0) = E_1^{1m}(\mathcal{M}_x)$ and $\mathrm{Sym}\mathcal{M}^0 = \mathrm{Sym}\mathcal{M}_x$).

Note that if $\mathcal{D} = \mathbb{I} \otimes \mathcal{D}_x$ then $\frac{\partial}{\partial t} \circ \mathcal{D} = \mathcal{D} \circ \frac{\partial}{\partial t}$, and $\int_0^t \circ \mathcal{D} = \mathcal{D} \circ \int_0^t$.

After these explanations the counterpart of Theorem 16.10 is formulated as follows.

Theorem 16.20. *Let* $\mathcal{D}=\mathbb{I}\otimes\mathcal{D}_x$ *be a Hamiltonian operator, and* $f=\mathcal{D}\circ d_1[H]$, $[H]\in E_1^{0m}(\mathcal{M}^{in})$. *If* $[T]\in E_1^{0m}(\mathcal{M}^{in})$ *is a conservation law of the system* $\{u_t=f\}$, *then* $\Phi=\mathcal{D}\circ d_1[T]$ *is a Hamiltonian symmetry of this system. Conversely, if* Φ *is a Hamiltonian symmetry of the system* $\{u_t=f\}$, *then there is a conservation law* $[T]$ *of this system such that* $\Phi=\mathcal{D}\circ d_1[T]$.

Proof. Let $[T]\in E_1^{0m}(\mathcal{M}^{in})$ be a conservation law of the system $\{u_t=f\}$, i.e., $[T_t]+\{[H],[T]\}=0$ (see Proposition 16.18). Then, by Theorem 16.19,

$$\Phi_t + \left\{f,\Phi\right\} = \frac{\partial}{\partial t}\left(\mathcal{D}\circ d_1[T]\right) + \left\{\mathcal{D}\circ d_1[H],\mathcal{D}\circ d_1[T]\right\}$$

$$= \mathcal{D}\circ d_1\left([T_t]+\left\{[H],[T]\right\}\right) = 0,$$

hence $\Phi=\mathcal{D}\circ d_1[T]$ is a symmetry. Conversely, Let $\Phi\in\mathrm{Sym}_0\mathcal{M}^{in}$ be a Hamiltonian symmetry, i.e., $\Phi_t+\{f,\Phi\}=0$, and $\Phi=\mathcal{D}\circ d_1[R]$, $[R]\in E_1^{0m}(\mathcal{M}^{in})$. Then, again by Theorem 16.19,

$$\mathcal{D}\circ d_1\left([R_t] + \left\{[H],[R]\right\}\right) = \Phi_t + \left\{f,\Phi\right\} = 0,$$

hence $[R_t]+\{[H],[R]\}=[L]\in\mathrm{Ker}\mathcal{D}\circ d_1$ is a distinguished functional. By the above arguments, the t-primitive $[\int_0^t L]\in\mathrm{Ker}\mathcal{D}\circ d_1$, and we denote $[T]=[R-\int_0^t L]$. Then

$$[T_t]+\left\{[H],[T]\right\} = [R_t] - [L] + \left\{[H],[R]\right\} - \left\{[H],[\int_0^t L]\right\} = 0,$$

because $\left\{[H],[\int_0^t L]\right\}=0$. Moreover, $\mathcal{D}\circ d_1[T]=\mathcal{D}\circ d_1[R]=\Phi$, and $[T]$ is the conservation law we need.

Thus we see that for Hamiltonian systems, just as for Lagrangian systems, there is a deep relation between conservation laws and symmetries. Note that Hamiltonian systems became popular not because of this property, but mostly because they usually have an infinite-dimensional spaces of conservation laws. As a matter of fact, the theory was initiated by the study of the Korteweg-de Vries equation (KdV), which got the international recognition for being the first evolution equation known to possess an infinite number of integrals (i.e., conservation laws, in our language), and only after that it was shown that KdV has a Hamiltonian structure. Also, though the Hamiltonian origin does not help one to calculate conservation laws, symmetries and other geometrical objects, associated with a given system, nevertheless Hamiltonian systems as a rule admit different methods (inverse scattering method, recursion formulas, Painleve test and so on), making these calculations an easy, elegant procedure. We refer an interested reader to the numerous literature on the subject.

17. Example: The Nambu's String

The Nambu's string is a simple system of PDE that governs a classical (not quantum) relativistic string in $(m+1)$-dimensional Minkowski space. It is a rare example of system playing an important role in physics and admitting the complete geometrical description. Our exposition here follows to Zharinov [5].

17.1. Determining Equations and Statements of Results. In the cone variables $\xi=\sigma+\tau$ and $\eta=\sigma-\tau$, Nambu's string has the form (see, for example, Scherk [1])

$$x^{\mu}_{\xi\eta} = 0, \quad \mu = 0,\dots,m,$$

$$x_{\xi}^{2} \equiv (x^{0}_{\xi})^{2} - \sum_{\mu=1}^{m} (x^{\mu}_{\xi})^{2} = 0, \tag{1.1}$$

$$x_{\eta}^{2} \equiv (x^{0}_{\eta})^{2} - \sum_{\mu=1}^{m} (x^{\mu}_{\eta})^{2} = 0,$$

here ξ,η are independent variables, $x=(x^{\mu};\mu=0,\dots,m)$ are dependent variables, subscripts $_{\xi},_{\eta},\dots$ denote the corresponding partial derivatives (we use the physical notation). We solve the last two equations in respect to x^{0}_{ξ}, x^{0}_{η} and rewrite (1.1) as

$$x^{\mu}_{\xi\eta} = 0, \quad \mu = 1,\dots,m,$$

$$x^{0}_{\xi} = \pm\sqrt{\sum_{\mu=1}^{m} (x^{\mu}_{\xi})^2} \equiv \pm f_0, \tag{1.2}$$

$$x^{0}_{\eta} = \pm\sqrt{\sum_{\mu=1}^{m} (x^{\mu}_{\eta})^2} \equiv \pm g_0.$$

For simplicity, we restrict ourselves to consideration of only the positive sector, corresponding to the + signs (the remaining sectors are treated in exactly the same way). Note that the equations (1.2) imply the equality $x^{0}_{\xi\eta}=0$. We add to this system all its differential prolongations obtained by successive differentiation with respect to ξ and η.

Thus, here the ambient differential manifold $\mathcal{M}=(M,D)$, where $M=\mathbb{R}^2\times\mathbb{R}^{m+1}_I$, $I=\mathbb{N}^2$, has global coordinates $\xi,\eta\in\mathbb{R}$, $x^{\mu}_{ij}\in\mathbb{R}^{m+1}_I$,

$$x^{\mu}_{ij} \longleftrightarrow \frac{\partial^{i+j}x^{\mu}}{\partial\xi^i\partial\eta^j}, \quad \mu=0,\dots,m, \;\; i,j\in\mathbb{N},$$

and basis Cartan fields

$$D_{\xi} = \frac{\partial}{\partial\xi} + \sum_{\mu=0}^{m}\sum_{i,j} x^{\mu}_{i+1,j}\frac{\partial}{\partial x^{\mu}_{ij}}, \quad D_{\eta} = \frac{\partial}{\partial\eta} + \sum_{\mu=0}^{m}\sum_{i,j} x^{\mu}_{i,j+1}\frac{\partial}{\partial x^{\mu}_{ij}},$$

(see Section 10). The system (1.2) yields the seed manifold

$$F_0 = \left\{\xi,\eta,x^{\mu}_{ij} \;:\; x^{\mu}_{11}=0,\; \mu=0,\dots,m,\; x^{0}_{10}=f_0,\; x^{0}_{01}=g_0\right\},$$

where $f_0=\sqrt{\Sigma^{m}_{\mu=1}(x^{\mu}_{10})^2}$ and $g_0=\sqrt{\Sigma^{m}_{\mu=1}(x^{\mu}_{01})^2}$. The prolongated manifold

$$F = \left\{\xi,\eta,x^{\mu}_{ij} \;:\; x^{\mu}_{ij}=0,\; x^{0}_{i0}=f_{i-1},\; x^{0}_{0j}=g_{j-1},\; \mu=0,\dots,m,\; i,j=1,2,\dots\right\},$$

where $f_i=D_\xi^i f_0$, $g_j=D_\eta^j g_0$, has global coordinates

$$\xi,\ \eta,\ x_{00}^0,\ \mathbf{x}_{00}=(x_{00}^\mu),\ \mathbf{x}_{i0}=(x_{i0}^\mu),\ \mathbf{x}_{0j}=(x_{0j}^\mu),$$

where $\mu=1,\ldots,m$, $i,j=1,2,\ldots$, and the restricted basis fields

$$\partial_\xi = \frac{\partial}{\partial\xi} = \sum_{\mu=1}^{m}\sum_{i=1}^{\infty} x_{i+1,0}^\mu \frac{\partial}{\partial x_{i0}^\mu} + |\mathbf{x}_{10}|\frac{\partial}{\partial x_{00}^0},$$

$$\partial_\eta = \frac{\partial}{\partial\eta} = \sum_{\mu=1}^{m}\sum_{i=1}^{\infty} x_{0,i+1}^\mu \frac{\partial}{\partial x_{0i}^\mu} + |\mathbf{x}_{01}|\frac{\partial}{\partial x_{00}^0}.$$

Here and below we use the notation

$$(\mathbf{a},\mathbf{b})=\sum_{\mu=1}^{m} a^\mu b^\mu,\quad |\mathbf{a}|=\sqrt{(\mathbf{a},\mathbf{a})},\quad \text{for any } \mathbf{a},\mathbf{b}\in\mathbb{R}^m.$$

We shall also write

$$f(\xi,\eta,x_{00}^0,x_{00},\mathbf{x}_{10},\ldots,\mathbf{x}_{p0},\mathbf{x}_{01},\ldots,\mathbf{x}_{0q}) = f(\mathbf{x}_{p0},\mathbf{x}_{0q},\ldots),$$

and we shall call the pair (p,q) *the bi-order* of a differential function f. For example,

$$f_1 = D_\xi|\mathbf{x}_{10}| = \frac{(\mathbf{x}_{20},\mathbf{x}_{10})}{|\mathbf{x}_{10}|^2},\quad g_1 = D_\eta|\mathbf{x}_{01}| = \frac{(\mathbf{x}_{02},\mathbf{x}_{01})}{|\mathbf{x}_{01}|^2},$$

and f_i has the bi-order $(i,0)$, g_j has the bi-order $(0,j)$.

Thus, we have constructed a reduction $\mathcal{F}=(F,\partial)$ of the ambient manifold $\mathcal{M}=(M,D)$.

According to Proposition 3.1, a vector field

$$X = \sum_{\mu=1}^{m}\sum_{i=1}^{\infty}\left(X_{i0}^\mu \frac{\partial}{\partial x_{i0}^\mu} + X_{0i}^\mu \frac{\partial}{\partial x_{0i}^\mu}\right) + \sum_{\mu=0}^{m} X_{00}^\mu \frac{\partial}{\partial x_{00}^\mu}$$

is a symmetry of the system (1.1) (i.e., $X\in\mathrm{Sym}\mathcal{F}$) iff

$$X^{\mu}_{io} = \partial^{i}_{\xi}X^{\mu}_{00}, \quad X^{\mu}_{oi} = \partial^{i}_{\eta}X^{\mu}_{00}, \quad \mu = 1,\dots,m, \quad i = 1,2,\dots,$$

where differential functions X^{μ}_{00}, $\mu=0,\dots,m$, satisfy the defining equations

$$\partial_{\xi}\partial_{\eta}X^{\mu}_{00} = 0, \quad \mu = 1,\dots,m, \tag{1.3}$$

$$|\mathbf{x}_{10}|\partial_{\xi}X^{0}_{00} - (\mathbf{x}_{10},\partial_{\xi}\mathbf{X}_{00}) = 0, \tag{1.4}$$

$$|\mathbf{x}_{01}|\partial_{\eta}X^{0}_{00} - (\mathbf{x}_{01},\partial_{\eta}\mathbf{X}_{00}) = 0, \tag{1.5}$$

where $\mathbf{X}_{00}=(X^{\mu}_{00})$, $\mu=1,\dots,m$. Note that the equations (1.3-5) imply the equality $\partial_{\xi}\partial_{\eta}X^{0}_{00}=0$.

According to Section 4, a pair $J=(F,G)$ is a conserved current of the system (1.1) (i.e.,$J\in\mathrm{Curr}^{1}\mathcal{F}$) iff functions $F(\mathbf{x}_{p0},\mathbf{x}_{0q},\dots)$, $G(\mathbf{x}_{k0},\mathbf{x}_{0l},\dots)$ satisfy the divergency equation

$$\partial_{\xi}F + \partial_{\eta}G = 0. \tag{1.6}$$

The number $r=\max\{p,q,k,l\}$ is the order of the current J. Trivial currents have here the form $F=-\partial_{\eta}\varphi$, $G=\partial_{\xi}\varphi$, where $\varphi(x_{p0},\mathbf{x}_{0q},\dots)$ is an arbitrary function.

We denote

$$\mathbf{e}_{10} = \frac{\mathbf{x}_{10}(\mathbf{x}_{20},\mathbf{x}_{20}) - \mathbf{x}_{20}(\mathbf{x}_{20},\mathbf{x}_{10})}{(\mathbf{x}_{20},\mathbf{x}_{20})(\mathbf{x}_{10},\mathbf{x}_{10}) - (\mathbf{x}_{20},\mathbf{x}_{10})^{2}},$$

$$\mathbf{e}_{20} = \frac{\mathbf{x}_{20}(\mathbf{x}_{10},\mathbf{x}_{10}) - \mathbf{x}_{10}(\mathbf{x}_{10},\mathbf{x}_{20})}{(\mathbf{x}_{20},\mathbf{x}_{20})(\mathbf{x}_{10},\mathbf{x}_{10}) - (\mathbf{x}_{20},\mathbf{x}_{10})^{2}}.$$

It is easy to verify the equalities

$$(\mathbf{e}_{p0},\mathbf{x}_{q0}) = \delta_{pq}, \quad p,q = 1,2,$$

where δ_{pq} is the Kronecker delta. The vectors $\mathbf{e}_{01}$ and $\mathbf{e}_{02}$ are determined similarly.

Theorem 1. *A vector field $X \in \mathrm{Sym}\mathcal{F}$ iff it has a representation*

$$X = \sum_{\mu=1}^{m} \sum_{i=1}^{\infty} \left(\partial_{\xi}^{i} X_{00}^{\mu} \frac{\partial}{\partial x_{i0}^{\mu}} + \partial_{\eta}^{i} X_{00}^{\mu} \frac{\partial}{\partial x_{0i}^{\mu}}\right) + \sum_{\mu=0}^{m} X_{00}^{\mu} \frac{\partial}{\partial x_{00}^{\mu}},$$

where $X_{00}^{\mu}=A^{\mu}+B^{\mu}+C^{\mu}$, $\mu=0,\dots,m$,

$$A^{\mu} = \sum_{\nu=0}^{m} a^{\mu\nu} x_{00}^{\nu}, \quad a^{\mu\nu} = \mathrm{const},$$

$$\mathbf{B} = f\mathbf{e}_{10} + \left[(\partial_{\xi} f) + (\mathbf{x}_{10},\partial_{\xi}\mathbf{A}) - |\mathbf{x}_{10}|\partial_{\xi}(A^{0}+B^{0})\right]\mathbf{e}_{20} + \mathbf{h},$$

$$\mathbf{C} = g\mathbf{e}_{01} + \left[(\partial_{\eta} g) + (\mathbf{x}_{01},\partial_{\eta}\mathbf{A}) - |\mathbf{x}_{01}|\partial_{\eta}(A^{0}+C^{0})\right]\mathbf{e}_{02} + \mathbf{k},$$

the functions $B^{0}, f(\xi,\mathbf{x}_{10},\dots,\mathbf{x}_{p0})$ *and* $C^{0}, g(\eta,\mathbf{x}_{01},\dots,\mathbf{x}_{0q})$ *are arbitrary, while the functions* $\mathbf{h}(\xi,\mathbf{x}_{10},\dots,\mathbf{x}_{p0})$ *and* $\mathbf{k}(\eta,\mathbf{x}_{01},\dots,\mathbf{x}_{0q})$ *satisfy the orthogonality conditions*

$$(\mathbf{h},\mathbf{x}_{10}) = (\mathbf{h},\mathbf{x}_{20}) = (\mathbf{k}.\mathbf{x}_{01}) = (\mathbf{k},\mathbf{x}_{02}) = 0.$$

For a given symmetry X, this representation is unique.

Here and below, the $(m+1)$-dimensional vector $a=(a^{0},a^{1},\dots,a^{m})$ is written in the form $a=(a^{0},\mathbf{a})$, $\mathbf{a}=(a^{1},\dots,a^{m})$.

Theorem 2. *A pair $J=(F,G) \in \mathrm{Curr}^{1}\mathcal{F}$ iff it has a representation*

$$F = \sum_{\mu,\nu=0}^{m} a^{\mu\nu}x^{\mu}_{01}x^{\nu}_{00} + \phi - \partial_{\eta}\Omega, \quad G = \sum_{\mu,\nu=0}^{m} a^{\mu\nu}x^{\mu}_{10}x^{\nu}_{00} + \psi + \partial_{\xi}\Omega,$$

where the smooth functions $\phi(\eta,\mathbf{x}_{01},\dots,\mathbf{x}_{0p})$, $\psi(\xi,\mathbf{x}_{10},\dots,\mathbf{x}_{q0})$, *and* $\Omega(\mathbf{x}_{k0},\mathbf{x}_{01},\dots)$ *are arbitrary,* $a^{\mu\nu}=-a^{\nu\mu}=\text{const}$ (for convenience we have introduced the notation $\overset{0}{\mathbf{x}}_{10}=|\mathbf{x}_{10}|$, $\overset{0}{\mathbf{x}}_{01}=|\mathbf{x}_{01}|$).

17.2. Auxiliary Lemmas. We start our proofs with several lemmas.

Lemma 1. *The general solution of the equation* $\partial_{\xi}\Phi=0$ *has the form* $\Phi=\phi(\eta,\mathbf{x}_{01},\dots,\mathbf{x}_{0q})$, *where* ϕ *is an arbitrary smooth function.*

Proof. Let us substitute a function $\Phi(\mathbf{x}_{p0},\mathbf{x}_{0q},\dots)$ into the equation, then we get

$$\partial_{\xi}\Phi = \sum_{\mu=1}^{m} x^{\mu}_{p+1,0}\Phi_{x^{\mu}_{p0}} + \dots = 0,$$

where here and in all similar situations below the dotes denote lower orders terms. Hence, $\Phi_{x^{\mu}_{p0}}=0$, so in fact $\Phi=\Phi(\mathbf{x}_{p-10},\mathbf{x}_{0q},\dots)$. By induction, $\Phi=\Phi(\xi,\eta,\overset{0}{x}_{00},\mathbf{x}_{00},\mathbf{x}_{01},\dots,\mathbf{x}_{0q})$, and

$$\partial_{\xi}\Phi = \sum_{\mu=1}^{m} x^{\mu}_{10}\Phi_{x^{\mu}_{00}} + |\mathbf{x}_{10}|\Phi_{\overset{0}{x}_{00}} + \Phi_{\xi} = 0.$$

It is readily verified that the last equation holds iff Φ has the required form.

Lemma 2. *The general solution* $\phi^{\alpha}(x,y),\psi^{\mu}(x,y)$ *of the system*

$$\phi^{\alpha}_{y^{\mu}} + \psi^{\mu}_{x^{\alpha}} = 0, \quad \alpha = 1,\dots,a, \quad \mu = 1,\dots,m, \qquad (2.1)$$

has the form

$$\phi^{\alpha} = A^{\alpha}(x) + B_{x^{\alpha}}(x,y), \quad \psi^{\mu} = C^{\mu}(y) - B_{y^{\mu}}, \tag{2.2}$$

where A^{α},B,C^{μ} *are arbitrary smooth functions.*

Proof. Functions ϕ^{α},ψ^{μ} of the form (2.2) obviously satisfy the system (2.1). On the other hand, let ϕ^{α},ψ^{μ} be a solution of the system (2.1). Then from $\phi^{\alpha}_{y^{\mu}}=-\psi^{\mu}_{x^{\alpha}}$ we have $\phi^{\alpha}_{y^{\mu}x^{\beta}}=-\psi^{\mu}_{x^{\alpha}x^{\beta}}=\phi^{\beta}_{y^{\mu}x^{\alpha}}$, i.e., $\left(\phi^{\alpha}_{x^{\beta}}-\phi^{\beta}_{x^{\alpha}}\right)_{y^{\mu}}=0$, whence $\phi^{\alpha}_{x^{\beta}}-\phi^{\beta}_{x^{\alpha}}=a^{\alpha\beta}(x)=-a^{\beta\alpha}(x)$. The verification shows that $a^{\alpha\beta}_{x^{\gamma}}+a^{\beta\gamma}_{x^{\alpha}}+a^{\gamma\alpha}_{x^{\beta}}=0$, and by the Poincare lemma, $a^{\alpha\beta}=\tilde{A}^{\alpha}_{x^{\beta}}-\tilde{A}^{\beta}_{x^{\alpha}}$, for some $\tilde{A}^{\alpha}(x)$. Thus, $\phi^{\alpha}_{x^{\beta}}-\phi^{\beta}_{x^{\alpha}}=\tilde{A}^{\alpha}_{x^{\beta}}-\tilde{A}^{\beta}_{x^{\alpha}}$, i.e., $\left(\phi^{\alpha}-\tilde{A}^{\alpha}\right)_{x^{\beta}}=\left(\phi^{\beta}-\tilde{A}^{\beta}\right)_{x^{\alpha}}$, and again by the Poincare lemma, there exists a function $\tilde{B}(x,y)$ such that $\phi^{\alpha}-\tilde{A}^{\alpha}=\tilde{B}_{x^{\alpha}}$. Similarly, $\psi^{\mu}-\tilde{C}^{\mu}=\tilde{D}_{y^{\mu}}$, where $\tilde{C}^{\mu}=\tilde{C}^{\mu}(y)$, $\tilde{D}=\tilde{D}(x,y)$. The substitution in (2.1) gives $\left(\tilde{B}+\tilde{D}\right)_{x^{\alpha}y^{\mu}}=0$, so $\tilde{B}+\tilde{D}=E(x)+F(y)$. Setting now $B=\tilde{B}-E=F-\tilde{D}, A^{\alpha}=\tilde{A}^{\alpha}+E_{x^{\alpha}}$, $C^{\mu}=\tilde{C}^{\mu}+F_{y^{\mu}}$, we obtain the representation (2.2).

Lemma 3. *The general solution* $F(x,y),\phi^{\alpha\mu}(y)$ *of the system*

$$F_{x^{\alpha}y^{\mu}} = \phi^{\alpha\mu}, \quad \alpha = 1,\ldots,a, \ \mu = 1,\ldots,m, \tag{2.3}$$

has the form

$$F = \sum_{\alpha=1}^{a} \phi^{\alpha}x^{\alpha} + \psi + \chi, \quad \phi^{\alpha\mu} = \phi^{\alpha}_{y^{\mu}}, \tag{2.4}$$

where $\phi^{\alpha}(y),\psi(x),\chi(y)$ *are arbitrary smooth functions.*

Proof. Functions of the form (2.4) obviously satisfy the system (2.3). On the other hand, let $F,\phi^{\alpha\mu}$ be a solution of the system (2.3). Then $\phi^{\alpha\mu}_{y^\nu}=-F_{x^\alpha y^\mu y^\nu}=\phi^{\alpha\nu}_{y^\mu}$, whence $\phi^{\alpha\mu}=\phi^{\alpha}_{y^\mu}$ (again by the Poincare lemma), where $\phi^\alpha=\phi^\alpha(y)$. Substituting this into (2.3), we obtain $\left(F_{x^\alpha}-\phi^\alpha\right)_{y^\mu}=0$, so $F_{x^\alpha}=\phi^\alpha+\psi^\alpha$, where $\psi^\alpha=\psi^\alpha(x)$. Moreover, $\psi^\alpha=\psi_{x^\alpha}$, where $\psi=\psi(x)$, since $\psi^\alpha_{x^\beta}=F_{x^\alpha x^\beta}=\psi^\beta_{x^\alpha}$. Thus, $\left(F-\psi\right)_{x^\alpha}=\phi^\alpha(y)$, and hence $F-\psi=\Sigma\phi^\alpha x^\alpha+\chi(x)$, as we claimed.

Lemma 4. *The general solution $f(x,y)$, $a^{\mu\alpha}(x)$, $b^\alpha(x)$, $c^\alpha(x)$ of the system*

$$f_{x^\alpha} = \sum_{\mu=1}^{m} y^\mu a^{\mu\alpha} + |y|b^\alpha + c^\alpha, \quad \alpha = 1,\dots,a, \tag{2.5}$$

has the form

$$f = \sum_{\mu=1}^{m} y^\mu A^\mu+|y|B+C+D, \; a^{\mu\alpha} = A^\mu_{x^\alpha}, \; b^\alpha = B_{x^\alpha}, \; c^\alpha = C_{x^\alpha}, \tag{2.6}$$

where $A^\mu(x)$, $B(x)$, $C(x)$, $D(y)$ are arbitrary smooth functions.

Proof. Functions of the form (2.6) obviously solve the system (2.5). On the other hand, let $f,a^{\mu\alpha},b^\alpha,c^\alpha$ be a solution of (2.5). Then

$$f_{x^\alpha x^\beta} = \sum_{\mu=1}^{m} y^\mu a^{\mu a}_{x^\beta} + |y|b^\alpha_{x^\beta} + c^\alpha_{x^\beta} = \sum_{\mu=1}^{m} y^\mu a^{\mu\beta}_{x^\alpha} + |y|b^\beta_{x^\alpha} + c^\beta_{x^\alpha}.$$

This is possible iff $a^{\mu\alpha}_{x^\beta}-a^{\mu\beta}_{x^\alpha}=b^\alpha_{x^\beta}-b^\beta_{x^\alpha}=c^\alpha_{x^\beta}-c^\beta_{x^\alpha}=0$. Therefore by the Poincare lemma, $a^{\mu\alpha}=A^\mu_{x^\alpha}$, $b^\alpha=B_{x^\alpha}$, $c^\alpha=C_{x^\alpha}$, where $A^\mu,B,C(x)$ are smooth functions. Substitution in (2.5) yields

$$f_{x^\alpha} = \left(\sum_{\mu=1}^{m} y^\mu A^\mu + |y|B + C \right)_{x^\alpha},$$

and hence, f has the representation we claimed.

Lemma 5. *A function $h(x,y)$ has the representations*

$$h = \sum_{\alpha=1}^{a} x^\alpha a^\alpha(y) + |x|b(y) + c(y) \tag{2.7}$$

$$= \sum_{\mu=1}^{m} y^\mu d^\mu(x) + |y|e(x) + f(x) \tag{2.8}$$

iff

$$h = \sum_{\alpha,\mu} x^\alpha y^\mu A^{\alpha\mu} + \sum_\alpha x^\alpha |y| B^\alpha + \sum_\mu |x| y^\mu C^\mu + |x||y|D$$

$$+ \sum_\alpha x^\alpha E^\alpha + \sum_\mu y^\mu F^\mu + |x|G + |y|H + K, \tag{2.9}$$

where $A^{\alpha\mu},\dots,K$ are arbitrary constants.

Proof. Clear that the representation (2.9) yields the equalities (2.7) and (2.8). Let us check that the converse is also true. For $x=0$, (2.7) and (2.8) give

$$c(y) = \sum_\mu y^\mu d^\mu(0) + |y|e(0) + f(0).$$

Substituting this into (2.7), we obtain

$$\sum_{\alpha=1}^{a} x^\alpha a^\alpha(y) + |x|b(y) = \sum_{\mu=1}^{m} y^\mu \tilde{d}^\mu(x) + |y|\tilde{e}(x) + \tilde{f}(x). \tag{2.10}$$

Let us fix an index $\alpha=1,\dots,a$, and set $x=(\alpha)\equiv(0,\dots,0,1,0,\dots,0)$, where the unit stays in the position α. Then (2.10) implies

$$a^{\alpha}(y) + b(y) = \sum_{\mu} y^{\mu}\tilde{d}^{\mu}((\alpha)) + |y|\tilde{e}((\alpha)) + \tilde{f}((\alpha)).$$

Setting $x=-(\alpha)\equiv(0,\ldots,0,-1,0,\ldots,0)$, we get

$$-a^{\alpha}(y) + b(y) = \sum_{\mu} y^{\mu}\tilde{d}^{\mu}(-(\alpha)) + |y|\tilde{e}(-(\alpha)) + \tilde{f}-((\alpha)).$$

Solving these two equations for a^{α} and b, and substituting the results in (2.7), we obtain the representation (2.9).

Lemma 6. *The general solution $\phi^{\mu},\psi^{\mu}(x)$ of the system*

$$\phi^{\mu}_{x^{\nu}} + \psi^{\nu}_{x^{\mu}} = 0, \quad \mu,\nu = 1,\ldots,m, \tag{2.11}$$

has the form

$$\phi^{\mu} = \Omega_{x^{\mu}} + \sum_{\nu} a^{\mu\nu}x^{\nu} + b^{\mu}, \quad \psi^{\mu} = -\Omega_{x^{\mu}} + \sum_{\nu} a^{\mu\nu}x^{\nu} + c^{\mu}, \tag{2.12}$$

where $\mu=1,\ldots,m$, $\Omega(x)$ is an arbitrary smooth function, $a^{\mu\nu}=-a^{\nu\mu}$, b^{μ}, c^{μ} are arbitrary constants.

Proof. The functions ϕ^{μ},ψ^{μ} of the form (2.12) clearly satisfy the system (2.11). Conversely, let functions ϕ^{μ},ψ^{μ} be a solution of (2.11). Then, $\phi^{\mu}_{x^{\nu}x^{\lambda}}=-\psi^{\nu}_{x^{\mu}x^{\lambda}}=\phi^{\lambda}_{x^{\nu}x^{\mu}}$, i.e., $\left(\phi^{\mu}_{x^{\lambda}}-\phi^{\lambda}_{x^{\mu}}\right)_{x^{\nu}}=0$, whence $\phi^{\mu}_{x^{\lambda}}-\phi^{\lambda}_{x^{\mu}}=2a^{\mu\lambda}$, where $a^{\mu\lambda}=-a^{\lambda\mu}=\text{const}$. Further,

$$\left(\phi^{\mu} - \sum_{\lambda}a^{\mu\lambda}x^{\lambda}\right)_{x^{\nu}} - \left(\phi^{\nu} - \sum_{\lambda}a^{\nu\lambda}x^{\lambda}\right)_{x^{\mu}} = \phi^{\mu}_{x^{\nu}} - \phi^{\nu}_{x^{\mu}} - a^{\mu\nu} + a^{\nu\mu} = 0,$$

so by the Poincare lemma,

$$\phi^{\mu} = \sum_{\nu} a^{\mu\nu}x^{\nu} + \Omega_{x^{\mu}}(x). \tag{2.13}$$

In the same way,

$$\psi^\mu = \sum_\nu \tilde{a}^{\mu\nu} x^\nu + \tilde{\Omega}_{x^\mu}(x). \tag{2.14}$$

The substitution of (2.13) and (2.14) into (2.11) yields

$$\left(\Omega + \tilde{\Omega}\right)_{x^\mu x^\nu} = -\left(a^{\mu\nu} + \tilde{a}^{\nu\mu}\right) = 0,$$

since the expression on the right-hand side is skew-symmetric with respect to the indices μ,ν. Thus, $\tilde{a}^{\mu\nu}=-a^{\mu\nu}=a^{\nu\mu}$, and

$$\tilde{\Omega} = -\Omega + \sum_\mu c^\mu x^\mu + d,$$

where c^μ and d are constants. Substituting the resulting representations into (2.13) and (2.14), we obtain (2.12).

17.3. Proof of Theorem 1. We must find the general solution of the system (1.3-5).

Proposition 1. *The general solution of a bi-order (p,q) of the equation $\partial_\xi \partial_\eta \chi=0$ has the form*

$$\chi = \sum_{\mu=0}^{m} a^\mu x^\mu_{00} + \phi(\xi, \mathbf{x}_{10}, \ldots, \mathbf{x}_{p0}) + \psi(\eta, \mathbf{x}_{01}, \ldots, \mathbf{x}_{0q}),$$

where ϕ,ψ are arbitrary smooth functions, a^μ=const (cf. Ibragimov [1; §18.4).

Proof. A function χ of the indicated form obviously satisfy the equation. On the other hand, let $\chi(\mathbf{x}_{p0}, \mathbf{x}_{0q}, \ldots)$ be a solution. Then

$$\partial_\xi \partial_\eta \chi = \sum_{\mu=1}^{m} x^\mu_{p+1,0} \partial_\eta \chi_{x^\mu_{p0}} + \ldots = 0,$$

and hence $\partial_\eta \chi_{x^\mu_{p0}}=0$, so by Lemma 1, we have $\chi_{x^\mu_{p0}}=A^\mu(\xi,\mathbf{x}_{10},\ldots,\mathbf{x}_{p0})$, whence $\chi=B(\xi,\mathbf{x}_{10},\ldots,\mathbf{x}_{p0})+C(\mathbf{x}_{p-1,0},\mathbf{x}_{0q},\ldots)$. By induction we verify $\chi=\phi(\xi,\mathbf{x}_{10},\ldots,\mathbf{x}_{p0})+f(\xi,\eta,x^0_{00},\mathbf{x}_{00},\mathbf{x}_{01},\ldots,\mathbf{x}_{0q})$, where $\partial_\xi\partial_\eta f=0$. Repeating the procedure, we obtain the representation

$$f = \psi(\eta,\mathbf{x}_{01},\ldots\mathbf{x}_{0q}) + g(\xi,\eta,x^0_{00},\mathbf{x}_{00}),$$

where $\partial_\xi\partial_\eta g=0$, again. In more detail,

$$\sum_{\mu=1}^{m} x^\mu_{10}\left(\partial_\eta\chi\right)_{x^\mu_{00}} + |\mathbf{x}_{10}|\left(\partial_\eta\chi\right)_{x^0_{00}} + \left(\partial_\eta\chi\right)_\xi = 0,$$

whence $\partial_\eta\chi=h(\eta,\mathbf{x}_{01})$, i.e.,

$$\sum_{\mu=1}^{m} x^\mu_{01} g_{x^\mu_{00}} + |\mathbf{x}_{01}| g_{x^0_{00}} + g_\eta = h(\eta,\mathbf{x}_{01}).$$

Therefore, $g_{x^\mu_{00}}=a^\mu(\eta)$, $\mu=0,\ldots,m$, and hence,

$$g = \sum_{\mu=0}^{m} a^\mu x^\mu_{00} + b(\xi) + c(\eta).$$

Redefining the functions ϕ and ψ, we can assume that $b=c=0$. Then

$$\partial_\xi\partial_\eta g = \sum_{\mu=1}^{m} a^\mu_\eta x^\mu_{10} + a^0_\eta|\mathbf{x}_{10}| = 0,$$

so that $a^\mu_\eta=0$, hence a^μ=const for all $\mu=0,\ldots,m$, and the proof is completed.

By Proposition 1, the general solution of subsystem (1.3) has the form

$$x^\mu_{00} = A^\mu + B^\mu(\xi,\mathbf{x}_{10},\ldots,\mathbf{x}_{p0}) + C^\mu(\eta,\mathbf{x}_{01},\ldots,\mathbf{x}_{0q}), \qquad (3.1)$$

where $\mu=0,\ldots,m$,

$$A^{\mu} = \sum_{\nu=0}^{m} a^{\mu\nu} x^{\mu}_{00}, \quad a^{\mu\nu} = \text{const}.$$

Substitution of (3.1) into (1.4) yields

$$(\mathbf{x}_{10}, \partial_{\xi}\mathbf{B}) = |\mathbf{x}_{10}| \partial_{\xi}(B^{0} + A^{0}) - (\mathbf{x}_{10}, \partial_{\xi}\mathbf{A}). \tag{3.2}$$

Proposition 2. *The general solution of the homogeneous equation*

$$(\mathbf{x}_{10}, \partial_{\xi}\mathbf{B}) = 0 \tag{3.3}$$

in the class of differential functions depending on $\xi, \mathbf{x}_{10}, \mathbf{x}_{20}, \ldots$ *has the form*

$$\mathbf{B} = f\mathbf{e}_{10} + (\partial_{\xi}f)\mathbf{e}_{20} + \mathbf{h}, \tag{3.4}$$

where the function f *is arbitrary, and* $\mathbf{h}$ *satisfies the orthogonality conditions* $(\mathbf{h},\mathbf{x}_{10})=(\mathbf{h},\mathbf{x}_{20})=0$. *For a given solution* $\mathbf{B}$ *the representation (3.4) is unique.*

Proof. Every vector function $\mathbf{B}$ has the unique decomposition

$$\mathbf{B} = f\mathbf{e}_{10} + g\mathbf{e}_{20} + \mathbf{h},$$

where $f=(\mathbf{B},\mathbf{x}_{10})$, $g=(\mathbf{B},\mathbf{x}_{20})$, $\mathbf{h}=\mathbf{B}-(\mathbf{B},\mathbf{x}_{10})\mathbf{e}_{10}-(\mathbf{B},\mathbf{x}_{20})\mathbf{e}_{20}$, so that $(\mathbf{h},\mathbf{x}_{10})=(\mathbf{h},\mathbf{x}_{20})=0$. Such function $\mathbf{B}$ will satisfy (3.3) iff

$$\begin{aligned}(\mathbf{x}_{10}, \partial_{\xi}\mathbf{B}) &= \partial_{\xi}(\mathbf{x}_{10}, \mathbf{B}) - (\mathbf{B}, \partial_{\xi}\mathbf{x}_{10}) \\ &= \partial_{\xi}(\mathbf{x}_{10}, \mathbf{B}) - (\mathbf{B}, \mathbf{x}_{20}) = \partial_{\xi}f - g = 0,\end{aligned}$$

i.e., when $g=\partial_{\xi}f$, as we claimed.

Proposition 3. *The inhomogeneous equation*

$$(\mathbf{x}_{10}, \partial_\xi \mathbf{B}) = \rho, \tag{3.5}$$

where $\rho(\xi, \mathbf{x}_{10}, \ldots, \mathbf{x}_{p0})$ *is a given function has the particular solution* $\mathbf{B}=-\mathbf{e}_{20}\rho$.

Proof. Indeed,

$$(\mathbf{x}_{10}, \partial_\xi \mathbf{e}_{20}\rho) = \partial_\xi(\mathbf{x}_{10}, \mathbf{e}_{20}\rho) - (\mathbf{e}_{20}\rho, \partial_\xi \mathbf{x}_{10})$$

$$= 0 - \rho(\mathbf{e}_{20}, \mathbf{x}_{20}) = -\rho.$$

By propositions 2 and 3, the functions $A^0, \mathbf{A}, B^0$ on the right-hand side of the equality (3.2) can be chosen arbitrary, while the vector function **B** must have the form, indicated in the theorem. The function **C** is treated in the same way.

17.4. Proof of Theorem 2. Functions F and G of the indicated form obviously satisfy satisfy the divergency equation (1.6). On the other hand, let functions $F, G(\mathbf{x}_{p0}, \mathbf{x}_{0p}, \ldots)$ solve (1.6). Then

$$\sum_{\mu=1}^{m} \left(x^\mu_{p+1,0} F_{x^\mu_{p0}} + x^\mu_{0,p+1} G_{x^\mu_{0p}} \right) + \ldots = 0,$$

and therefore $F_{x^\mu_{p0}} = G_{x^\mu_{0p}} = 0$, so that in fact $F=F(\mathbf{x}_{p-1,0}, \mathbf{x}_{0p}, \ldots)$ and $G=G(\mathbf{x}_{p0}, \mathbf{x}_{0,p-1}, \ldots)$.

We assume that for some $n \le p-2$ the representation

$$F = a(\mathbf{x}_{p-1,0}, \mathbf{x}_{0,p-n}, \ldots) + b(\mathbf{x}_{p-2,0}, \mathbf{x}_{0,p-1}, \ldots)$$

$$+ c(\mathbf{x}_{p-n-1,0}, \mathbf{x}_{0p}, \ldots) - \partial_\eta \omega, \tag{4.1}$$

$$G = d(\mathbf{x}_{p-n,0}, \mathbf{x}_{0,p-1}, \ldots) + e(\mathbf{x}_{p-1,0}, \mathbf{x}_{0,p-2}, \ldots)$$

$$+ f(\mathbf{x}_{p0}, \mathbf{x}_{0,p-n-1}, \ldots) + \partial_\xi \omega, \qquad (4.2)$$

(which is trivial when $n=0$) is valid, and we show that such representation the is valid for $n+1$ too. The substitution of (4.1) and (4.2) into (1.6) yields

$$\sum_{\mu=1}^{m} x^\mu_{p0} a_{x^\mu_{p-1,0}} + \partial_\eta f + \ldots = 0,$$

so that

$$a_{x^\mu_{p-1,0}} + \partial_\eta f_{x^\mu_{p0}} = a_{x^\mu_{p-1,0}} + \sum_{\nu=1}^{m} x^\nu_{0,p-n} f_{x^\mu_{p0} x^\nu_{0,p-n-1}} + \ldots = 0,$$

and hence $a_{x^\mu_{p-1,0} x^\nu_{0,p-n}} + f_{x^\mu_{p0} x^\nu_{0,p-n-1}} = 0$. In other words,

$$a_{x^\mu_{p-1,0} x^\nu_{0,p-n}} = -f_{x^\mu_{p0} x^\nu_{0,p-n-1}} = \phi^{\mu\nu}(\mathbf{x}_{p-1,0}, \mathbf{x}_{0,p-n-1}, \ldots).$$

By Lemma 3,

$$a = \sum_{\nu=1}^{m} \alpha^\nu(\mathbf{x}_{p-1,0}, \mathbf{x}_{0,p-n-1}, \ldots) x^\nu_{0,p-n} + \beta(\mathbf{x}_{p-1,0}, \mathbf{x}_{0,p-n-1}, \ldots)$$

$$+ \gamma(\mathbf{x}_{p-2,0}, \mathbf{x}_{0,p-n}, \ldots),$$

$$f = \sum_{\mu=1}^{m} \delta^\mu(\mathbf{x}_{p-1,0}, \mathbf{x}_{0,p-n-1}, \ldots) x^\mu_{p0} + \varepsilon(\mathbf{x}_{p,0}, \mathbf{x}_{0,p-n-2}, \ldots)$$

$$+ \lambda(\mathbf{x}_{p-1,0}, \mathbf{x}_{0,p-n-1}, \ldots),$$

with $\alpha^\nu_{x^\mu_{p-1,0}} + \delta^\mu_{x_{0,p-n-1}} = 0$. Now, by Lemma 2,

$$\alpha^{\nu} = A^{\nu}(\mathbf{x}_{p-2,0}, \mathbf{x}_{0,p-n-1}, \ldots) + B_{x^{\nu}_{0,p-n-1}}(\mathbf{x}_{p-1,0}, \mathbf{x}_{0,p-n-1}, \ldots),$$

$$\delta^{\mu} = C^{\mu}(\mathbf{x}_{p-1,0}, \mathbf{x}_{0,p-n-2}, \ldots) - B_{x^{\mu}_{p-1,0}}(\mathbf{x}_{p-1,0}, \mathbf{x}_{0,p-n-1}, \ldots),$$

whence,

$$a = \sum_{\nu=1}^{m} B_{x^{\nu}_{0,p-n-1}} x^{\nu}_{0,p-n} + \beta + \tilde{\gamma}(\mathbf{x}_{p-2,0}, \mathbf{x}_{0,p-n}, \ldots)$$

$$= \tilde{\beta}(\mathbf{x}_{p-1,0}, \mathbf{x}_{0,p-n-1}, \ldots) + \tilde{\gamma} + \partial_{\eta} B, \qquad (4.3)$$

since

$$\partial_{\eta} B = \sum_{\nu=1}^{m} B_{x^{\nu}_{0,p-n-1}} x^{\nu}_{0,p-n} + \beta'(\mathbf{x}_{p-1,0}, \mathbf{x}_{0,p-n-1}, \ldots).$$

In the same manner,

$$f = \tilde{\varepsilon}(\mathbf{x}_{p0}, \mathbf{x}_{0,p-n-2}, \ldots) + \tilde{\lambda}(\mathbf{x}_{p-1,0}, \mathbf{x}_{0,p-n-1}, \ldots) - \partial_{\xi} B. \qquad (4.4)$$

Substituting (4.3), (4.4) and the similar expressions for c and d into (4.1), we obtain the representation we claimed with $n+1$. By induction we get ($n=p-1$)

$$F = a(\mathbf{x}_{p-1,0}, \mathbf{x}_{01}, \ldots) + b(\mathbf{x}_{p-2,0}, \mathbf{x}_{0,p-1}, \ldots)$$

$$+ c(\mathbf{x}_{00}, \mathbf{x}_{0p}, \ldots) - \partial_{\eta}\omega, \qquad (4.5)$$

$$G = d(\mathbf{x}_{10}, \mathbf{x}_{0,p-1}, \ldots) + e(\mathbf{x}_{p-1,0}, \mathbf{x}_{0,p-2}, \ldots)$$

$$+ f(\mathbf{x}_{p0}, \mathbf{x}_{00}, \ldots) + \partial_{\xi}\omega. \qquad (4.6)$$

The substitution of these representations into (1.6) yields

$$\sum_{\mu=1}^{m} x^{\mu}_{p0} a_{x^{\mu}_{p-1,0}} + \partial_{\eta} f + \ldots = 0,$$

hence

$$a_{x^{\mu}_{p-1,0}} + \partial_{\eta} f_{x^{\mu}_{p0}}$$

$$= a_{x^{\mu}_{p-1,0}} + \sum_{\nu=1}^{m} x^{\nu}_{01} f_{x^{\mu}_{p0} x^{\nu}_{00}} + |\mathbf{x}_{01}| f_{x^{\mu}_{p0} x^{0}_{00}} + f_{x^{\mu}_{p0}\eta} = 0. \qquad (4.7)$$

Now, by Lemma 4,

$$a = \sum_{\nu=1}^{m} x^{\nu}_{01} \alpha^{\nu} + |\mathbf{x}_{01}|\beta + \gamma + \phi,$$

where $\alpha^{\nu}, \beta, \gamma(\mathbf{x}_{p-1,0}, \mathbf{x}_{00}, \ldots), \phi(\mathbf{x}_{p-2,0}, \mathbf{x}_{01}, \ldots)$ are smooth functions. Redefining the function $b(\mathbf{x}_{p-2,0}, \mathbf{x}_{0,p-1}, \ldots)$ in (4.5), we may assume $\phi=0$. The substitution of the function a into (4.7) yields the system

$$\alpha^{\nu}_{x^{\mu}_{p-1,0}} + f_{x^{\mu}_{p0} x^{\nu}_{00}} = \beta_{x^{\mu}_{p-1,0}} + f_{x^{\mu}_{p0} x^{0}_{00}} = \gamma_{x^{\mu}_{p-1,0}} + f_{x^{\mu}_{p0}\eta} = 0,$$

whence by Lemma 3,

$$f = \sum_{\mu=1}^{m} x^{\mu}_{p0} \phi^{\mu}(\mathbf{x}_{p-1,0}, \mathbf{x}_{00}, \ldots) + \psi(\xi, \mathbf{x}_{10}, \ldots, \mathbf{x}_{p0})$$
$$+ \chi(\mathbf{x}_{p-1,0}, \mathbf{x}_{00}, \ldots).$$

Redefining the function e in (4.6), we may assume $\chi=0$. Substituting f into the last system, we obtain

$$\alpha^{\nu}_{x^{\mu}_{p-1,0}} + \phi^{\mu}_{x^{\nu}_{00}} = \beta_{x^{\mu}_{p-1,0}} + \phi^{\mu}_{x^{0}_{00}} = \gamma_{x^{\mu}_{p-1,0}} + \phi^{\mu}_{\eta} = 0,$$

and by Lemma 2,

$$\alpha^{\nu} = \tilde{\alpha}^{\nu} + \tilde{\Omega}_{x^{\nu}_{00}}, \quad \beta = \tilde{\beta} + \tilde{\Omega}_{x^{0}_{00}}, \quad \gamma = \tilde{\gamma} + \tilde{\Omega}_{\eta}, \quad \phi^{\mu} = \tilde{\phi}^{\mu} - \tilde{\Omega}_{x^{\mu}_{p-1,0}},$$

where $\tilde{\alpha}^{\nu}, \tilde{\beta}, \tilde{\gamma}(\mathbf{x}_{p-2,0}, \mathbf{x}_{00}, \ldots)$, $\tilde{\Omega}(\mathbf{x}_{p-1,0}, \mathbf{x}_{00}, \ldots)$, $\tilde{\phi}^{\mu}(\xi, \mathbf{x}_{10}, \ldots, \mathbf{x}_{p0})$ are smooth functions. Substituting thus obtained representations for a and f and similar representations for c and d into (4.5) and (4.6) we get

$$F = \tilde{F}(\mathbf{x}_{p-2,0}, \mathbf{x}_{0,p-1}, \ldots) + \tilde{\phi}(\eta, \mathbf{x}_{01}, \ldots, \mathbf{x}_{0p}) - \partial_{\eta}\tilde{\omega},$$

$$G = \tilde{G}(\mathbf{x}_{p-1,0}, \mathbf{x}_{0,p-2}, \ldots) + \tilde{\psi}(\xi, \mathbf{x}_{10}, \ldots, \mathbf{x}_{p0}) + \partial_{\xi}\tilde{\omega},$$

the term wit $\tilde{\Omega}$ and the analogous term from c and d are included into $\tilde{\omega}$. Repeating the arguments the required number of times, We obtain the representation

$$F = f(\mathbf{x}_{00}, \mathbf{x}_{01}, \ldots) + \phi(\eta, \mathbf{x}_{01}, \ldots, \mathbf{x}_{0p}) - \partial_{\eta}\Omega, \tag{4.8}$$

$$G = g(\mathbf{x}_{10}, \mathbf{x}_{00}, \ldots) + \psi(\xi, \mathbf{x}_{10}, \ldots, \mathbf{x}_{p0}) + \partial_{\xi}\Omega. \tag{4.9}$$

The substitution into (1.6) yields the divergency equation for functions f and g.

$$\partial_{\xi} f + \partial_{\eta} g = 0. \tag{4.10}$$

Thus, it remains to solve (4.10) with $f=f(\xi, \eta, x^{0}_{00}, \mathbf{x}_{00}, \mathbf{x}_{01})$ and $g=g(\xi, \eta, x^{0}_{00}, \mathbf{x}_{00}, \mathbf{x}_{10})$. To this aim we rewrite (4.10) in the form

$$h = \partial_{\xi} f = \sum_{\mu=1}^{m} x^{\mu}_{10} f_{x^{\mu}_{00}} + |\mathbf{x}_{10}| f_{x^{0}_{00}} + f_{\xi}$$

$$= -\partial_\eta g = -\sum_{\mu=1}^{m} x_{01}^{\mu} g_{x_{00}^{\mu}} - |\mathbf{x}_{01}| g_{x_{00}^{0}} - g_\eta .$$

Hence, by Lemmas 3 and 4,

$$f = \sum_{\mu=1}^{m} x_{01}^{\mu} A^{\mu} + |\mathbf{x}_{01}| A^{0} + B, \tag{4.11}$$

$$g = \sum_{\mu=1}^{m} x_{10}^{\mu} C^{\mu} + |\mathbf{x}_{10}| C^{0} + D, \tag{4.12}$$

where $A^{\mu}, A^{0} B, C^{\mu}, C^{0}, D(\xi, \eta, x_{00}^{0}, \mathbf{x}_{00})$ are smooth functions. Substituting (4.11) and (4.12) into (4.10), we obtain the system

$$A^{\mu}_{x_{00}^{\nu}} + C^{\nu}_{x_{00}^{\mu}} = A^{\mu}_{\xi} + D_{x_{00}^{\mu}} = B_{x_{00}^{\mu}} + C^{\mu}_{\eta} = B_{\xi} + D_{\eta} = 0, \tag{4.13}$$

$\mu, \nu = 0, 1, \ldots, m$. Lemma 6 and the subsystem $A^{\mu}_{x_{00}^{\nu}} + C^{\nu}_{x_{00}^{\mu}} = 0$ yield

$$A^{\mu} = \chi_{x_{00}^{\mu}} + \sum_{\nu=0}^{m} a^{\mu\nu} x_{00}^{\nu} + b^{\mu}, \quad \mu = 0, \ldots, m, \tag{4.14}$$

$$C^{\mu} = -\chi_{x_{00}^{\mu}} + \sum_{\nu=0}^{m} a^{\mu\nu} x_{00}^{\nu} + c^{\mu}, \quad \mu = 0, \ldots, m, \tag{4.15}$$

where $\chi = \chi(\xi, \eta, x_{00}^{0}, \mathbf{x}_{00})$, $a^{\mu\nu} = -a^{\nu\mu}, b^{\mu}, c^{\mu}(\xi, \eta)$. Redefining the function Ω in (4.8)-(4.9), we may assume $\chi = 0$. Substituting A^{μ} into the subsystem $A^{\mu}_{\xi} + D_{x_{00}^{\mu}} = 0$, we get

$$D_{x_{00}^{\mu}} + \sum_{\nu=0}^{m} a^{\mu\nu}_{\xi} x_{00}^{\nu} + b^{\mu}_{\xi}, \quad \mu = 0, \ldots, m,$$

whence, $D_{x_{00}^{\mu} x_{00}^{\nu}} = -a^{\mu\nu}_{\xi} = 0$, since the matrix $a^{\mu\nu}$ is skew-symmetric.

Similarly, $B_{x_{00}^{\mu} x_{00}^{\nu}} = -a^{\mu\nu}_{\eta} = 0$. Therefore, $a^{\mu\nu} = \text{const}$, and

$$D = -\sum_{\mu=0}^{m} b^{\mu}_{\xi} x^{\mu}_{00} + d(\xi,\eta), \quad B = -\sum_{\mu=0}^{m} c^{\mu}_{\eta} x^{\mu}_{00} + e(\xi,\eta),$$

where

$$-\sum_{\mu=0}^{m} \left(b^{\mu}_{\xi\eta} + c^{\mu}_{\xi\eta}\right) x^{\mu}_{00} + e_{\xi} + d_{\eta} = 0,$$

by virtue of the remaining equation $B_{\xi}+D_{\eta}=0$. Hence,

$$\left(b^{\mu} + c^{\mu}\right) = 0, \quad \mu = 0,\dots,m, \tag{4.16}$$

$$e_{\xi} + d_{\eta} = 0. \tag{4.17}$$

From (4.17) we derive that $e=-h_{\eta}$, $d=h_{\xi}$, where $h(\xi,\eta)$ is a smooth function. Redefining Ω in (4.8)-(4.9), we may assume $h=e=d=0$. The equation (4.16) give

$$b^{\mu} + c^{\mu} = \alpha^{\mu}(\xi) + \beta^{\mu}(\eta), \quad \mu = 0,\dots,m,$$

whence $b^{\mu}-\beta^{\mu}=-c^{\mu}+\alpha^{\mu}=\rho^{\mu}(\xi,\eta)$, so $b^{\mu}=\rho^{\mu}+\beta^{\mu}$, $c^{\mu}=-\rho^{\mu}+\alpha^{\mu}$. Including the functions $\Sigma\beta^{\mu}x^{\mu}_{01}$, $\Sigma\alpha^{\mu}x^{\mu}_{10}$, $\Sigma\rho^{\mu}x^{\mu}_{00}$ into the functions ϕ,ψ,Ω from (4.8)-(4.9), we obtain the representation (4.14)-(4.15) with $\chi=b^{\mu}=c^{\mu}=0$. In this case, functions $B=D=0$ too. Thus, in the representations (4.11)-(4.12) the functions f and g are reduced to the required form, and Theorem 2 is proved.

APPENDIX

DIFFERENTIAL ALGEBRAS

Let us remind the standard multi-index notation:

$\mathbb{N}=\{0,1,2,\dots\}$ is the set of all natural numbers, including zero;

a multi-index $i=(i_1,\dots,i_m)\in\mathbb{N}^m$;

$|i|=i_1+\dots+i_m$ and $i!=i_1!\dots i_m!$ for $i\in\mathbb{N}^m$;

the multi-index $(\mu)=(0,\dots,0,1,0,\dots,0)$, where the unity stands in the μ-th position, $\mu=1,\dots,m$;

$i+k=(i_1+k_1,\dots,i_m+k_m)$ for $i,k\in\mathbb{N}^m$;

$i\le k$ means $i_1\le k_1,\dots,i_m\le k_m$ for $i,k\in\mathbb{N}^m$;

polynomial coefficients $C(i)=\dfrac{|i|!}{i!}$ for $i\in\mathbb{N}^m$;

multinomial coefficient $\begin{pmatrix}k\\ i\end{pmatrix}=\dfrac{k!}{i!(k-i)!}$ for $i,k\in\mathbb{N}^m$, $i\le k$.

We shall also use another multi-index notation. Namely, every multi-index $i=(i_1,\dots,i_m)\in\mathbb{N}^m$, $|i|=k$, has the unique representation $i=(\mu_1)+\dots+(\mu_k)$, where the unordered family $\mu_1\dots\mu_k\in\{1,\dots,m\}^k$, and we shall write $i=(i_1,\dots,i_m)=\mu_1\dots\mu_k$ in this case.

Let us introduce some special notations. For a module M and natural numbers $m,n=1,2,\dots$ and $p=0,\dots,m$ we denote by

$$M_n = \left\{ f = \left(f_\sigma \right) \ : \ f_\sigma \in M \right\}, \quad M^n = \left\{ f = \left(f^\sigma \right) \ : \ f^\sigma \in M \right\},$$

$$M^{\wedge p} = \left\{ f = \left(f^{\mu_1 \ldots \mu_p} \right) \ : \ f^{\mu_1 \ldots \mu_p} \in M \right\},$$

where $\sigma=1,\ldots,n$, $\mu_1,\ldots,\mu_p=1,\ldots,m$, and

$$f^{\pi(\mu_1 \ldots \mu_p)} = \text{sign}\ \pi\ f^{\mu_1 \ldots \mu_p},$$

sign π is the sign of a permutation π of indices $\mu_1,\ldots,\mu_p$ (we distinguish the modules M_n and M^n to make formulas more readable). In particular, $M_1=M^1=M^{\wedge 0}=M$ and

$$\begin{aligned} M_n^{\wedge p} &= \left\{ f = \left(f_\sigma \right) \ : \ f_\sigma \in M^{\wedge p} \right\} \\ &= \left\{ f = \left(f^{\mu_1 \ldots \mu_p} \right) \ : \ f^{\mu_1 \ldots \mu_p} \in M_n \right\} \\ &= \left\{ f = \left(f_\sigma^{\mu_1 \ldots \mu_p} \right) \ : \ f_\sigma^{\mu_1 \ldots \mu_p} \in M \right\}. \end{aligned}$$

Let $\mathcal{A}$ be a (real or complex) algebra and let $D_1,\ldots,D_m$ be pairwise commuting differentiations of $\mathcal{A}$ (a linear mapping $\zeta:\mathcal{X}\longrightarrow\mathcal{X}$ is called the *differentiation* of an algebra $\mathcal{X}$ if the Leibniz' rule

$$\zeta(fg) = \zeta(f)g + f\zeta(g)$$

is valid for all $f,g\in\mathcal{X}$). The algebra $\mathcal{A}$ with the differentiations $D=(D_1,\ldots,D_m)$ is called *the differential algebra*. Let us denote by $\mathcal{P}=\mathcal{A}[D]$ the set of all polynomials

$$P = \sum_i f^i D_i,$$

in D's with coefficients $f^i\in\mathcal{A}$, where

$$D_i = D_1^{i_1}\dots D_m^{i_m} \quad \text{for } i = (i_1,\dots,i_m) \in \mathbb{N}^m$$

(we don't write the summation limits if there is no danger for a mistake). The degree of a polynomial P is the biggest of all numbers $|i|$ such that $f^i \neq 0$. Every polynomial $P \in \mathcal{P}$ defines the linear differential operator $P: \mathcal{A} \longrightarrow \mathcal{A}$ and the order of this operator coincides with the degree of the polynomial. The composition rule of operators defines the multiplication in the set $\mathcal{P}$, so $\mathcal{P}$ is the algebra. The algebra $\mathcal{A}$ is identified with subalgebra of $\mathcal{P}$ of all polynomials of zero order. We assume that the algebra $\mathcal{A}$ contains the identity element (otherwise we formally add it), so the differentiations $D_1,\dots,D_m$ are elements of the algebra $\mathcal{P}$.

An element $c \in \mathcal{A}$ is called a *constant* if

$$D_\nu c = 0 \text{ for all } \nu = 1,\dots,m.$$

The set of all constants defines the subalgebra $\mathbb{K} \subset \mathcal{A}$. In particular, the identity element belongs to $\mathbb{K}$.

The characteristic of a polynomial $P \in \mathcal{P}$ is the element

$$\chi(P) = \sum_i (-1)^{|i|} D_i f^i \in \mathcal{A}.$$

In particular,

$$\chi(D_\mu) = 0, \qquad \text{for all } \mu = 1,\dots,m,$$

$$\chi(P \circ R) = R(\chi(P)) \qquad \text{for all } P, R \in \mathcal{P},$$

so

$$\chi(D_\mu \circ P) = 0 \text{ for all } \mu = 1,\dots,m \text{ and } P \in \mathcal{P}.$$

The rule $P \mapsto \chi(P)$ defines the mapping $\chi: \mathcal{P} \longrightarrow \mathcal{A}$.

For $p=0,\ldots,m$ let us denote by $\mathcal{P}^{\wedge p}$ the two-sided $\mathcal{P}$-modules, defined by the above rules. For an element $P\in\mathcal{P}^{\wedge p}$ with components

$$P^{\mu_1\ldots\mu_p} \in \mathcal{P}, \quad \mu_1,\ldots,\mu_p = 1,\ldots,m$$

we denote by $D\circ P$ the element of $\mathcal{P}^{\wedge(p-1)}$ with the components

$$(D\circ P)^{\mu_1\ldots\mu_{p-1}} = \sum_{\mu} D_{\mu}\circ P^{\mu\mu_1\ldots\mu_{p-1}} \in \mathcal{P}.$$

In particular, $D\circ P=0$ if $P\in\mathcal{P}^{\wedge 0}$. The rule $P\mapsto D\circ P$ defines the linear mappings

$$D\circ \, : \, \mathcal{P}^{\wedge p} \longrightarrow \mathcal{P}^{\wedge(p-1)}, \quad p = 1,\ldots,m.$$

Clear the composition $D\circ D\circ=0$.

To formulate our first theorem it is convenient to use another multi-index notation. Namely, a polynomial $P\in\mathcal{P}$ we shall write in two forms

$$P = \sum_i f^i D_i = \sum_q \sum_{\nu_1\ldots\nu_q} f^{\nu_1\ldots\nu_q} D_{\nu_1}\ldots D_{\nu_q},$$

where $q\in\{0,\ldots,Q\}$, Q is the degree of the polynomial P,

$$f^{\pi(\nu_1\ldots\nu_m)} = f^{\nu_1\ldots\nu_m}$$

for any permutation π of indices $\nu_1\ldots\nu_q\in\{1,\ldots,m\}$.

For an element $P\in\mathcal{P}^{\wedge p}$ with components

$$P^{\mu_1\ldots\mu_p} = \sum_q \sum_{\nu_1\ldots\nu_q} f^{\nu_1\ldots\nu_q;\mu_1\ldots\mu_p} D_{\nu_1}\ldots D_{\nu_q}$$

we denote by $h(P)$ an element from $\mathcal{P}^{\wedge(p+1)}$ with the components

$$h(P)^{\mu_1\ldots\mu_{p+1}}$$

$$= \sum_{r,s} (-1)^r c(p,r,s) \sum_{(\nu)} \left(\sum_{(\lambda)} D_{(\lambda)} f^{(\lambda)(\nu)[\mu_1;\mu_2\ldots\mu_{p+1}]} \right) D_{(\nu)},$$

where $r+s\in\{0,\ldots,Q-1\}$, the coefficients

$$c(p,r,s) = \frac{(p+1)(s+1)\ldots(s+r+1)}{(p+s+1)\ldots(p+s+r+1)}$$

$$= (p+1)\binom{p+s}{p} \Big/ \binom{p+s+r+1}{p},$$

$(\nu)=\nu_1\ldots\nu_s$, $(\lambda)=\lambda_1\ldots\lambda_r$, the brackets $[\ldots]$ denote the skew-symmetrization in indices $\mu_1,\ldots,\mu_{p+1}$ (for example,

$$f^{[\mu_1;\mu_2\ldots\mu_{p+1}]} = \frac{1}{p+1} \sum_{k=1}^{p+1} (-1)^{k-1} f^{\mu_k;\mu_1\ldots\hat{\mu}_k\ldots\mu_{p+1}},$$

where the index $\hat{\mu}_k$ is omitted). In particular, $h(P)=0$, if $P\in\mathcal{P}^{\wedge m}$ or the order Q of P is equal to zero. The rule $P\mapsto h(P)$ defines the linear mappings

$$h : \mathcal{P}^{\wedge p} \longrightarrow \mathcal{P}^{\wedge(p+1)}, \quad p = 0,\ldots,m-1.$$

Theorem A.1. *The following homotopy formula*

$$D\circ h(P) + h(D\circ P) = P - \delta(p)\chi(P)$$

is valid for any $P\in\mathcal{P}^{\wedge p}$, $p=0,\ldots,m$, where $\delta(p)=1$ if $p=0$ and $\delta(p)=0$ if $p\geq 1$.

In more details, for $p=0$ we have

$$D\circ h(P) = P - \chi(P)$$

the Green's formula for linear differential operators (see, for example, Vladimirov, Zharinov [1], Zharinov [3], Vinogradov [4]). Further,

$$D\circ h(P) + h(D\circ P) = P \quad \text{if } p = 1,\dots,m-1,$$

$$h(D\circ P) = P \quad \text{if } p=m.$$

Proof. Let us consider the mappings h of the above form with unknown coefficients $c(p,r,s)$ and let us try to define these coefficients in a way that the homotopy formula would be valid. In doing so, one should bear in mind that

$$\sum_{(\nu)} f^{\nu_1\dots\nu_s;\nu_{s+1}\mu_1\dots\mu_{p-1}} D_{\nu_1}\dots D_{\nu_s} D_{\nu_{s+1}}$$

$$= \sum_{(\nu)} f^{\{\nu_1\dots\nu_s;\nu_{s+1}\}\mu_1\dots\mu_{p-1}} D_{\nu_1}\dots D_{\nu_s} D_{\nu_{s+1}},$$

where $(\nu)=\nu_1\dots\nu_{s+1}$, the curly braces $\{\dots\}$ denote the symmetrization in indices $\nu_1\dots\nu_{s+1}$, for instance,

$$f^{\{\nu_1\dots\nu_s;\nu_{s+1}\}} = \frac{1}{s+1}\sum_{i=1}^{s+1} f^{\nu_1\dots\hat{\nu}_i\dots\nu_{s+1};\nu_i},$$

the index $\hat{\nu}_i$ is omitted. After rather tiresome computations one gets the following system of equations for coefficients $c(p,r,s)$:

$$\bar{\delta}(s)\frac{c(p,r,s-1)}{p+1} - \bar{\delta}(r)\frac{c(p,r-1,s)}{p+1} + \frac{c(p-1,r,s)}{r+s+1} = \delta(r),$$

$$\bar{\delta}(s)\left(\frac{c(p,r,s-1)}{(p+1)s} - \frac{c(p-1,r,s)}{(r+s+1)p}\right) = 0,$$

$$\bar{\delta}(r)\left(\frac{c(p,r-1,s)}{(p+1)r} + \frac{c(p-1,r,s)}{(r+s+1)p} - \frac{c(p-1,r-1,s)}{pr}\right) = 0,$$

where $p\geq 1$, $\bar{\delta}(s)=1-\delta(s)=0$ if $s=0$ and $\bar{\delta}(s)=1$ if $s\geq 1$. Now, to finish the proof it is enough to check that the proposed coefficients $c(p,r,s)$ satisfy to this system.

Let us consider the important case $p=0$ more closely. Here the coefficients $c(0,r,s)=1$, so the following proposition is valid.

Proposition A.2. *Let a differential polynomial*

$$P = \sum_i f^i D_i \in \mathcal{P}.$$

Then the following formula is true

$$\sum_{\mu=1}^{m} D_\mu \circ h^\mu(P) = P - \chi(P),$$

where the differential polynomials

$$h^\mu(P) = \sum_{r,s} (-1)^r \sum_{(\nu)} \left(\sum_{(\lambda)} D_{(\lambda)} f^{(\nu)(\lambda)\mu} \right) D_{(\nu)}, \quad \mu = 1,\dots,m,$$

or

$$h^\mu(P) = \sum_i \left(\sum_k (-1)^{|k|} M(i,k,\mu) D_k f^{i+k+(\mu)} \right) D_i, \quad \mu = 1,\dots,m,$$

if we use the original multi-index notation, where factors

$$M(i,k,\mu) = \frac{C(i)C(k)}{C(i+k+(\mu))}.$$

Let $P\in\mathcal{P}$ be a differential polynomial. Consider P as an operator $P:\mathcal{A}\longrightarrow\mathcal{A}$ and let us define the *adjoint* operator $P^*:\mathcal{A}\longrightarrow\mathcal{A}$ by the rule

$$\mathcal{A} \ni \varphi \mapsto P^*(\varphi) = \chi(\varphi P) = \sum_i (-1)^{|i|} D_i\left(\varphi f^i\right) \in \mathrm{A},$$

where $\chi(\varphi P)$ is the characteristic of the polynomial $\varphi P\in\mathcal{P}$. One can easily check that if an algebra $\mathcal{A}$ is commutative then

$$P^* = \sum_i f^{*i} D_i, \quad \text{where } f^{*i} = \sum_k (-1)^{|i+k|} \binom{i+k}{k} D_k f^{i+k}, \quad i \in \mathbb{N}^m.$$

Proposition A.2 can be rewritten now as follows.

Lemma A.3. *Let an operator $P \in \mathcal{P}$ and elements $\varphi, \psi \in \mathcal{A}$. Then*

$$\varphi P(\psi) - P^*(\varphi)\psi = \sum_\mu D_\mu J^\mu,$$

where

$$J^\mu = h^\mu(\varphi P)\psi$$

$$= \sum_{i,k} (-1)^{|i|} M(i,k,\mu) \left(D_i \varphi f^{i+k+(\mu)}\right) \left(D_k \psi\right), \quad \mu = 1,\dots,m.$$

The lemma has obvious extensions to matrix cases. For example, let a matrix operator $P: \mathcal{A}^b \longrightarrow \mathcal{A}_a$ act by the rule

$$\mathcal{A}^b \ni \psi \mapsto P(\psi) \in \mathcal{A}_a, \quad \text{where } P(\psi)_\alpha = \sum_{\beta=1}^{b} \sum_i f^i_{\alpha\beta} D_i \psi^\beta,$$

$\alpha = 1,\dots,a$. Then the adjoint operator $P^*: \mathcal{A}^a \longrightarrow \mathcal{A}_b$ acts by the rule

$$\mathcal{A}^a \ni \varphi \mapsto P^*(\varphi) \in \mathcal{A}_b, \quad \text{where } P^*(\varphi)_\beta = \sum_{\alpha=1}^{a} \sum_i (-1)^{|i|} D_i \left(\varphi^\alpha f^i_{\alpha\beta}\right),$$

$\alpha = 1,\dots,a$, and we have

Proposition A.4 (Green's formula). *Let an operator $P: \mathcal{A}^a \longrightarrow \mathcal{A}_b$ and elements $\varphi \in \mathcal{A}^a$, $\psi \in \mathcal{A}^b$. Then*

$$\sum_\alpha \varphi^\alpha P(\psi)_\alpha - \sum_\beta P^*(\varphi)_\beta \psi^\beta = \sum_\mu D_\mu J^\mu,$$

where

$$J^\mu = \sum_{\alpha,\beta} \sum_{i,k} (-1)^{|i|} M(i,k,\mu) \left(D_i \varphi^\alpha f^{i+k+(\mu)}_{\alpha\beta}\right) \left(D_k \psi^\beta\right),$$

$\mu = 1,\dots,m.$

Theorem A.5.. *The complex of the linear spaces*

$$0 \longrightarrow \mathcal{P}^{\wedge m} \xrightarrow{D\circ} \mathcal{P}^{\wedge(m-1)} \xrightarrow{D\circ} \ldots \xrightarrow{D\circ} \mathcal{P}^{\wedge 1} \xrightarrow{D\circ} \mathcal{P}^{\wedge 0} \xrightarrow{\chi} \mathcal{A} \longrightarrow_{\mathbb{T}} 0$$

is exact.

We shall denote this complex by $\left\{\mathcal{P}^{\wedge p}, \mathcal{A}\right\}$.

Proof. To prove the exactness of the complex it is enough to note that for any element $f\in\mathcal{A}$ the characteristic $\chi(P)$ of the zero-order polynomial $P=f\in\mathcal{P}$ is equal to the element f $(\chi(f)=f)$ and to use Theorem A.1.

Remark A.6. Consider the modules $\mathcal{P}_n^{\wedge p}$, $p\in\{0,\ldots,m\}$, and define linear mappings $D\circ:\mathcal{P}_n^{\wedge p} \longrightarrow \mathcal{P}_n^{\wedge(p-1)}$ by the rule $(D\circ P)_\sigma=D\circ P_\sigma$ for $\sigma\in\{1,\ldots,n\}$. In the same way let us define the characteristic $\chi(P)\in\mathcal{A}_n$ of an element $P\in\mathcal{P}_n$ by the rule $\chi(P)_\sigma=\chi(P_\sigma)$. Then the homotopy formula will be valid and the complex $\left\{\mathcal{P}_n^{\wedge p}, \mathcal{A}_n\right\}$ will be exact.

Remark A.7. Consider the modules $\mathcal{A}^{\wedge p}$, $p=0,\ldots,m$. The rule

$$\Phi = \left(\Phi^{\mu_1\ldots\mu_p}\right) \mapsto D\Phi = \left((D\Phi)^{\mu_1\ldots\mu_{p-1}} = \sum_\mu D_\mu\Phi^{\mu\mu_1\ldots\mu_{p-1}}\right)$$

defines linear mappings $D:\mathcal{A}^{\wedge p}\longrightarrow\mathcal{A}^{\wedge(p-1)}$, $p=1,\ldots,m$. Clear that the composition $DD=0$, so we have the complex

$$0 \longrightarrow \mathbb{K} \longrightarrow \mathcal{A}^{\wedge m} \xrightarrow{D} \mathcal{A}^{\wedge(m-1)} \xrightarrow{D} \ldots \xrightarrow{D} \mathcal{A}^{\wedge 1} \xrightarrow{D} \mathcal{A}^{\wedge 0} \longrightarrow 0.$$

We denote this complex by $\left\{\mathcal{A}^{\wedge p}\right\}$. In the general case this complex should not be exact and we denote by $\mathcal{H}^p(\mathcal{A})$ its homology spaces, $p=0,\ldots,m$.

An ideal $\mathcal{Y}$ of a differential algebra $\mathcal{A}$ is called *differential* if it is closed with respect to differentiations D_μ, i.e. if

$$D_\mu \Phi \in \mathcal{Y} \quad \text{for all } \mu = 1,\dots,m \text{ and } \Phi \in \mathcal{Y}.$$

Let $\mathcal{A}$ be a differential algebra and $\mathcal{Y}$ be its two-sided differential ideal. The differentiations D_μ of the algebra $\mathcal{A}$ define the differentiations ∂_μ of the quotient algebra $\bar{\mathcal{A}}=\mathcal{A}/\mathcal{Y}$ by the rule: $\partial_\mu \varphi=[D_\mu \Phi]=D_\mu \Phi+\mathcal{Y}$ for an equivalence class $\varphi=[\Phi]=\Phi+\mathcal{Y}\in\bar{\mathcal{A}}$. Hence, the differential algebra $\bar{\mathcal{A}}$ with the differentiations ∂_μ, $\mu=1,\dots,m$, is defined. Let us denote by $\bar{\mathcal{P}}=\bar{\mathcal{A}}[\partial]$ the algebra of all polynomials in $\partial_1,\dots,\partial_m$ with coefficients in $\bar{\mathcal{A}}$.

Every polynomial $P=\sum f^i D_i\in\mathcal{P}$ defines the quotient polynomial $\bar{P}=\sum [f^i]\partial_i\in\bar{\mathcal{P}}$. Clear that the corresponding mapping $\mathcal{P}\longrightarrow\bar{\mathcal{P}}$ is surjective. The kernel of this mapping is two-sided ideal $Q=\mathcal{Y}[D]\subset\mathcal{P}$ of all polynomials with coefficients in $\mathcal{Y}$. The mapping $\mathcal{P}\longrightarrow\bar{\mathcal{P}}$ is compatible with the operator action, i.e., $[P\Phi]=\bar{P}[\Phi]$ for all $P\in\mathcal{P}$ and $\Phi\in\mathcal{A}$. Moreover, $P(\mathcal{Y})\subset\mathcal{Y}$ for all $P\in\mathcal{P}$, and $P(\mathcal{A})\subset\mathcal{Y}$ for all $P\in Q$. It is also clear that Theorems A.1 and A.5 are valid for the complexes $\left\{\bar{\mathcal{P}}^{\wedge p},\bar{\mathcal{A}}\right\}$ and $\left\{Q^{\wedge p},\mathcal{Y}\right\}$.

$$
\begin{array}{ccccccccccc}
 & & 0 & & 0 & & 0 & & 0 & & \\
 & & \uparrow & & \uparrow & & \uparrow & & \uparrow & & \\
0 & \longrightarrow & \overline{\mathbb{K}} & \longrightarrow & \overline{\mathcal{A}}^{\wedge m} & \xrightarrow{\partial} \dots \xrightarrow{\partial} & \overline{\mathcal{A}}^{\wedge 1} & \xrightarrow{\partial} & \overline{\mathcal{A}}^{\wedge 0} & \longrightarrow & 0 \\
 & & \uparrow & & \uparrow & & \uparrow & & \uparrow & & \\
0 & \longrightarrow & \mathbb{K} & \longrightarrow & \mathcal{A}^{\wedge m} & \xrightarrow{D} \dots \xrightarrow{D} & \mathcal{A}^{\wedge 1} & \xrightarrow{D} & \mathcal{A}^{\wedge 0} & \longrightarrow & 0 \\
 & & \uparrow & & \uparrow & & \uparrow & & \uparrow & & \\
0 & \longrightarrow & \mathbb{L} & \longrightarrow & \mathcal{Y}^{\wedge m} & \xrightarrow{D} \dots \xrightarrow{D} & \mathcal{Y}^{\wedge 1} & \xrightarrow{D} & \mathcal{Y}^{\wedge 0} & \longrightarrow & 0 \\
 & & \uparrow & & \uparrow & & \uparrow & & \uparrow & & \\
 & & 0 & & 0 & & 0 & & 0 & &
\end{array}
$$

Let us denote by $\mathbb{L}$ the subalgebra of all constants in the ideal $\mathcal{Y}$, i.e., $\mathbb{L}=\mathbb{K}\cap\mathcal{Y}$, and let $\overline{\mathbb{K}}=\mathbb{K}/\mathbb{L}$. Let us consider the commutative diagram

Let us denote by $\mathcal{H}^p(\overline{\mathcal{A}})$, $\mathcal{H}^p(\mathcal{A})$ and $\mathcal{H}^p(\mathcal{Y})$, $p=0,\dots,m$, the homology spaces of the rows. Since all columns are exact the connecting linear mappings $\delta:\mathcal{H}^p(\overline{\mathcal{A}})\to\mathcal{H}^{p-1}(\mathcal{Y})$ are defined, where

$$\mathcal{H}^p(\overline{\mathcal{A}}) \ni \varphi = [\Phi],\ D\Phi \in \mathcal{Y}^{\wedge(p-1)},\ \mapsto \delta\varphi = [D\Phi] \in \mathcal{H}^{p-1}(\mathcal{Y}),$$

$p=1,\dots,m$, and the homological sequence

$$0 \to \mathcal{H}^m(\mathcal{Y}) \to \mathcal{H}^m(\mathcal{A}) \to \mathcal{H}^m(\overline{\mathcal{A}}) \xrightarrow{\delta} \mathcal{H}^{m-1}(\mathcal{Y}) \to \dots$$

$$\dots \to \mathcal{H}^1(\overline{\mathcal{A}}) \xrightarrow{\delta} \mathcal{H}^0(\mathcal{Y}) \to \mathcal{H}^0(\mathcal{A}) \to \mathcal{H}^0(\overline{\mathcal{A}}) \to 0$$

is exact (see, for example, Cartan, Eilenberg [1]). Note that here $\mathcal{H}^m(\mathcal{Y})=\mathcal{H}^m(\mathcal{A})=0$.

Lemma A.8. *If* $\mathcal{H}^p(\mathcal{A})=0$ *for all* $p=1,\dots,m$, *then*

$$\mathcal{H}^p(\bar{\mathcal{A}}) \simeq \mathcal{H}^{p-1}(\mathcal{Y}), \quad p = 2,\dots,m,$$

$$0 \longrightarrow \mathcal{H}^1(\bar{\mathcal{A}}) \xrightarrow{\delta} \mathcal{H}^0(\mathcal{Y}) \longrightarrow \mathcal{H}^0(\mathcal{A}) \longrightarrow \mathcal{H}^0(\bar{\mathcal{A}}) \longrightarrow 0.$$

In particular,

$$\delta : \mathcal{H}^1(\bar{\mathcal{A}}) \simeq \mathcal{Y} \cap D(\mathcal{A}^{\wedge 1}) \Big/ D(\mathcal{Y}^{\wedge 1}).$$

Proof. It is enough to use the exactness of the above sequence.

Every element $F \in \mathcal{A}^n$ defines the left differential ideal

$$\mathcal{Y}_F = \left\{ \Phi = \sum_{\sigma=1}^{n} \sum_i f^i_\sigma D_i F^\sigma, \ f^i_\sigma \in \mathcal{A} \right\}$$

$$= \left\{ \Phi = P(F) = \sum_\sigma P_\sigma(F^\sigma), \ P \in \mathcal{P}_n \right\}.$$

Lemma A.9. *The representation*

$$\mathcal{Y}_F = \left\{ \Phi = DJ + \sum_\sigma \Gamma_\sigma F^\sigma, \ J \in \mathcal{Y}_F^{\wedge 1}, \ \Gamma \in \mathcal{A}_n \right\}$$

is valid.

Proof. Let $\Phi \in \mathcal{Y}_F$, i.e., $\Phi = P(F)$ for some $P \in \mathcal{P}_n$. Then by Theorem A.1,

$$\Phi = P(F) = D \circ Q(F) + \sum_\sigma \chi(P)_\sigma F^\sigma,$$

where $Q=h(P)$, so the representation is valid with $J=Q(F)$ and $\Gamma=\chi(P)$. The inverse inclusion is obvious.

Let us consider the linear mapping $F : \mathcal{P}_n \longrightarrow \mathcal{A}$ defined by the rule $P \mapsto P(F)$. Its image is the ideal $\mathcal{Y}_F$ and its kernel is the left $\mathcal{P}$-module

$$\mathcal{N}_F = \left\{P \in \mathcal{P}_n,\ P(F) = 0\right\}.$$

Let us put $\mathcal{R}_F = \chi(\mathcal{N}_F) \subset \mathcal{A}_n$.

Lemma A.10. *The representation*

$$\mathcal{R}_F = \left\{\Gamma \in \mathcal{A}_n,\ \sum_\sigma \Gamma_\sigma F^\sigma \in D\left(\mathcal{Y}_F^{\wedge 1}\right)\right\}$$

is valid.

Proof. Let $\Gamma \in \mathcal{R}_F$, i.e., $\Gamma = \chi(P)$, $P(F)=0$. According to Theorem A.1, $P = D\circ Q + \chi(P) = D\circ Q + \Gamma$, so

$$\sum_\sigma \Gamma_\sigma F^\sigma = P(F) - D\circ Q(F) = D\left(-Q(F)\right) \in D\left(\mathcal{Y}_F^{\wedge 1}\right).$$

On the other hand, let $\Gamma \in \mathcal{A}_n$, $\sum_\sigma \Gamma_\sigma F^\sigma \in D\left(\mathcal{Y}_F^{\wedge 1}\right)$, i.e.,

$$\sum_\sigma \Gamma_\sigma F^\sigma = DJ,\ J = Q(F),\ Q \in \mathcal{P}_n.$$

Let us put $P = \Gamma - D\circ Q$, then $P \in \mathcal{N}_F$ and $\chi(P) = \Gamma - \chi(D\circ Q) = \Gamma$, so the inverse inclusion is also valid.

Let us consider the following commutative diagram

$$\begin{array}{ccccccccccc}
 & & 0 & & & & 0 & & 0 & & 0 \\
 & & \uparrow & & & & \uparrow & & \uparrow & & \uparrow \\
0 & \longrightarrow & \mathcal{Y}_F^{\wedge m} & \xrightarrow{D} & \ldots & \xrightarrow{D} & \mathcal{Y}_F^{\wedge 1} & \xrightarrow{D} & \mathcal{Y}_F^{\wedge 0} & \xrightarrow{\bar{\chi}} & \mathcal{A}_F & \longrightarrow 0 \\
 & & F\uparrow & & & & F\uparrow & & F\uparrow & & \uparrow \\
0 & \longrightarrow & \mathcal{P}_n^{\wedge m} & \xrightarrow{D\circ} & \ldots & \xrightarrow{D\circ} & \mathcal{P}_n^{\wedge 1} & \xrightarrow{D\circ} & \mathcal{P}_n^{\wedge 0} & \xrightarrow{\chi} & \mathcal{A}_n & \longrightarrow 0 \\
 & & \uparrow & & & & \uparrow & & \uparrow & & \uparrow \\
0 & \longrightarrow & \mathcal{N}_F^{\wedge m} & \xrightarrow{D\circ} & \ldots & \xrightarrow{D\circ} & \mathcal{N}_F^{\wedge 1} & \xrightarrow{D\circ} & \mathcal{N}_F^{\wedge 0} & \xrightarrow{\chi} & \mathcal{R}_F & \longrightarrow 0 \\
 & & \uparrow & & & & \uparrow & & \uparrow & & \uparrow \\
 & & 0 & & & & 0 & & 0 & & 0
\end{array}$$

where $\mathcal{A}_F=\mathcal{A}_n/\mathcal{R}_F$ and the linear mapping $\bar{\chi}$ is induced by χ. Let us denote by $\tilde{\mathcal{H}}^p(\mathcal{Y}_F)$, $\mathcal{H}^p(\mathcal{P}_n)$ and $\mathcal{H}^p(\mathcal{N}_F)$ the homology spaces of the rows, $p=-1,0,1,\ldots,m$. Since all columns are exact the connecting linear mappings $\delta:\tilde{\mathcal{H}}^p(\mathcal{Y}_F)\longrightarrow\mathcal{H}^{p-1}(\mathcal{N}_F)$ are defined, where

$$\tilde{\mathcal{H}}^p(\mathcal{Y}_F) \ni \varphi = [P(F)],\ D\circ P \in \mathcal{N}_F^{\wedge(p-1)},\ \mapsto \delta\varphi = [D\circ P] \in \mathcal{H}^{p-1}(\mathcal{N}_F),$$

$p=1,\ldots,m$, and the homological sequence

$$0 \longrightarrow \mathcal{H}^m(\mathcal{N}_F) \longrightarrow \mathcal{H}^m(\mathcal{P}_n) \longrightarrow \tilde{\mathcal{H}}^m(\mathcal{Y}_F) \xrightarrow{\delta} \mathcal{H}^{m-1}(\mathcal{N}_F) \longrightarrow\ldots$$

$$\ldots\xrightarrow{\delta} \mathcal{H}^0(\mathcal{N}_F) \longrightarrow \mathcal{H}^0(\mathcal{P}_n) \longrightarrow \tilde{\mathcal{H}}^0(\mathcal{Y}_F)$$

$$\xrightarrow{\delta} \mathcal{H}^{-1}(\mathcal{N}_F) \longrightarrow \mathcal{H}^{-1}(\mathcal{P}_n) \longrightarrow \tilde{\mathcal{H}}^{-1}(\mathcal{Y}_F) \longrightarrow 0$$

is exact. Here, $\mathcal{H}^m(\mathcal{N}_F)=0$, $\tilde{\mathcal{H}}^m(\mathcal{Y}_F)=\mathbb{L}$, $\tilde{\mathcal{H}}^p(\mathcal{Y}_F)=\mathcal{H}^p(\mathcal{Y}_F)$ for $p=1,\ldots,m-1$, $\mathcal{H}^{-1}(\mathcal{N}_F)=0$. By Theorem A.5 (see Remark A.6), in our case $\mathcal{H}^p(\mathcal{P}_n)=0$ for $p=-1,0,\ldots,m$. Hence, the following lemma is valid.

Lemma A.11. *We have*

$$\mathcal{H}^p(\mathcal{Y}_F) \simeq \mathcal{H}^{p-1}(\mathcal{N}_F),\ p = 1,\ldots,m-1,$$

$$\mathcal{H}^m(\mathcal{N}_F) \simeq \mathbb{L},\ \tilde{\mathcal{H}}^0(\mathcal{Y}_F) = \tilde{\mathcal{H}}^{-1}(\mathcal{Y}_F) = 0.$$

BIBLIOGRAPHY

Abellanas L., Galindo A.J.

[1] Conserved densities for nonlinear evolution equations. I. Even order case. J. Math. Phys. **20**(1979), 1239-1243.

Bourbaki N.

[1] Variétés différentielles et analytiques. Fascicule de résults. Fasc. XXXIII, XXXVI. Hermann 1967, 1971.

Cartan H.

[1] Calcul différentiel. Formes différentielles. Hermann, Paris 1967.

Cartan H., Eilenberg S.

[1] Homological algebra. Princeton university press, Princeton 1956.

Dubrovin B.A., Novikov S.P., Fomenko A.T.

[1] Modern geometry. Methods and applications (Russian). Nauka, Moscow 1979. Engl. transl., Springer-Verlag, New York - Berlin - Heidelberg - Tokio 1984-1985-1990.

Duzhin S.U., Tsujishita T.

[1] Conservation laws of BBM equation. J. Phys. A **17**(1984), 3267-3276.

Galindo A., Martinez L.

[1] Kernels and ranges in the variational formalism. Lett. Math. Phys. **2**(1978), 385-390.

Gelfand I.M., Manin Yu.I., Shubin M.A.

[1] Poisson brackets and the kernel of the variational derivative in the formal calculus of variations. Funk. Anal. i Prilozen. **10**(4)(1976), 274-278.

Godeman R.

[1] Topologie algébrique et théorie des faiseaux. Hermann, Paris 1964.

Ibragimov N.Kh.

[1] Transformation groups applied to mathematical physics (Russian). Nauka, Moscow 1983. Engl. transl., Reidel, Dordrecht 1985.

Kobayashi S., Nomizu K.

[1] Foundations of differential geometry. Vol. I, II. Interscience Publishers, New York - London - Sydney 1963, 1969.

Krasil'shchik I.S., Vinogradov A.M.

[1] Nonlocal symmetries and the theory of coverings: an addendum to A.M.Vinogradov's "Local symmetries and conservation laws". Acta Appl. Math. **2**(1984), 79-96.

Lang S.

[1] Introduction to differentiable manifolds. Interscience Publishers, New York - London 1962.

Lax P.D.

[1] Periodic solutions of the KdV equation. Comm. Pure Appl. Math. **28**(1975), 141-188.

Martinez L.A.

[1] On the Noether map. Lett. Math. Phys. **3**(1979), 419-424.

Nirenberg L.

[1] Lectures on linear partial differential equations. CBMS Regional Conf. Series in Math., **17**, Amer. Math. Soc., Providence, R.I., 1973.

Olver P.J.

[1] Applications of Lie groups to differential equations. Springer-Verlag, New York - Berlin - Heidelberg - Tokio 1986.

Ovsyannikov L.V.

[1] Group analysis of differential equations (Russian). Nauka, Moscow 1978. Engl. transl., Academic Press, New York 1982.

Pommaret J.F.

[1] Systems of partial differential equations and Lie pseudogroups. Gordon and Breach, New York - London - Paris 1978.

Ritt J.F.

[1] Differential algebra. Dover, New York 1966.

Scherk J.

[1] An introduction to the theory of dual models and strings. Rev. Mod. Phys. **47**(1975), 123-164.

Sternberg S.

[1] Lectures on differential geometry. Prentice Hall, Inc. Engle-

wood Cliffs, N.J. 1964.

Tsujishita T.

[1] On variation bicomplexes associated to differential equations. Osaka J. Math. **19**(1982), 311-363.

Vinogradov A.M.

[1] On the algebra-geometric foundations of Lagrangian field theory (Russian). Dokl. Akad. Nauk SSSR **236**(1977), 284-287; Engl. transl. in Soviet Math. Dokl.**18**(1977), 1200-1204.

[2] A spectral sequence associated with a nonlinear differential equation and algebra-geometric foundations of Lagrangian field theory with constraints (Russian). Dokl. Akad. Nauk SSSR **238**(1978), 1028-1031; Engl. transl. in Soviet Math. Dokl. **19**(1978), 144-148.

[3] The theory of higher infinitesimal symmetries of nonlinear partial differential equations (Russian). Dokl. Akad. Nauk SSSR **248**(1979), 274-278; Engl. transl. in Soviet Math. Dokl. **20**(1979), 985-990.

[4] The C-spectral sequence, Lagrangian formalism, and conservation laws. J. Math. Anal. Appl. **100**(1984), 1-129.

[5] Local symmetries and conservation laws. Acta. Appl. **2**(1984), 21-78.

Vinogradov A.M., Krasil'shchik I.S.

[1] A method of computing higher symmetries of nonlinear evolution equations and nonlocal symmetries (Russian). Dokl. Akad. Nauk

SSSR **253**(1980), 1289-1293; Engl. transl. in Soviet Math. Dokl. **22**(1980), 235-239.

Vinogradov A.M., Krasil'shchik I.S., Lychagin V.V.

[1] Introduction to geometry of nonlinear differential equations (Russian). Nauka, Moscow 1986; Engl. transl., Geometry of jet spaces and nonlinear partial differential equations. Gordon and Breach, New York 1986.

Vladimirov V.S., Zharinov V.V.

[1] Closed forms associated with linear differential operators (Russian). Differentsial'nye Uravneniya **16**(1980), 845-867; Engl. transl. in Differential Equations **16**(1980).

Wahlquist H.D., Eastabrook F.B.

[1] Prolongation structures of nonlinear evolution equations. J. Math. Phys. **16**(1975), 1-7.

Whitham G.B.

[1] Linear and nonlinear waves. John Willey and Sons, New York - London - Sydney - Toronto 1974.

Zharinov V.V.

[1] The geometry of analytic jets (Russian). Mat. Sb. **125**(1984), 364-383; Engl. transl. in Math. USSR-Sb. **53**(1986), 367-383.

[2] Conservation laws of evolution systems (Russian). Teor. Mat. Fiz. **68**(1986), 163-171; Engl. transl. in Theor. Math. Phys. **68**(1987), 745-751.

[3] Green's formula for nonlinear systems of partial differential

equations (Russian). Dokl. Akad. Nauk SSSR **296**(1987), 785-787; Engl.transl. in Soviet Math. Dokl. **36**(1988), 314-316.

[4] On Bäcklund correspondences (Russian). Mat. Sb. **136**(1988), 274-291; Engl. transl. in Math. USSR-Sb. **64**(1989), 277-293.

[5] Symmetries and conservation laws of the string (Russian). Teor. Mat. Fiz. **81**(1989), 163-174; Engl. transl. in Theor. Math. Phys. **81**(1990).

[6] External geometry of differential equations and Green's formula (Russian). Izv. Akad. Nauk. SSSR Ser. Mat. **53**(1989), 708-730; Engl. transl. in Math. USSR-Izv.

[7] Differential algebras and low-dimensional conservation laws (Russian). Mat. Sb.

SUBJECT INDEX

E

F

G

H

I

J

L

M

Minkowski space 247

N

O

P

Q

S

T

U

V

W

Y